TROPHIC AND GUILD INTERACTIONS
IN BIOLOGICAL CONTROL

Progress in Biological Control

Volume 3

Trophic and Guild Interactions in Biological Control

Edited by

Jacques Brodeur
Institut de Recherche en Biologie Végétale,
Université de Montréal, Montréal, Canada

and

Guy Boivin
CRDH, Agriculture and Agri-Food Canada,
Saint-Jean-sur-Richelieu, Québec, Canada

Springer

réseau biocontrôle biocontrol network

A C.I.P. Catalogue record for this book is available from the Library of Congress.

ISBN-10 1-4020-4766-5 (HB)
ISBN-13 978-1-4020-4766-4 (HB)
ISBN-10 1-4020-4767-3 (e-book)
ISBN-13 978-1-4020-4767-1 (e-book)

Published by Springer,
P.O. Box 17, 3300 AA Dordrecht, The Netherlands.

www.springer.com

Cover illustration:
Example of intra-guild competition. A female *Trichogramma turkestanica* parasitizes eggs of *Ephestia kuehniella* (top picture) and a larva of *Harmonia axyridis* predates both unparasitized (white) and parasitized (black) eggs of *Ephestia kuehniella* (bottom picture).

Pictures by D. Thibodeau and G. Boivin, Agriculture and Agri-Food Canada.

Printed on acid-free paper

Contents

Contributing Authors

Steve P. Arthurs, NERC Centre for Population Biology, Imperial College London, Silwood Park Campus, Ascot, Berks. SL5 7PY, UK

Guy Boivin, CRDH, Agriculture and Agrifood Canada, Saint-Jean-sur-Richelieu, Québec, Canada, J3B 3E6; email: boiving@agr.gc.ca

Jacques Brodeur, Institut de Recherche en Biologie Végétale, Université de Montréal, 4101, rue Sherbrooke Est, Montréal, Québec, Canada H1X 2B2; email: jacques.brodeur@umontreal.ca

Robert Denno, Department of Entomology, Plant Sciences Building, University of Maryland, College park MD 20742, USA; email: rdenno@umd.edu

Micky Eubanks, Department of Entomology and Plant Pathology, 301 Funchess Hall, Auburn University, Auburn, AL 36849, USA; email: meubanks@acesag.auburn.edu

Deborah Finke, Department of Entomology, Plant Sciences Building, University of Maryland, College park MD 20742; Department of Entomology, Washington State University, Pullman, Washington, 99164, USA; email: dfinke@wsu.edu

Dave Gillespie, Pacific Agri-Food Research Centre, Agriculture and Agri-Food Canada, POB 1000, Agassiz, British Columbia, Canada, V0M 1A0; Department of Biology, Simon Fraser University, Burnaby, British Columbia, Canada, V5A 1S6; email: GillespieD@agr.gc.ca

Charles J. Godfray, NERC Centre for Population Biology and Division of Biology, Imperial College London, Silwood Park Campus, Ascot, Berkshire, SL5 7PY, United Kingdom; email: c.godfray@imperial.ac.uk

Jason P. Harmon, Department of Entomology, University of California, Davis, CA 95616, USA; email: jharmon@ucdavis.edu

Reinier HilleRisLambers, CSIRO Entomology, Indooroopilly, Australia

Arne Janssen, IBED, section Population Biology, University of Amsterdam, PO Box 94084, 1090 GB Amsterdam, The Netherlands; Department of Animal Biology, Section Entomology, Federal University of Viçosa, Brazil; email: janssen@science.uva.nl

Jane Memmott, School of Biological Sciences, University of Bristol, Woodland Road, Bristol, BS8 1UG, United Kingdom

Nick J. Mills, Department of Environmental Science, Policy and Management, Mulford Hall, University of California, Berkeley, CA 94720-3114, USA; email: nmills@nature.berkeley.edu

Martha Montserrat, IBED, section Population Biology, University of Amsterdam, PO Box 94084, 1090 GB Amsterdam, The Netherlands; email: M.MontserratLarrosa@uva.nl

Angelo Pallini, Department of Animal Biology, Section Entomology, Federal University of Viçosa, Brazil

Bernard D. Roitberg, Department of Biology, Simon Fraser University, Burnaby, BC, Canada V5A 1S6; email: bernard_roitberg@sfu.ca

André M. de Roos, IBED, section Population Biology, University of Amsterdam, PO Box 94084, 1090 GB Amsterdam, The Netherlands

Jay A. Rosenheim, Department of Entomology, University of California, Davis, CA 95616, USA; email: jarosenheim@ucdavis.edu

Maurice W. Sabelis, IBED, section Population Biology, University of Amsterdam, PO Box 94084, 1090 GB Amsterdam, The Netherlands email: M.W.Sabelis@uva.nl

William E. Snyder, Department of Entomology, Washington State University, Pullman, WA 99164, USA; email: wesnyder@wsu.edu

Cory S. Straub, Department of Entomology, Washington State University, Pullman, WA 99164, USA; email: corystraub@wsu.edu

John D. Styrsky, Department of Entomology and Plant Pathology, 301 Funchess Hall, Auburn University, Auburn, AL 36849, USA; email: styrsjd@auburn.edu

Matt Thomas, CSIRO Entomology, GPO Box 1700, Canberra ACT 2601, Australia; email: Matthew.Thomas@csiro.au

Frank Van Veen, NERC Centre for Population Biology and Division of Biology, Imperial College London, Silwood Park Campus, Ascot, Berkshire, SL5 7PY, United Kingdom; email: f.vanveen@imperial.ac.uk

Emma L. Watson, NERC Centre for Population Biology, Imperial College London, Silwood Park Campus, Ascot, Berks. SL5 7PY, UK

Preface

This book origins from a symposium we organized in May 2005 at a joint meeting of the Biocontrol Network of Canada and the International Organization for Biological Control in Magog, Québec, Canada. During this symposium, we discussed concepts of direct and indirect interactions among natural enemies of herbivores in natural and agricultural ecosystems — a field of growing interest in ecology and biological control.

Natural enemies of herbivores exist in nature as an assemblage of species that interact with one another and may transcend trophic levels. The community embracing a natural enemy can be complex and includes taxonomically dissimilar species of pathogens, parasitoids, and predators. Trophic and guild interactions involve predation and competition processes and share the typical characteristics of resource-consumer relationships where the resource species is killed and consumed by the other. Although they are mostly viewed as primary carnivores (developing on herbivores), natural enemies can also be secondary carnivores (when they attack other natural enemies), hosts, prey, or even herbivores, as several species may also feed on and acquire energy from plant resources.

Historically, research on biological control has been conducted using a vertical approach, focusing on simple trophic interactions between plants, herbivores and natural enemies. In the 1980s, this approach has been extended to multitrophic interactions to include contributions of the fourth trophic level. More recently, organisms that live below ground and may interact with higher trophic levels have been added to our knowledge. This trophic approach has been very successful as we now have a much better understanding of the role of bottom-up and top-down effects in both natural and managed ecosystems.

Guild interactions occur among species within the same trophic level. For example, complex relationships might evolve between a fungus, a parasitoid and a predator that exploit a herbivore. In our opinion, a key paper published in 1995 by Jay Rosenheim and his colleagues from the University of California, Davis has largely contributed to change our perception of the importance of guild interactions in biological control (Rosenheim, J.A., H.K. Kaya, L.E. Ehler, J.J. Marois and B.A. Jaffee, 1995. Intraguild predation among biological control agents: theory and evidence. Biological Control 5: 303-335). They concluded that intraguild interactions are widespread within communities of biological control agents of arthropod pests and that they are likely to influence the efficacy of biological control. The publication of this review has immediately stimulated original research on the nature and outcome of interactions among natural enemies. The literature is now filled with fascinating evidence of positive, negative, or neutral interactions between different types of biocontrol agents.

The study of trophic and guild interactions has led to a reconsideration of many paradigms in ecology, such as community structure, species exclusion, trophic cascades in food webs, management of endangered species and biological control. From an applied perspective in biological control, understanding trophic and guild interactions is important to determine (i) the number and composition of species to release, (ii) the most suitable developmental stages of arthropods (or doses of pathogens) to be used, (iii) the timing of multiple species releases, and (iv) the impact of biological control agents on non-target species.

This book explores a broad range of ecological and evolutionary issues in animal species interactions, mostly in the context of biological control. From the beginning of this project we were seeking original viewpoints on a growing field. All authors have used ecological theory to better interpret emerging patterns of interactions in biological control. The core of the book is a series of chapters that examine how species interactions, such as competition, predation, parasitism, disease, mutualism, and omnivory affect population dynamics of natural enemies. Chapters include critical discussions of the current status of research in the field, comparative and meta-analyses, case studies, new data, models, and approaches to measure trophic and guild interactions. Drawing on a diversity of plant, herbivore and natural enemy examples from different ecosystems, each contribution illustrates how trophic and guild interactions, whether they be direct or indirect, simple or complex, strongly affect the efficiency of natural enemies and, over time, determine the outcome of biological control. The contributing authors were selected because they have each added significantly to our understanding of trophic and guild interactions in biological control. This book constitutes a survey of their knowledge, fresh perspective, efforts and enthusiasm.

We are therefore indebted to all authors for their dedication to the symposium and the book, including their help in reviewing chapters. Our deepest gratitude goes to Claude Godin for editorial assistance, compilation of the index and careful attention to detail throughout the book's production. We extend our appreciation to Zuzana Bernhart and Ineke Ravesloot, our editors at Springer, and to Heikki Hokkanen, editor of the book series 'Progress in Biological Control' for their helpful guidance in the production of this book. Finally, we wish to acknowledge the generous contributions of the Biocontrol Network of Canada and the International Organization for Biological Control towards the organization of the symposium and the publication of the book. In particular, these projects have required the coordinated efforts and encouragement of Raynald Laprade, Jean-Louis Schwartz, Robert Wiedenmann, Lucie Lévesque and Stéphane Dupont.

Jacques Brodeur and Guy Boivin

THE INFLUENCE OF INTRAGUILD PREDATION ON THE SUPPRESSION OF A SHARED PREY POPULATION: AN EMPIRICAL REASSESSMENT

Jay A. Rosenheim and Jason P. Harmon

Department of Entomology, University of California, Davis

Abstract. The experimental literature on the effects of intraguild predation on population growth rates of herbivorous arthropod prey has expanded substantially in the last decade, creating a body of results that can be used to test hypotheses relevant to biological control. Here we present a formal meta-analysis of the published experimental literature to assess two hypotheses: (1) intraguild predation causes an increase in the density of the shared herbivore prey, and (2) 'coincidental intraguild predation', in which a predatory arthropod (the 'intraguild predator') consumes a herbivore that harbors a developing parasitoid (the 'intermediate predator'), is less likely to disrupt biological control than is 'omnivorous intraguild predation', in which the intermediate predator is consumed directly. The meta-analysis reveals that intraguild predation does not universally cause an increase in the density of the shared prey; instead, the mean effect size viewed across all studies is not significantly different from zero, and there is strong variability in effects across studies. The meta-analysis also reveals a marginally significant difference between the effects of coincidental and omnivorous intraguild predation: inclusion of a coincidental intraguild predator significantly enhances biological control, at least in the short-term trials included in our database, whereas inclusion of an omnivorous intraguild predator has little overall effect. Thus, our analysis highlights the diversity of effects generated by intraguild predators within arthropod communities. The discrepancy between theory and empirics appears likely to stem from their different time-frames, with theory often emphasizing equilibria and experimentation examining instead short-term transients, and also with the artificial simplification of arthropod communities depicted in theoretical treatments. More work, both theoretical and empirical, is needed to bridge the gap between theory and observation and to develop a deeper understanding of factors generating the observed diversity of intraguild predator effects.

1. INTRODUCTION

The last fifteen years have witnessed a revolution in our understanding of the trophic structure of animal communities. From a view that emphasized functionally discrete trophic levels, as most forcefully enunciated by Hairston *et al.* (1960; see also Hairston and Hairston 1993, 1997), we have now moved to a new understanding of animal communities as often being replete with omnivory, defined as feeding at a variety of trophic levels (Hurd and Eisenberg 1990, Polis 1991, Wise 1993, Polis and Strong 1996, McCann *et al.* 1998, Rosenheim 1998, Halaj and Wise 2001, Finke and Denno 2004). For arthropod communities, two types of omnivory have been especially highlighted. First, many arthropods feed both as herbivores and as predators (Coll and Guershon 2002; see also Gillespie and Roitberg, this volume).

J. Brodeur and G. Boivin (eds.), Trophic and Guild Interactions in Biological Control, 1–20.

Second, many predatory arthropods feed on prey that are found at diverse locations within the food web, including insects that are detritivores, herbivores, and other predators (Polis *et al.* 1989, Rosenheim *et al.* 1995).

Intraguild predation is a type of omnivory that may have particular relevance to the suppression of herbivorous insects by their natural enemies. It occurs when two consumers that share a resource, and which therefore are potential competitors, also engage in predator-prey interactions with each other (Polis *et al.* 1989, Arim and Marquet 2004). Intraguild predation appears to be widespread in communities of biological control agents (Rosenheim *et al.* 1995, Sunderland *et al.* 1997, Janssen *et al.* 1998, Brodeur and Rosenheim 2000, Snyder *et al.* 2005).

The late Gary Polis and his colleagues produced a series of seminal papers that shaped both our theoretical and empirical understanding of intraguild predation (Polis *et al.* 1989, Polis 1991, Polis and Holt 1992, Polis and Strong 1996, Holt and Polis 1997). Although Polis himself did not work in applied insect ecology, he recognized the implications of his work for biological control, and he introduced the idea that intraguild predation might cause biological control programs to fail (Polis and Holt 1992). This idea was bolstered by two types of evidence. The first was a series of analytical models of a three-species module of interacting species: an intraguild predator, an intermediate predator, and a shared prey (which we will henceforth call the "herbivore") (Polis *et al.* 1989, Polis and Holt 1992, Holt and Polis 1997). A salient prediction emanating from these models was that the presence of the intraguild predator increased the equilibrium density of the shared herbivore prey. The logic underlying this prediction was simple and compelling: in the simplest three-species module of intraguild predation, the two predators are in 'perfect' competition for the one species of prey that they both consume (the herbivore). Were there no direct trophic interactions occurring between the two predators (i.e., if they were only competitors and not engaged in intraguild predation), only one of the predators, the superior competitor, would be able to persist at equilibrium, while the inferior competitor would be competitively excluded. Theory demonstrates that the superior predator is the one that can sustain its population at the lowest equilibrium density of the herbivore prey population; that is, the superior competitor is always the best biological control agent. This is, of course, a happy outcome for biological control, because it means that the community would 'self select' the predator that produces the best pest control.

If, however, the two predators are not only competitors, but also engage in trophic interactions with one another, the outcome is quite different. The only way to achieve a stable three-species equilibrium is for the intraguild predator's special advantage (it's ability to consume the intermediate predator) to be balanced against some advantage possessed by the intermediate predator; the only possible advantage that the intermediate predator can have is a superior competitive ability. That is, the intermediate predator when present singly must be a better biological control agent than is the intraguild predator when present singly. In this case, for the intraguild predator to be present in a stable 3-species equilibrium, it must always elevate the equilibrium density of the herbivore, because it suppresses the population of the superior biological control agent: the intermediate predator. This, of course, is an unhappy outcome for biological control.

Polis and Holt's earliest models have been extended in a variety of ways, to include variable ecosystem productivity, stage structure, saturating functional responses, and immigration (Diehl and Feissel 2000, Mylius *et al.* 2001, Revilla 2002, Briggs and Borer 2005, Ives *et al.* 2005). Most of these elaborations of the basic theory have upheld the basic prediction that intraguild predators should elevate the density of the herbivore. However, as explored in more detail below (see Discussion), recent work by Briggs and Borer (2005) has begun to demonstrate scenarios that broaden the possible range of expected outcomes. Thus, although theoretical treatments of intraguild predation are the first and primary basis for the expectation that intraguild predation has a negative influence on biological control, ongoing theoretical developments are beginning to cast doubt on how robust this prediction really is.

The second reason to think that intraguild predation might be a problem for biological control came from some of the earliest experimental studies of multi-predator biological control systems, which demonstrated exactly this unwanted outcome (Hoy *et al.* 1972, Press *et al.* 1974, Spiller 1986). Some of these studies were discussed by Polis and Holt in their early reviews. There were, however, actually very few studies that had involved experimental manipulations of intraguild predators, and thus the literature provided only the scantiest basis for a critical assessment of the prediction that intraguild predation always disrupts biological control (see Janssen *et al.* this volume).

The literature was still too immature to provide a definitive test when Ehler (1995) and Rosenheim *et al.* (1995) attempted to review what was known about intraguild predators in biological control systems. Their reviews did, however, produce a second and more nuanced prediction regarding the differential impact of two types of intraguild predation. The first type, called 'coincidental' intraguild predation by Polis *et al.* (1989), occurs only when an intraguild predator and an intermediate predator both attack the same herbivore individual. Coincidental intraguild predation occurs most often when a predator (the 'intraguild predator') attacks a herbivore that has previously been attacked by a parasitoid (or a pathogen; see Thomas *et al.* this volume), and which therefore harbors a developing offspring of the parasitoid (the 'intermediate predator'). In this case, intraguild predation will impose mortality on the intermediate predator population that is often similar in magnitude to the mortality that it imposes on the herbivore population – the two will generally be linked (and they will be equal in cases where the predator does not distinguish between parasitized and unparasitized hosts). In contrast, the second type of intraguild predation, termed 'omnivorous' intraguild predation by Polis *et al.* (1989), occurs without joint attack on the herbivore. Instead, intraguild predation occurs when one predator encounters and consumes another predator. In this case, the intraguild predator may impose mortality on the intermediate predator that is independent of any direct effects on the herbivore population – it is easier to decouple mortality that is being imposed on the intermediate predator population and the herbivore population. This simple verbal argument, supported only weakly by the earliest hints of patterns emerging from the empirical literature, led to the suggestion that coincidental intraguild predation would have less potential to disrupt biological control than would omnivorous intraguild predation.

The intraguild predation literature has expanded dramatically in the last decade, providing the first opportunity to produce a quantitative synthetic test of the effect of adding an intraguild predator to a system made up of an intermediate predator and its herbivore prey. Here we present a formal meta-analysis of the literature on intraguild predation among natural enemies of terrestrial herbivorous arthropods to address two questions: (1) Does intraguild predation consistently lead to an elevated density of the shared herbivore population? (2) Is coincidental intraguild predation less disruptive of biological control than omnivorous intraguild predation?

2. METHODS

2.1. Survey of studies

We attempted to conduct an exhaustive survey of the peer-reviewed literature published through May 2005 to construct a dataset for analysis. Our criteria for including a study in the dataset were as follows. (1) The interacting species must have been terrestrial arthropods, including at least one herbivore and two natural enemies. (2) The trophic interactions must have been demonstrated to include intraguild predation; that is, both of the predators must have been known to feed on the shared herbivore population, and one predator (the 'intraguild predator') must have been demonstrated to feed upon the other (the 'intermediate predator'). In all cases the intraguild predation was exclusively or primarily unidirectional. (3) The study must have included at least two experimental treatments, applied through manipulation of the system by the investigator: a 'control', in which the herbivore was present with only the intermediate predator, and an 'intraguild predator treatment' that was identical to the 'control' treatment in all respects except for the inclusion of one or more species of intraguild predators, thus completing the 3-species module. (4) The experiment must have employed an additive design, in which the intermediate predator was maintained at a constant density in the treatments with and without the intraguild predator present, as opposed to a replacement series, in which the total number of predators is maintained constant as the number of predator species present is increased. The relative merits of these two designs have been extensively discussed in the literature (Connolly 1988, Jolliffe 2000, see Straub and Snyder this volume); whereas the replacement series design may be particularly appropriate for some aspects of the study of how predator diversity *per se* influences biological control (e.g., Snyder *et al.* 2005), it is inappropriate for testing the effect of adding an intraguild predator, because it confounds an interspecific effect (the effect of adding the intraguild predator) with an intraspecific effect (the effect of halving the density of the intermediate predator). (5) The experimental organisms must have been studied in the setting where they are normally found: in the field for arthropods found in natural ecosystems or attacking crops grown outdoors; in greenhouses for arthropods associated with greenhouse-grown crops; and indoors for arthropods that are economically significant in stored products settings. (6) The studies must have provided measurements of herbivore population densities and a measure of variance across replicates. Using these criteria, we identified 25 published articles that contributed a total of 30 observations concerning the effect of adding an intraguild predator (Table 1). Often, the published articles did not report all the information that we needed for the meta-analysis. Thus, we wrote to the authors and solicited the needed data. We received extremely generous assistance from the contacted authors; thus, this study reflects the collective efforts of the community of ecologists studying intraguild predation, both in the conduct of the original work and in the preparation of the raw dataset. In the few cases where the original data files were no longer available, means and variance estimates were estimated digitally from published figures.

A key assumption underlying formal meta-analysis is that the observations included in the dataset are statistically independent. In some cases, meta-analyses are now being conducted using explicit phylogenetic hypotheses of the relationships among the taxa being studied (e.g., Verdú and Traveset 2005). However, our study concerned interactions in 3-species modules, and techniques have not yet been developed to correct for phylogenetic effects in this case. In choosing an approach for defining independent observations, we also wanted to consider a second major concern for meta-analysis: the tendency for non-significant results to remain unpublished, generating substantial biases in the published literature. We reasoned that studies that reported comparative assessments of more than a single intraguild predator, intermediate predator, or herbivore were more likely to produce an unbiased measure of the distribution of possible intraguild predator effects, because if at least one of the tested predators produced a significant effect it would support the publication of the entire study. Thus, we extracted multiple data points from a single study when the study measured more than one target herbivore (Lang 2003), when the study measured more than one intraguild predator (Rosenheim *et al.* 1993, Rosenheim 2001, Colfer *et al.* 2003), or when the study included two qualitatively different environmental contexts (short plants versus tall plants, Snyder and Ives 2001). In all other cases, including cases where there were multiple herbivore density treatments and in which multiple experiments were performed on the same 3-species module, results were averaged across the multiple observations to produce a single entry in our final dataset.

2.2 Meta-analysis

For each study, the magnitude of the effect of the intraguild predator on the density of the herbivore was measured as a ln-transformed response ratio:

$$\ln R = \ln\left(\frac{\overline{X}^{E}}{\overline{X}^{C}}\right) \tag{1}$$

where R is the response ratio, $\overline{X}^{E}$ is the mean herbivore density in the presence of the intraguild predator, and $\overline{X}^{C}$ is the mean herbivore density in the absence of the intraguild predator. Negative values for ln$R$ indicate that herbivore densities are decreased in the presence of the intraguild predator (improved biological control), whereas positive values indicate elevated herbivore densities in the presence of the intraguild predator (disrupted biological control). The variance of this response measure was calculated as:

$$v_{\ln R} = \frac{\left(s^{E}\right)^{2}}{N^{E}\left(\overline{X}^{E}\right)^{2}} + \frac{\left(s^{C}\right)^{2}}{N^{C}\left(\overline{X}^{C}\right)^{2}} \tag{2}$$

where s^E and s^C are the standard deviation for the treatments with and without the intraguild predator, respectively, and N^E and N^C are the respective sample sizes. Osenberg *et al.* (1999) have emphasized the importance of choosing a metric for effect size that is appropriate for the underlying dynamics of the system being studied. The response ratio used here is ideal for studying instantaneous rates of exponential population growth. However, it does not account for variation in the duration of different experiments, which can be a key source of variation for meta-analyses. Osenberg *et al.* (1999) suggest that the response ratio be divided by the duration of each experiment to eliminate this important source of variation. We explored this approach, but found it to have a perhaps unexpected and undesirable outcome. If experiments are comparable in all respects except for their durations, then we might expect longer-duration studies to report larger response ratios, simply because the populations subjected to different treatments have had more time to diverge. However, when we examined the relationship between the absolute value of the response ratio and the duration of the experiment, we did not observe this expected positive relationship; instead we found exactly the desired result that experimental duration was not correlated with effect size (r = 0.000, N = 30, P = 0.99). When we divided the response ratio by the duration of each experiment, we obtained a non-significant but still undesirable trend towards a negative relationship between the duration of the experiment and the absolute value of the effect size (r = -0.29, N = 30, P = 0.12). We reason that different researchers choose a duration for their experimentation that matches the underlying tempo of the system being studied. While short experiments are appropriate for some arthropods that reproduce very rapidly (e.g., aphids and mites), longer experiments are appropriate for arthropods that reproduce more slowly (e.g., some Lepidoptera). Thus, we chose to retain the time-dependent measure of the response ratio for our analyses. We did, however, correct our response ratio measure for variable experiment duration in those cases where the same 3-species module was tested in multiple experiments, and where we then averaged the replicate response ratios to produce a single datum for inclusion in the meta-analysis.

For most of the studies, we used herbivore densities measured at the end of the trial to calculate the response ratio (often, this was the only measure taken by the original workers). In cases where herbivore densities were measured repeatedly as a time-series and in which the relative values of the control and intraguild predator treatments changed markedly over the course of the experiment, we used the arithmetic mean herbivore densities across the duration of the experiment to calculate the response ratio. Variance estimates for mean herbivore densities across the duration of the studies were calculated for us *de novo* from the raw data by the authors of the original studies.

The meta-analysis was conducted using the MetaWin2.0 statistical program (Rosenberg *et al.* 2000). After checking for the existence of significant heterogeneity among studies, we discarded the fixed-effect model (the fixed-effect model provides a test of the hypothesis that intraguild predators generated the same effect size across all studies included in the data set), and fitted only random-effect models, which assume that different studies are estimating different effect sizes. To test whether or not the overall dataset was consistent with the hypothesis that the presence of an intraguild predator elevates herbivore densities, we asked whether a bootstrapped 95% confidence interval, constructed by re-sampling the dataset 999 times, overlapped zero. We also examined the effect of a categorical variable coding the type of intraguild predation (coincidental versus omnivorous). In an approach directly analogous to a standard ANOVA, MetaWin2.0 partitions the total variance in effect sizes into variance explained by the categorical variable and the residual, error variance. The significance of the variance explained by the categorical variable can then be tested using either a parametric test, using the X^2 distribution, or a more conservative non-parametric randomization test; here, we report both test results. We used a 1-tailed test to reflect our *a priori* hypothesis that coincidental intraguild predation would be less disruptive of biological control.

Table 1. The full dataset of studies included in the meta-analysis examining the influence of intraguild predation on the density of the shared herbivore prey.

Herbivore	Intermediate predator	Intraguild predator	Type of intraguild predation	Expt length (days)	Effect size	Effect variance	Study
Plodia interpunctella	*Bracon hebetor*	*Xylocoris flavipes*	coincid	28	0.594	0.215	Press *et al.* 1974
Melanoplus sanguinipes	*Blaesoxipha hunteri*	Asilidae spp.	omniv	13	0.301	0.00840	Rees and Onsager 1982
Aphis gossypii	*Chrysoperla carnea*	*Orius tristicolor*	omniv	8	-1.128	0.198	Rosenheim *et al.* 1993, Rosenheim 2001
Aphis gossypii	*Chrysoperla carnea*	*Geocoris* spp.	omniv	7.8	0.281	0.150	Rosenheim *et al.* 1993, Rosenheim 2001
Aphis gossypii	*Chrysoperla carnea*	*Nabis* spp.	omniv	7.8	0.641	0.113	Rosenheim *et al.* 1993, Rosenheim 2001
Aphis gossypii	*Chrysoperla carnea*	*Zelus renardii*	omniv	8.7	1.700	0.080	Rosenheim *et al.* 1993, Cisneros and Rosenheim 1997, Rosenheim 2001
Dactynotus sp.	*Aphidius floridaensis*	*Cycloneda sanguinea*	coincid	18	0.270	0.177	Ferguson and Stiling 1996
Diuraphis noxia	*Aphelinus asychis*	*Paecilomyces fumosoroseus*	omniv	13	-0.119	0.0298	Mesquita *et al.* 1997
Aphis gossypii	*Lysiphlebus testaceipes*	*Hippodamia convergens*	coincid	24	-8.268	0.353	Colfer and Rosenheim 2001

Table 1. continued

Acyrthosiphon pisum	*Aphidius ervi*	*Pterostichus melanarius*	coincid	15	-0.364	0.044	Snyder & Ives 2001 (short plants)
Acyrthosiphon pisum	*Aphidius ervi*	*Pterostichus melanarius*	coincid	17	0.371	0.049	Snyder & Ives 2001 (tall plants)
Anasa tristis	*Nabis* sp.	Lycosids	omniv	83.5	0.871	0.120	Snyder & Wise 2001
Tetranychus urticae	*Phytoseiulus persimilis*	*Orius laevigatus*	omniv	28.7	-0.144	0.0546	Venzon *et al.* 2001
Trialeurodes vaporariorum	*Macrolophus caliginosus*	*Dicyphus tamaninii*	omniv	4	-0.285	0.00358	Lucas and Alomar 2002
Acyrthosiphon pisum	*Aphidius ervi*	*Harmonia axyridis* and *Nabis* sp.	coincid	31	-1.374	0.131	Cardinale *et al.* 2003
Tetranychus spp.	*Galendromus occidentalis*	*Orius tristicolor*	omniv	7	-4.376	0.149	Colfer *et al.* 2003
Tetranychus spp.	*Galendromus occidentalis*	*Geocoris* spp.	omniv	7	-0.528	0.130	Colfer *et al.* 2003
Prokelisia	*Tytthus vagus*	*Pardosa littoralis*	omniv	38	0.897	0.120	Finke and Denno 2003
Aphids	Lycosids and Linyphiids	Carabids	omniv	37	0.0274	0.0539	Lang 2003
Thysanoptera	Lycosids and Linyphiids	Carabids	omniv	37	-0.332	0.0409	Lang 2003

Table 1. continued

Herbivore	Intermediate predator	Intraguild predator	Type of intraguild predation	Expt length (days)	Effect size	Effect variance	Study
Cicadellidae & Delphacidae	Lycosids and Linyphiids	Carabids	omniv	37	-0.0953	0.0473	Lang 2003
Sitobion avenae	Aphidiidae	Ground-dwelling predators	coincid	61	-0.165	0.0318	Schmidt *et al.* 2003
Acyrthosiphon pisum	*Aphidius ervi*	Generalist predators	coincid	21	-1.378	0.166	Snyder and Ives 2003
Prokelisia	*Grammonota trivittata*	*Pardosa littoralis*	omniv	49	-1.025	0.339	Denno *et al.* 2004
Plutella xylostella	*Cotesia plutellae* and others	*Solenopsis invicta*	coincid	70.5	-0.092	0.0635	Harvey & Eubanks 2004
Tetranychus cinnabarinus	*Phytoseiulus macropilis*	*Nesticodes rufipes*	omniv	11.5	-0.102	0.0444	Rosenheim *et al.* 2004a
Tetranychus cinnabarinus	*Stethorus siphonulus*	*Nesticodes rufipes*	omniv	10.33	0.545	0.0532	Rosenheim *et al.* 2004b
Macrosiphum euphorbiae	*Aphelinus asychis*	*Harmonia axyridis*	coincid	45	-0.593	0.114	Snyder *et al.* 2004
Trialeurodes vaporariorum	*Encarsia formosa*	*Dicyphus hesperus*	coincid	19	-0.0294	0.1150	McGregor & Gillespie 2005
Musca domestica	Small carabids	*Pterostichus melanarius*	omniv	7	0.051	0.384	Prasad and Snyder in press

3. RESULTS

3.1. Effects of intraguild predation on herbivore density

The overall effect of moving from a 2-species module (herbivore + intermediate predator) to a 3-species module (herbivore + intermediate predator + intraguild predator) was not as predicted by theory. Instead of producing a significant increase in herbivore densities, the mean effect size was –0.38, with a bootstrapped 95% confidence interval that overlapped zero (Fig. 1). Thus, across all the studies, adding an intraguild predator had no significant effect on herbivore densities, with a non-significant trend towards a slight improvement of biological control (the –0.38 value represents a 31% decrease in herbivore densities when the intraguild predator is present).

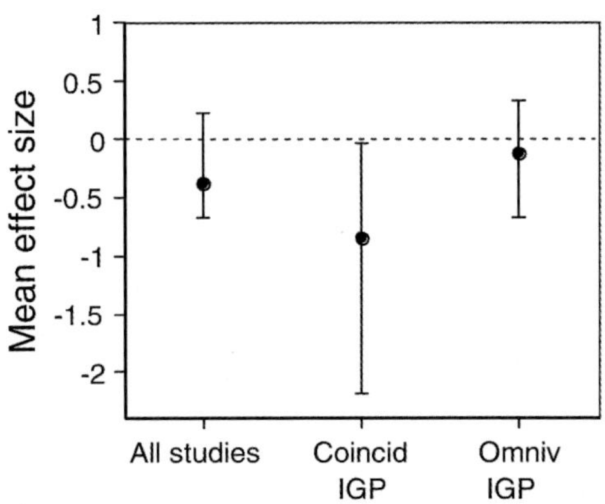

Figure 1. Mean influence of intraguild predation on the density of the shared herbivore population, measured as $ln(\overline{X}^E / \overline{X}^C)$, observed across all studies in the meta-analysis. Shown are the means and bootstrapped 95% confidence intervals for all studies (N = 30), for studies of coincidental intraguild predation (Coincid IGP; n = 11), and for omnivorous intraguild predation (Omniv IGP; n = 19). Negative values indicate that the presence of the intraguild predator results in a decrease in herbivore density.

There was significant heterogeneity across the 30 studies in the magnitude of the effect size ($Q_T = 100.3$, df = 29, $P < 0.00001$). Thus, although the overall mean was not significantly different from zero, the variance among effect sizes was greater than expected simply by sampling error. This suggests that different systems exhibit significantly different outcomes: in some cases, adding an intraguild predator elevates herbivore density, and in other cases it depresses herbivore density. This result was reinforced by the observation that at least in some cases, the response ratios observed in replicate observations of a given 3-species module were repeatable. For example, the meta-analysis included a single datum for the 3-species module of *Aphis gossypii* (Aphididae) (herbivore), *Chrysoperla carnea* (Chrysopidae) (intermediate predator), and *Zelus renardii* (Reduviidae) (intraguild predator), which represented the mean of five independent experimental trials; the trials produced quite concordant response ratio measures (1.93, 1.35, 1.71, 1.71, 1.83). The distribution of effect sizes included many studies that produced only small effects on herbivore densities, some studies that produced large increases in herbivore densities, and other studies that produced dramatic decreases in herbivore densities (Fig. 2). The meta-analysis therefore allows us to reject emphatically the notion that intraguild predation produces one consistent type of effect on herbivore densities. Variability is instead the dominant result.

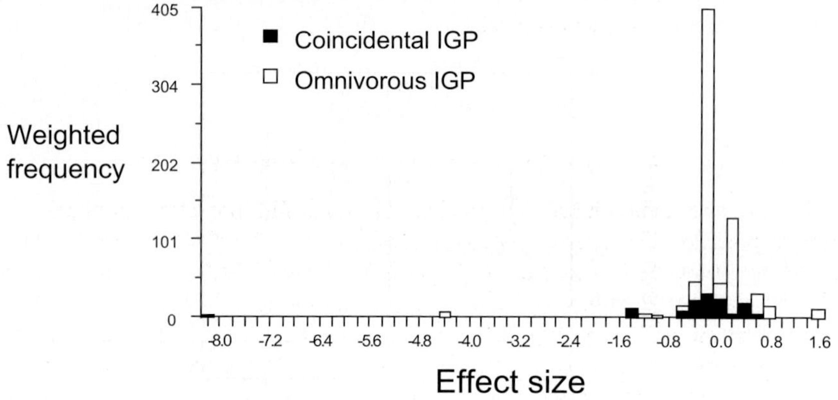

Figure 2. Distribution of effect sizes of intraguild predation on the density of the shared herbivore population observed for coincidental and omnivorous intraguild predation. Effect sizes are measured as $ln(\overline{X}^{E}/\overline{X}^{C})$*; thus positive values indicate that the intraguild predator elevated herbivore density, and negative values indicate that the intraguild predator suppressed herbivore density. The vertical axis shows the weighted frequency (weights = 1/variance), and thus the histogram shows the relative contribution of the data to each effect class (coincidental versus omnivorous intraguild predation) rather than the sample size for that effect class.*

3.2. Coincidental versus omnivorous intraguild predation

Eleven of the 30 observations in the dataset involved coincidental intraguild predation, where predators attacked parasitized herbivore prey, thereby consuming both the herbivore and the developing parasitoid (the intermediate predator), with the remaining 19 cases examples of omnivorous intraguild predation. The analyses provided only mixed support for the hypothesis that coincidental intraguild predators were less likely to elevate herbivore densities than were omnivorous intraguild predators: the less conservative parametric test was significant ($Q = 4.4$, df = 1, $P =$ 0.018), whereas the more conservative randomization test was not ($P = 0.12$). The significance of the parametric test was lost if (a) the analysis was conducted without weighting the observations by their associated variance estimates ($Q = 1.6$, df = 1, $P = 0.10$) or (b) the strongest outlier (Colfer and Rosenheim 2001) was excluded ($Q = 1.5$, df = 1, $P = 0.11$). The mean effect size for coincidental intraguild predation was –0.84 (a 57% decrease in herbivore density), and the bootstrapped confidence interval did not overlap zero (Fig. 1). Thus, our analysis suggests that adding a coincidental intraguild predator to a herbivore/intermediate predator system results in an overall improvement in herbivore suppression, at least in the short-term trials included in our data set. In contrast, the mean effect size for omnivorous intraguild predation was close to zero (mean effect size = -0.12); these studies included cases of predators eating other predators, predators eating adult parasitoids, and a pathogen infecting an adult parasitoid (Table 1). Our analysis provides the first empirical support for the idea that coincidental intraguild predation has a reduced potential to disrupt biological control.

4. DISCUSSION

4.1. Intraguild predation and the density of the shared prey population

The earliest and most influential models of intraguild predation predicted that intraguild predators increase the equilibrium density of the shared prey population (Polis *et al.* 1989, Polis and Holt 1992, Holt and Polis 1997). Translated into the context of applied insect ecology, this suggests that intraguild predators will always act to disrupt biological control, suppressing populations of intermediate predators and thereby allowing herbivore populations to expand. Nevertheless, our analysis of the experimental literature on intraguild predation in terrestrial arthropod communities does not support any single, dominant effect of intraguild predators. Instead, our primary result is that intraguild predators have widely variable effects on herbivore populations. Some intraguild predators produce dramatic decreases in herbivore populations, whereas others have the reverse effect, triggering herbivore outbreaks. Strikingly, the overall effect size, viewed across all studies, was not significantly different from zero.

How can we explain the discrepancy between the empirical record and the theoretical predictions? We suggest two primary possibilities. The first is that there is a profound disconnect between the time frames of the empirical and the theoretical work: while the theory has been almost universally concerned with long-term equilibrium conditions, the empirical work has been almost entirely very short-term (Venzon *et al.* 2001, Hastings 2004, Briggs and Borer 2005). Indeed, the mean duration of the experiments included in our dataset was a mere 25.5 days (range: 4-83.5 d). Even for arthropods, these are very short experiments, in at least some cases too short to incorporate reproductive recruitment of the intermediate and intraguild predators. As incisively demonstrated by Briggs and Borer (2005), short-term transient effects of intraguild predators may be very different from the long-term equilibrium effects. As noted earlier, to achieve a stable 3-species equilibrium, the intraguild predator must be an inferior biological control agent of the herbivore population in comparison to the better competitor, the intermediate predator. In contrast, in a non-equilibrial setting, there is no reason to exclude the possibility that the intraguild predator has all the advantages over the intermediate predator – that is, it may be the best competitor (and therefore the best biological control agent) and also be able to exploit the intermediate predator as a food resource. In this case, it is not all surprising that adding the intraguild predator to the system would improve herbivore control.

This may be exactly the scenario that underlies some of the most dramatic cases of improved herbivore control following introduction of an intraguild predator. For example, *Geocoris* spp. (Hemiptera: Lygaeidae), *Orius tristicolor* (Anthocoridae), and *Galendromus occidentalis* (Phytoseiidae) are three important predators of herbivorous spider mites *Tetranychus* spp. (Tetranychidae) in cotton (Colfer *et al.* 2003). These predators engage in a size-structured ladder of intraguild predation, in which the larger species feed on progressively smaller species: so, *Geocoris* feeds on *Orius*, and both feed heavily on *Galendromus* (Rosenheim 2005). In at least some short-term trials, *Geocoris* and *Orius* are better at controlling spider mites than is *Galendromus* (R. G. Colfer, pers. comm.). The observation, then, that adding *Geocoris* or *Orius* to a 2-species module of *Galendromus* + spider mites produces strong improvements in biological control is not surprising (Colfer *et al.* 2003). Because *Geocoris* and *Orius* may be better short-term competitors and intraguild predators of *Galendromus*, they might be expected to exclude *Galendromus* from cotton. And, indeed, *Galendromus* does appear to be excluded from cotton by intraguild predation, even when their populations are augmented experimentally with large releases (Colfer *et al.* 2004). But, what does this tell us about the longer-term dynamics? Perhaps not much. Spider mites exhibit irruptive dynamics in cotton fields, even in organically-managed cotton fields that are not subject to pesticide-generated disruptions (unpubl. data). *Galendromus* may not be the best predator of spider mites in the shortest-term trials in cotton, but it has been demonstrated to be a highly effective long-term biological control agent in more stable perennial cropping systems (Nyrop *et al.* 1998), and even in longer-term cage trials in cotton it may outperform *Orius* and *Geocoris* (R. G. Colfer, pers. comm.). We do not know why, in the long term, *Geocoris* and *Orius* cannot control spider mite populations in cotton, but other natural enemies (predators, parasitoids, and pathogens) that suppress their populations are the leading candidate explanations (Rosenheim 2005). Thus, in this system, and probably in many others, short-term and long-term effects of interactions within communities of natural enemies may be very different indeed.

There is a second disconnect between the theory and the real setting of the empirical work: the theory has been developed for a closed 3-species module, with perfect competition between the intraguild predator and the intermediate predator for a single shared prey, whereas in nature most systems are open and predators have diverse prey resources. This is particularly true for intraguild predators. A predator that has a diet broad enough to include a herbivore and another predator is very likely to consume multiple species of herbivore prey, and often plant-based resources as well (Arim and Marquet 2004). Holt and Polis (1997) suggested that their prediction that intraguild predators universally increase the density of the shared prey was likely to be sensitive to the addition of alternate prey. Our unpublished simulations have shown that this is the case: moving from a 3-species module to a 4-species module incorporating a second species of herbivore prey, immediately opens up the possibility that the lowest density of the target herbivore is achieved with both predators present in a 4-species equilibrium (unpubl. data). Briggs and Borer (2005) have also shown that immigration of the intraguild and intermediate predators, which can be thought of as another form of subsidy from an alternate resource, can have exactly this effect. While some intraguild predation systems may conform to the simplest 3-species configuration modelled by Holt and Polis, we suspect that most will not.

4.2. Coincidental versus omnivorous intraguild predation

Our analysis has provided the first empirical support for the hypothesis that coincidental intraguild predation may have less potential to disrupt biological control than omnivorous intraguild predation. Why might this be? There is no published theoretical work comparing the dynamical significance of these two forms of intraguild predation, so we are left with verbal argumentation. We tentatively suggest the following line of reasoning. For both coincidental and omnivorous intraguild predation, a key determinant of overall implications for biological control is the intraguild predator's preference for consuming the intermediate predator versus the herbivore. We suggest that omnivorous intraguild predators may be more likely to exhibit a preference for consuming the intermediate predator than are coincidental omnivorous predators. Consider first an omnivorous intraguild predator, which chooses between two potential prey: an intermediate predator, which in this case is generally another freely-foraging predator, or an herbivore. These two prey types may often occupy different microhabitats on the plant, differ significantly in body size, and differ in foraging or defensive behavior. These differences create abundant opportunities for different encounter probabilities, willingness to initiate attacks, and likelihood of subduing a prey given an attack. Strong preferences for one type of prey over another can decouple the mortality imposed by the intraguild predator on the intermediate predator population from that imposed on the herbivore population. In many cases, the intermediate predator may be more mobile than the herbivore, and thus may be at greater risk of attack (Rosenheim *et al.* 2004b). Biological control may be disrupted when an omnivorous intraguild predator generates strong mortality of the intermediate predator population but minimal mortality of the herbivore population.

Consider now a coincidental intraguild predator, which chooses between two potential prey: an intermediate predator, which in this case is an immature parasitoid developing within the herbivore, or the herbivore itself (now in an unparasitized state). We suggest that these two prey types are much more likely to occupy similar microhabitats, be similar in size, and be similar in behavior, because we are now simply comparing parasitized versus unparasitized individuals of the same species. Although the empirical record does show that predators may in some cases have strong preferences for consuming either parasitized (e.g., Jones 1987, Snyder and Ives 2001) or unparasitized individuals (e.g., Ruberson and Kring 1991), the most common result is that predators do not distinguish between parasitized and unparasitized individuals early during the parasitoid's development, and later develop an increasing preference for unparasitized hosts (reviewed in Rosenheim *et al.* 1995, Brodeur and Rosenheim 2000). Thus, in general, we expect coincidental intraguild predators to impose similar or somewhat lower levels of mortality on the intermediate predator population (the parasitoid) than on the herbivore population. Opportunities to release herbivores from control thus seem likely to be diminished.

4.3. Where do we do from here?

Ecologists in search of broad generalizations are often disappointed at the end of the day. The view that all predators operate from the third trophic level, acting reliably to suppress populations of herbivores and thereby freeing plants from strong effects of herbivory (Hairston *et al.* 1960) is no longer tenable. Our synthesis of the experimental literature suggests that the view that all intraguild predators act to disrupt the control of their shared herbivore prey is equally untenable. Instead, different intraguild predators seem to have very different effects on herbivore population suppression, at least in the short term.

What sort of work is now most needed? We suggest two approaches. First, we need to bring the theory and the real world tests of theory closer together. This will require renewed work by both the empiricists and the theoreticians. Empiricists will need to overcome the logistical barriers to conducting longer-term experiments to observe the dynamics of intraguild predators, intermediate predators, and their prey over multiple generations. Whether empiricists working with arthropods will ever be able to run experiments long enough to observe real equilibria is open to question, but it seems both reasonable and important to include at least several generations of all the key players to reach conclusions about longer-term effects of intraguild predation. This may require careful choices about which systems to study. Theoreticians will need to devote additional attention to the transient effects of intraguild predators (e.g., Venzon *et al.* 2001, Snyder and Ives 2003) and build models with more realistic representations of the array of resources available to most predators in nature. Work on transient dynamics is important because many real world systems are inherently non-equilibrial; this is especially true for agroecosystems, where regular disturbances are the norm, and management decisions may hinge entirely on events occurring during a short growing season or a narrow window of crop vulnerability to herbivore damage.

Second, we need to embrace the variability in intraguild predator effects and try to understand why different predators have such radically different influences in their communities. We are already making some progress in this area. Recent work has demonstrated that predator-predator interactions are influenced by features of the physical environmental within which they interact (MacRae and Croft 1996, Roda *et al.* 2000, Norton *et al.* 2001, Snyder and Ives 2001, Finke and Denno 2002), the relative body sizes of predators and other aspects of predator defensive behavior (Lucas *et al.* 1998, Rosenheim *et al.* 2004a), predator foraging behavior (Rosenheim and Corbett 2003, Rosenheim *et al.* 2004b), predator preferences (Venzon *et al.* 2001, Snyder and Ives 2003), and behaviorally-mediated effects of predators on their prey (Janssen *et al.* 1998, Magalhães *et al.* 2004). The current study has added to this growing list of potentially important influences on intraguild predator ecology: coincidental intraguild predation that occurs when predators consume parasitized herbivores may be more likely to enhance biological control than to disrupt it, at least in the short-term. The borderline level of statistical significance obtained in the meta-analysis suggests that a definitive conclusion on this point must await further growth of the empirical literature. In the meantime, however, this is another area that remains unexplored with models. Our understanding of intraguild predation is still very much in the process of unfolding.

5. ACKNOWLEDGEMENTS

We extend our deepest appreciation to our colleagues who provided data that were essential for the meta-analysis: Brad Cardinale, Ramy Colfer, Bob Denno, Debbie Finke, Chad Harvey, Rob McGregor, Martin Schmidt, and Bill Snyder. We also thank John Press and Terry Arbogast for making an effort to uncover old original data sets. Arne Janssen, Charles Godfray, and an anonymous reviewer made helpful suggestions on the manuscript. Finally, we thank Jacques Brodeur and Guy Boivin for inviting us to participate in the IOBC/Biocontrol Network symposium 'Trophic and guild interactions in biological control.'

6. REFERENCES

Arim, M, and Marquet, P. A., 2004, Intraguild predation: a widespread interaction related to species biology, Ecol. Lett. 7: 557-564.

Briggs, C. J., and Borer, E. T., 2005, Why short-term experiments may not allow long-term predictions about intraguild predation, Ecol. Appl. 15: 1111-1117.

Brodeur, J., and Rosenheim, J. A., 2000, Intraguild interactions in aphid parasitoids. Entomol. Exper. Appl. 97: 93-108.

Cardinale, B. J., Harvey, C. T., Gross, K., and Ives, A. R., 2003, Biodiversity and biocontrol: emergent impacts of a multi-enemy assemblage on pest suppression and crop yield in an agroecosystem, Ecol. Lett. 6: 857-865.

Cisneros, J. J., and Rosenheim, J. A., 1997, Ontogenetic change of prey preference in a generalist predator, *Zelus renardii*, and its influence on the intensity of predator-predator interactions, Ecol. Entomol. 22: 399-407.

Colfer, R. G., and Rosenheim, J. A., 2001, Predation on immature parasitoids and its influence of aphid population suppression, Oecologia 126: 292-304.
Colfer, R. G., Rosenheim, J. A., Godfrey, L. D., and Hsu, C. L., 2003, Interactions between the augmentatively released predaceous mite *Galendromus occidentalis* (Acari: Phytoseiidae) and naturally occurring generalist predators, Environ. Entomol. 32: 840-852.
Colfer, R. G., Rosenheim, J. A., Godfrey, L. D., and Hsu, C. L., 2004, Evaluation of large-scale releases of western predatory mite for spider mite control in cotton, Biol. Cont. 30: 1-10.
Coll, M., and Guershon, M., 2002, Omnivory in terrestrial arthropods: mixing plant and prey diets, Annu. Rev. Entomol. 47: 267-297.
Connolly, J., 1988, What is wrong with replacement series? Trends Ecol. Evol. 3: 24-26.
Denno, R. F., Mitter, M. S., Langellotto, G. A., Gratton, C., and Finke, D. L., 2004, Interactions between a hunting spider and a web-builder: consequences of intraguild predation and cannibalism for prey suppression, Ecol. Entomol. 29: 566-577.
Diehl, S., and Feissel, M., 2000, Effects of enrichment on three level food chains with omnivory, Am. Nat. 155: 200-218.
Ehler, L. E., 1995, Structure and impact of natural-enemy guilds in biological control of insect pests, *in*: Food Webs: Integration of Patterns and Dynamics, G. Polis and K. Winemiller, eds., Chapman and Hall, New York, pp. 337-342.
Ferguson, K. I., and Stiling, P., 1996, Non-additive effects of multiple natural enemies on aphid populations, Oecologia 108: 375-379.
Finke, D. L., and Denno, R. F., 2002, Intraguild predation diminished in complex-structured vegetation: implications for prey suppression, Ecology 83: 643-652.
Finke, D. L., and Denno, R. F., 2003, Intra-guild predation relaxes natural enemy impacts on herbivore populations, Ecol. Entomol. 28: 67-73.
Finke, D. L., and Denno, R. F., 2004, Predator diversity dampens trophic cascades, Nature 429: 407-410.
Hairston, N. G., Jr., and Hairston, N. G., Sr., 1993, Cause-effect relationships in energy flow, trophic structure and interspecific interactions, Am. Nat. 142: 379-411.
Hairston, N. G., Jr., and Hairston, N. G., Sr., 1997, Does food web complexity eliminate trophic-level dynamics? Am. Nat. 149: 1001-1007.
Hairston, N. G., Smith, F. E., and Slobodkin, L. B., 1960, Community structure, population control, and competition, Am. Nat. 94: 421-425.
Halaj, J., and Wise, D. H., 2001, Terrestrial trophic cascades: how much do they trickle? Am. Nat. 157: 262-281.
Harvey, C. T., and Eubanks, M. D., 2004, Effect of habitat complexity on biological control by the red imported fire ant (Hymenoptera: Formicidae) in collards, Biol. Cont. 29: 348-358.
Hastings, A., 2004, Transients: the key to long-term ecological understanding? Trends Ecol. Evol. 19: 39-45.
Holt, R. D., and Polis, G. A., 1997, A theoretical framework for intraguild predation, Am. Nat. 149: 745-764.
Hoy, J. G., Kauffman, E. E., and O'Berg, A. G., 1972, A large-scale field test of *Gambusia affinis* and chlorpyrifos for mosquito control, Mosquito News 32: 161-171.
Hurd, L. E., and Eisenberg, R. M., 1990, Arthropod community responses to manipulation of a bitrophic predator guild, Ecology 71: 2107-2114.
Ives, A. R., Cardinale, B. J., and Snyder, W. E., 2005, A synthesis of subdisciplines: predator-prey interactions, and biodiversity and ecosystem functioning, Ecol. Lett. 8: 102-116.
Janssen, A., Pallini, A., Venzon, M., and Sabelis, M. W., 1998, Behaviour and indirect interactions in food webs of plant-inhabiting arthropods, Exper. Appl. Acarol. 22: 497-521.
Jolliffe, P. A., 2000, The replacement series, J. Ecol. 88: 371-385.
Jones, R. E., 1987, Ants, parasitoids, and the cabbage butterfly *Pieris rapae*, J. Appl. Ecol. 56: 739-749.
Lang, A., 2003, Intraguild interference and biocontrol effects of generalist predators in a winter wheat field, Oecologia 134: 144-153.

Lucas, E., and Alomar, O., 2002, Impact of the presence of Dicyphus tamaninii Wagner (Heteroptera : Miridae) on whitefly (Homoptera : Aleyrodidae) predation by Macrolophus caliginosus (Wagner) (Heteroptera : Miridae), Biol. Cont. 25: 123-128.

Lucas, É., Coderre, D., and Brodeur, J., 1998, Intraguild predation among aphid predators: characterization and influence of extraguild prey density, Ecology 79: 1084-1092.

MacRae, I. V., and Croft, B. A., 1996, Differential impact of egg predation by *Zetzellia mali* (Acari: Stigmaeidae) on *Metaseiulus occidentalis* and *Typhlodromus pyri* (Acari: Phytoseiidae), Exp. Appl. Acarol. 20: 143-154.

Magalhães, S., Tudorache, C., Montserrat, M., van Maanen, R., Sabelis, M. W., and Janssen, A., 2004, Diet of intraguild predators affects antipredator behavior in intraguild prey, Behav. Ecol. 16: 364-370.

McCann, K., Hastings, A., and Huxel, G. R., 1998, Weak trophic interactions and the balance of nature, Nature 395: 794-798.

McGregor, R. R., and Gillespie, D. R., 2005, Intraguild predation by the generalist predator *Dicyphus hesperus* on the parasitoid *Encarsia formosa*, Biocont. Sci. Technol. 15: 219-227.

Mesquita, A. L. M., Lacey, L. A., and Leclant, F., 1997, Individual and combined effects of the fungus, *Paecilomyces fumosoroseus* and parasitoid, *Aphelinus asychis* Walker (Hym., Aphelinidae) on confined populations of Russian wheat aphid, *Diuraphis noxia* (Mordvilko) (Hom., Aphididae) under field conditions, J. Appl. Entomol. 121: 155-163.

Mylius, S. D., Klumpers, K., de Roos, A. M., and Persson, L., 2001, Impact of intraguild predation and stage structure on simple communities along a productivity gradient, Am. Nat. 158: 259-276.

Norton, A. P., English-Loeb, G., and Belden, E., 2001, Host plant manipulation of natural enemies: leaf domatia protect beneficial mites from insect predators. Oecologia 126: 535-542.

Nyrop, J., English-Loeb, G., and Roda, A, 1998, Conservation biological control of spider mites in perennial cropping systems, *in*: Perspectives on the Conservation of Natural Enemies of Pest Species, P. Barbosa, ed., Academic Press, San Diego, pp. 307-333.

Osenberg, C. W., Sarnelle, O., Cooper, S. D., and Holt, R. D., 1999, Resolving ecological questions through meta-analysis: goals, metrics, and models, Ecology 80: 1105-1117.

Polis, G. A, 1991, Complex trophic interactions in deserts: an empirical critique of food-web theory, Am. Nat. 138: 123-155.

Polis, G. A., and Holt, R. D., 1992, Intraguild predation: the dynamics of complex trophic interactions. Trends Ecol. Evol., 7: 151-155.

Polis, G. A., and Strong, D. R., 1996, Food web complexity and community dynamics. Am. Nat. 147: 813-846.

Polis, G., A., Myers, C. A., and Holt, R. D., 1989, The ecology and evolution of intraguild predation: potential competitors that eat each other, Annu. Rev. Ecol. Syst. 20: 297-330.

Prasad, R.P., and Snyder, W.E., Polyphagy complicates conservation biocontrol targeting generalist predators, J. Appl. Ecol. (in press).

Press, J. W., Flaherty, B. R., and Arbogast, R. T.,1974, Interactions among *Plodia interpunctella*, *Bracon hebetor*, and *Xylocoris flavipes*, Environ. Entomol. 3: 183-184.

Rees, N. E., and Onsager, J. A., 1982, Influence of predators on the efficiency of the *Blaesoxipha* spp. parasites of the migratory grasshopper, Environ. Entomol. 11: 426-428.

Revilla, T., 2002, Effects of intraguild predation on resource competition. J. Theor. Biol., 214: 49-62.

Roda, A., Nyrop, J., Dicke, M., and English-Loeb, G., 2000, Trichomes and spider-mite webbing protect predatory mite eggs from intraguild predation, Oecologia 125: 428-435.

Rosenberg, M. S., Adams, D. C., and Gurevitch, J., 2000, MetaWin: statistical software for meta-analysis with resampling tests, ver. 2.0, Sinauer, Sunderland, Mass..

Rosenheim, J. A., 1998, Higher-order predators and the regulation of insect herbivore populations, Annu. Rev. Entomol. 43: 421-447.

Rosenheim, J. A, 2001, Source-sink dynamics for a generalist insect predator in a habitat with strong higher-order predation, Ecol. Monogr. 71: 93-116.

Rosenheim, J. A., 2005, Intraguild predation on *Orius tristicolor* by *Geocoris* spp. and the paradox of irruptive spider mite dynamics in California cotton, Biol. Cont. 32: 172-179.

Rosenheim, J. A., and Corbett, A., 2003, Omnivory and the indeterminacy of predator function: can a knowledge of foraging behavior help? Ecology 84: 2538-2548.

Rosenheim, J. A., Glik, T. E., Goeriz, R. E., and Rämert, B., 2004b, Linking a predator's foraging behaviour with its effects on herbivore population suppression, Ecology 85: 3362-3372.

Rosenheim, J. A., Kaya, H. K., Ehler, L. E., Marois, J. J., and Jaffee, B. A., 1995, Intraguild predation among biological control agents: theory and evidence, Biol. Cont. 5: 303-335.

Rosenheim, J. A., Limburg, D. D., Colfer, R. G., Fournier, V., Hsu, C. L., Leonardo, T. E., and Nelson, E. H., 2004a, Herbivore population suppression by an intermediate predator, *Phytoseiulus macropilis*, is insensitive to the presence of an intraguild predator: an advantage of small body size? *Oecologia* 140: 577-585.

Rosenheim, J. A., Wilhoit, L. R., and Armer, C. A., 1993, Influence of intraguild predation among generalist insect predators on the suppression of an herbivore population, Oecologia 96: 439-449.

Ruberson, J. R., and Kring, J. T., 1991, Predation of *Trichogramma pretiosum* by the anthocorid *Orius insidiosus*. *in*: *Trichogramma* and Other Egg Parasitoids, Ed. INRA, Les Colloques no. 56, Paris, pp. 41-43.

Schmidt, M. H., Lauer, A., Purtauf, T., Thies, C., Schaefer, M., and Tscharntke, T., 2003, Relative importance of predators and parasitoids for cereal aphid control, Proc. Roy. Soc. Lond. B 270: 1905-1909.

Snyder, W. E., and Ives, A. R., 2001, Generalist predators disrupt biological control by a specialist parasitoid, Ecology 82: 705-716.

Snyder, W. E., and Ives, A. R., 2003, Interactions between specialist and generalist natural enemies: parasitoids, predators, and pea aphid biocontrol, Ecology 84: 91-107.

Snyder, W. E., and Wise, D. H., 2001, Contrasting trophic cascades generated by a community of generalist predators, Ecology 82: 1571-1583.

Snyder, W. E., Ballard, S. N., Yang, S., Clevenger, G. M., Miller, T. D., Ahn, J. J., Hatten, T. D., and Berryman, A. A., 2004, Complementary biocontrol of aphids by the ladybird beetle *Harmonia axyridis* and the parasitoid *Aphelinus asychis* on greenhouse roses, Biol. Cont. 30: 229-235.

Snyder, W.E., Chang, G.C., and Prasad, R.P., 2005, Biodiversity and successful conservation biological control: is there a relationship? *In:* Ecology of Predator-Prey Interactions, P. Barbosa and I. Castellanos, eds., Oxford University Press, Oxford, pp. 324-343.

Spiller, D. A., 1986, Interspecific competition between spiders and its relevance to biological control by general predators, Environ. Entomol. 15: 177-181.

Sunderland, K. D., Axelsen, J. A., Dromph, K., Freier, B., Hemptinne, J.-L., Holst, N. H., Mols, P. J. M., Petersen, M. K., Powell, W., Ruggle, P., Triltsch, H., and Winder, L., 1997, Pest control by a community of natural enemies, Acta Jutlandica 72: 271-326.

Venzon, M., Janssen, A., and Sabelis, M. W., 2001, Prey preference, intraguild predation and population dynamics of an arthropod food web on plants, Exper. Appl. Acarol. 25: 785-808.

Verdú, M., and Traveset, A., 2005, Early emergence enhances plant fitness: a phylogenetically controlled meta-analysis, Ecology 86: 1385-1394.

Wise, D. H., 1993, Spiders in Ecological Webs, Cambridge University Press, Cambridge.

INTRAGUILD PREDATION USUALLY DOES NOT DISRUPT BIOLOGICAL CONTROL

Arne Janssen[1,2], Marta Montserrat[1], Reinier HilleRisLambers[3], André M. de Roos[1], Angelo Pallini[2] and Maurice W. Sabelis[1]

[1]IBED, section Population Biology, University of Amsterdam, PO Box 94084, 1090 GB Amsterdam, the Netherlands; [2]Department of Animal Biology, Section Entomology, Federal University of Viçosa, Brazil; [3]Current address: CSIRO Entomology, Indooroopilly, Australia

Abstract: Intraguild predation is claimed to be ubiquitous in nature. It also occurs among natural enemies in biological control systems, where one natural enemy (the intraguild predator) attacks another species of natural enemy (the intraguild prey), whereas they also compete for the same pest. We review the theory of intraguild predation and its consequences for biological control for two different scenarios. 1. The intraguild predator is the superior natural enemy (*i.e.* reduces the pest population the most). In this case, the intraguild predator will exclude the intraguild prey, thus there will be no intraguild predation in the long term. 2. The intraguild prey is the superior natural enemy. In this case, the intraguild predator and intraguild prey may coexist or the intraguild predator can exclude the intraguild prey. Theory predicts for this scenario that pest numbers will always be lowest when only the intraguild prey is present. Hence, the occurrence of intraguild predation in cropping systems would never result in increased control, but can result in decreased control. We subsequently review experimental tests of the effect of intraguild predation among natural enemies on the population dynamics of pests. Contrary to expectations, we find that intraguild predation often did not result in an increase of pest populations, even when the intraguild predator was the inferior natural enemy. Often, the presence of the intraguild predator had no effect or even resulted in a decrease of pest populations. Although the number of studies was limited, we scanned the literature to identify possible causes for the discrepancy of experimental results with theoretical predictions. We specifically evaluated trends in the effects with respect to the length of the study period, the spatial scale at which experiments were carried out, the number of species involved in the studies and the spatial complexity of the experimental arenas. There was a slight trend towards experiments of longer duration showing less positive effects on pest densities, but no clear effect of spatial scale. All studies that showed positive effects on pest densities were studies with 3 species, but the number of studies with more than 3 species was small. Spatial complexity had mixed effects on experimental results. In conclusion, it is clear that intraguild predation most often does not increase pest densities as was predicted from theory, but more research is needed to reveal why theory does not meet practice.

1. INTRODUCTION

It has been sixteen years since the influential paper of Polis *et al.* (1989) on intraguild predation was published and a decade since the appearance of the paper by Rosenheim *et al.* (1995) on intraguild predation among biological control agents. Although earlier papers did address interactions among multiparasitizing parasitoids that are now often classified as intraguild predators (May and Hassell, 1981; Kakehashi *et al.*, 1984), it was only after the publication of Rosenheim *et al.*'s

J. Brodeur and G. Boivin (eds.), Trophic and Guild Interactions in Biological Control, 21–44.

(1995) paper that many biological control researchers concentrated on the possibility of adverse interactions among natural enemies and effects of intraguild predation on biological control. In the meantime, there have also been theoretical developments on the effects of intraguild predation on population dynamics and persistence of species. It is therefore timely to take stock. We will do so by addressing theoretical developments first. Because any theory is as strong as its support, we then review experimental tests of the theory. Subsequently, we address the guidelines for biological control that transpire from theory, and review experiments on intraguild predation in biological control systems. We subsequently discuss causes for the differences between theory and biological control.

2. THE THEORY OF INTRAGUILD PREDATION

Intraguild predation, the killing and eating of species that otherwise use similar resources (Polis *et al.*, 1989), was initially viewed as a rare interaction with little effect on food web dynamics (Pimm and Lawton, 1978). This view was partly caused by the theoretical prediction that omnivory and intraguild predation destabilize food web dynamics and could therefore not be important, and this led to a paucity of studies of this interaction. Since it has become clear that intraguild predation is by no means rare in food webs (Polis *et al.*, 1989; Polis and Holt, 1992; Rosenheim *et al.*, 1995; Polis and Winemiller, 1996), it has received considerable theoretical attention (Polis and Holt, 1992; Holt and Polis, 1997; Morin, 1999; Diehl and Feissel, 2000, 2001; Mylius *et al.*, 2001; Kuijper *et al.*, 2003). Most of these papers model populations of three species: an intraguild predator, an intraguild prey and a shared resource or prey that is attacked by intraguild predator and intraguild prey. They investigate the equilibrium densities of the three populations as a function of the productivity of the environment and show the following results. At low productivity levels, neither the intraguild prey nor the intraguild predator can persist with the resource because resource levels are too low (Fig. 1, region A). At higher productivity, the species that is the most efficient competitor for the shared resource can persist with the resource (Fig. 1B). If this species is the intraguild predator, there is no opportunity for the intraguild prey to persist because it will suffer both from competition and from intraguild predation. A necessary condition for coexistence of intraguild predators and intraguild prey is therefore that the intraguild prey is the superior competitor for the shared resource (Polis and Holt, 1992; Holt and Polis, 1997). If this condition is met, further increases in productivity give rise to several regions of coexistence or exclusion. The first of these regions is one in which all 3 species coexist; there is sufficient resource and intraguild prey to sustain a population of intraguild predators (Fig. 1C).

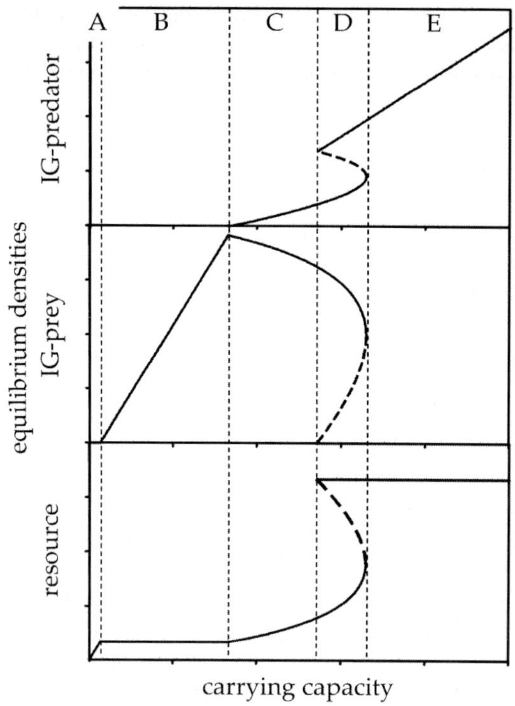

Figure 1. The equilibria of an intraguild predator (P), intraguild prey (C) and shared prey (R) as a function of productivity of the habitat (expressed as the carrying capacity of the shared prey). At very low productivity (region A), only the resource (R) can persist, at higher productivity the intraguild prey (C) coexists with the resource (region B). Then follows a region of productivity (C) in which all three species coexist. At yet higher productivity (region D), there are two alternative stable states, one with all three species and one with the resource and the intraguild predator (P). The equilibrium that will be reached depends on the initial conditions; starting above the dashed curve, the highest of the two equilibria will be reached, when starting below the dashed curve, the system will settle in the lowest equilibrium. At high productivity (region E), only the intraguild predator and the resource coexist. Redrawn after Mylius et al. *(2001).*

Higher productivity leads to a region with 2 stable equilibria in some (Fig. 1D), but not all models (Polis and Holt, 1992; Holt and Polis, 1997; Mylius *et al.*, 2001; Kuijper *et al.*, 2003). The first equilibrium is a 3-species equilibrium as was seen for lower productivity (as in Fig. 1C), whereas there is a second equilibrium in which the intraguild prey is excluded and the resource and intraguild predator persist. In this case, there is sufficient resource to sustain a population of intraguild predators even in absence of intraguild prey and the intraguild prey is excluded by a combination of competition and intraguild predation (Polis *et al.*, 1989). The system can end up in either of the two equilibria, depending on initial conditions or perturbation away from one of the equilibria. For example, when the intraguild prey and the resource are at equilibrium, a small population of intraguild predators will be able to invade, resulting in a system with all 3 species present. When the intraguild predator is at equilibrium with the resource, a small population of intraguild prey would suffer from intraguild predation and competition and would go extinct. Lastly, at highest productivity levels, all models predict exclusion of the intermediate prey through a combination of intraguild predation and competition, independent of initial conditions (Fig. 1E).

Based on theory, there is general agreement that the occurrence of intraguild predation in natural systems can only be understood when intraguild predation is a relatively weak interaction and productivity is not too high (May, 1973; McCann *et al.*, 1998; Emmerson and Yearsley, 2004), or when the effects of strong intraguild predation are overruled by other factors.

Theoretical models are necessarily simplifications of reality and it is well possible that factors that decrease the impact of intraguild predation on persistence of food webs have been overlooked. Holt and Polis (1997) suggested several such factors: age structure with invulnerable age classes; switching predators; antipredator behaviour of the prey; spatial heterogeneity; and food webs that are more complex. Some of these factors have been included in models, which we will now discuss.

Mylius *et al.* (2001) included *stage structure*, with either a stage of the intraguild prey being invulnerable to intraguild predation, or a stage of the intraguild predator being incapable of intraguild predation. This resulted in similar predictions as those of models without stage structure; elimination of the intermediate species by the top species is more likely than coexistence (Mylius *et al.* 2001). Possibly, the presence of stage structure in both intraguild prey and intraguild predator would lead to different predictions.

The effect of *switching intraguild predators* on persistence of a food web with intraguild predation was studied by Krivan (2000). He assumed that intraguild predators switched from feeding exclusively on one prey to feeding exclusively on the other prey, depending on prey abundance. This was found to increase persistence, especially at high productivity levels. This kind of switching is expected especially when predators cannot search for both prey at the same time, for example because both prey types occur in separate patches (Krivan and Diehl, 2005). Intraguild prey, however, are often found in patches with the shared prey, hence, intraguild predators encounter both prey in the same patch and such switching is not likely to be important. Another type of diet switching of predators was studied by Krivan and Diehl (2005). Contrary to the above study, intraguild predators were now assumed always to include the most profitable prey into their diet. The most profitable prey could be either the intraguild prey or the shared resource, and the less profitable prey was included only when the encounter rate with the profitable prey was sufficiently low. Such switching increased the parameter space for coexistence of intraguild prey and predator (Krivan and Diehl, 2005).

Heithaus (2001) studied the effect of *antipredator behaviour* of the intraguild prey in a model of habitat selection by intraguild predators and intraguild prey, whereas the distribution of the shared resource over the two habitats was fixed. He found that habitat segregation of intraguild prey and intraguild predators could occur when productivity was high. Although such habitat segregation does not lead to coexistence of intraguild predators and intraguild prey in the same habitat, coexistence at a larger scale, including several habitats, is possible.

Slightly more *complex food webs* were studied by HilleRisLambers and de Roos, in prep., who added reciprocal intraguild predation to a three-species food web. Reciprocal intraguild predation is quite common in nature (Polis and Holt, 1992; Janssen *et al.*, 2002; Janssen *et al.*, 2003). In this case, the terms intraguild prey and intraguild predator become somewhat confusing. We refer to the species that is the superior competitor but a relatively weak intraguild predator as the intraguild prey, in parallel to the terminology of models without reciprocal intraguild predation, in which the intraguild prey is assumed the superior competitor. At low productivities, model predictions do not differ from models with unilateral intraguild predation; the intraguild prey coexists with the resource only (*cf.* Fig. 1B). The parameter space for coexistence of all three species, however, becomes smaller with increasing reciprocal intraguild predation. At higher productivity, it is no longer the intraguild predator that excludes the intraguild prey (as in Fig. 1E), but either species can exclude the other, depending on initial conditions (priority effect). HilleRisLambers and Dieckmann (2003) studied the effects of a trade-off between the intraguild predator being a better competitor (hence, preying efficiently on the shared prey) and being a better predator (preying more efficiently on the intraguild prey). They find that coexistence is most likely with intermediately strong trade-offs between feeding on the resource and feeding on the competitor.

In summary, most of the extensions of the basic intraguild model suggested by Holt and Polis, 1997 have been studied to some extent. None of these resulted in increased persistence in food webs with intraguild predation, except perhaps for antipredator behaviour. The conclusion so far is that the omnipresence of intraguild predation in natural food webs can only be understood when it is a relatively weak interaction. Hence, if theoretical predictions are correct, strong intraguild predation will rarely occur in natural systems because it will result in exclusion of the intraguild prey. The occurrence of intraguild predation will then be limited to the period prior to exclusion.

Yet another factor might explain the common occurrence of intraguild predation in nature, which is the type of dynamics. Most models consider equilibria, whereas coexistence of species does not necessarily require such stable dynamics (Law and Blackford, 1992). Complex dynamics have been shown to create opportunity for coexistence for systems of competitors where stable dynamics would lead to exclusion of one of the species (Armstrong and McGehee, 1980; Huisman and Weissing, 1999). Perhaps similar mechanisms operate in systems with intraguild predation. This is confirmed by the study of McCann *et al.* (1998), who showed that adding intraguild predation to a system with complex dynamics resulted in well-bounded limit cycles with higher minimum numbers than in a system without intraguild predation. They conclude that intraguild predation resulted in increased coexistence of intraguild prey and intraguild predators. The difference of the study of McCann *et al.* (1998) with other studies was the presence of a top predator that attacked the intraguild predator and the intraguild prey. As such, this is another study on *more complex food webs* (see above). Hence, the increased persistence of the intraguild predator and the intraguild prey could have been caused by complex dynamics, by the shared predator, or by a combination of these two.

3. EXPERIMENTAL TESTS OF THEORY

A critical test of the model predictions outlined above would be to verify the existence of the various equilibria at different levels of productivity. If the intraguild predator is the superior competitor for the shared prey, the only equilibrium is that of the intraguild predator with the shared prey. This equilibrium resembles that of two competing species without intraguild predation, of which one is excluded through competition for a resource. In this case, the importance of intraguild predation for exclusion needs to be verified.

If the intraguild predator is the inferior competitor, there are several alternative equilibria (Fig. 1). Most of these equilibria, however, are similar to those of simple tritrophic systems without intraguild predation: at low productivity only the resource persists, resource and consumer persist at somewhat higher productivity, and resource, consumer and predator coexist at intermediate productivity (Fig. 1 A-C). In order to discriminate between dynamics of a tritrophic system and a system with intraguild predation, it is therefore necessary to show that the intraguild predator excludes the intraguild prey at high productivity levels, whereas the intraguild prey excludes the intraguild predator at low productivity levels. Alternatively, it could be demonstrated that there are levels of productivity where there is either coexistence of all three species or of the intraguild predator with the resource (Fig. 1D), but experimental demonstration of such alternative stable states is fraught with difficulties (Schröder *et al.*, 2005).

Although several studies tested various predictions of the theory of intraguild predation (Holyoak and Sachdev, 1998; Morin, 1999; Diehl and Feissel, 2000, 2001; Borer *et al.*, 2003; Price and Morin, 2004), only few studies tested whether intraguild predators can exclude intraguild prey at high productivity levels. Diehl and Feissel (2000) found that their intraguild predator excluded the intraguild prey at high productivity levels, but the prey excluded the intraguild predator again at even higher productivity. Lawler and Morin (1993) found exclusion of the intraguild prey by the intraguild predator in three out of five replicates, but did not vary productivity levels. A problem with many of these experimental studies is that the shared resource consisted of several species (usually bacteria), and intraguild prey and intraguild predators may have fed on different species or the species composition of the resource may have changed with productivity (Diehl and Feissel, 2000).

A further test of the theory of intraguild predation comes from a system of two predatory mites that both feed on pollen and prey, but are involved in reciprocal intraguild predation. As outlined above, such mutual intraguild predation leads to different predictions (HilleRisLambers and de Roos, in prep.); at low productivity, only the best competitor of the two intraguild predators can persist with the resource, whereas at higher productivity levels either of the two species can drive the other species to extinction, depending on initial conditions. This was exactly what happened in the experimental system, even with low levels of predation of the intraguild prey on the intraguild predator (Montserrat *et al.* in prep). This shows that reciprocal intraguild predation further limits possibilities for local coexistence in systems with intraguild predation.

In conclusion, there are only few critical tests of current theory of intraguild predation and most experimental systems consisted of more species than theory has taken into account so far. Although most natural food webs are also clearly more complex than the model food webs, the first step in further development of the theory of intraguild predation should be to test the theory as it stands. It is therefore essential that critical experiments be done with simple experimental systems.

4. THE THEORY OF INTRAGUILD PREDATION AND BIOLOGICAL CONTROL

Biological control systems are more complex than the 3-species systems considered in the theory of intraguild predation because there is at least a fourth species involved: the host plant. If we assume that the crop does not affect the dynamics of the intraguild system, several theoretical predictions can be used for biological control systems. First, theory predicts that the intraguild prey will always be excluded if the intraguild predator is the superior competitor. The exclusion of the intraguild prey implies that the intraguild predator reduces the pest densities to lower levels than the intraguild prey. Hence, in this case, there is no negative effect of intraguild predation on biological control, but there is also no advantage whatsoever in using the intraguild prey. Second, theory predicts that intraguild predators and intraguild prey can only coexist at intermediate levels of productivity when intraguild predators are the inferior competitor (Fig. 1C-D). At these levels, the intraguild prey alone would suppress pest densities to lower levels than both predators together (Fig. 2C, D); hence, intraguild predation has positive effects on pest densities. Finally, at high levels of productivity, the intraguild predator excludes the intraguild prey and pest densities are higher than they would be with the intraguild prey and the pest (Fig. 2E).

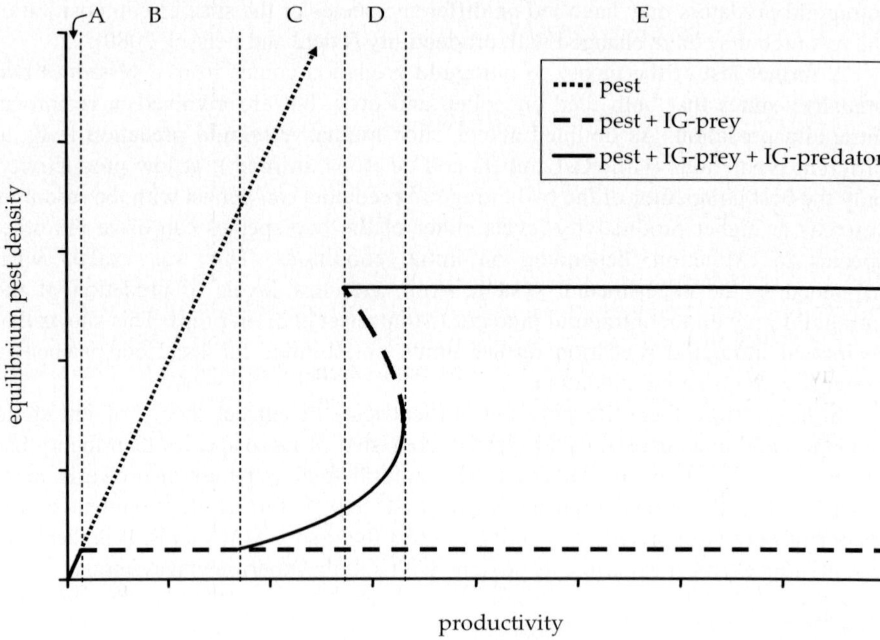

Figure 2. The equilibria of a pest species as a function of productivity (expressed as the carrying capacity of the pest). Shown are equilibrium densities of the pest without predators (dotted line), of the pest with the intraguild prey (broken line) and of the pest with the intraguild prey and the intraguild predator (black curve, see also Fig. 1). It is assumed that the intraguild predator is the inferior competitor for the pest. At very low productivity (region A), the equilibrium levels of all three systems are the same because the predators cannot persist and all systems consist of the pest only. At somewhat higher productivity (B), the intraguild prey (a specialist predator) controls the numbers of the pest (cf. broken line and the dotted line) and the intraguild predator is excluded by the intraguild prey through competition for the pest. At intermediate productivity levels (C), the intraguild predator, the intraguild prey and the pest can persist. At somewhat higher productivity levels (D) there is an area of bistability, in which either the three species persist or the intraguild prey is excluded. Finally, the intraguild prey is always excluded at the highest productivity levels (E). Note that a system with the intraguild predator always results in higher equilibrium pest densities than a system consisting of the pest and the intraguild prey alone, hence, intraguild predation results in increased pest densities, thus reduced control of the pest. The areas A – E correspond to the areas in Fig. 1.

Based on these predictions, the advice for use of intraguild predators for biological control would be a sound "no", unless the intraguild predator is the superior competitor for the pest, in which case the use of the intraguild prey would be redundant. Hence, there would be no advantage in using pairs of natural enemy species that are engaged in intraguild predation.

In practice, intraguild predation does occur frequently in biological control systems. Does this then result in reduced control? We reviewed the current literature on this topic and found 25 studies of food webs of plant-inhabiting arthropods in which the effects of intraguild predation on the intraguild prey as well as on the shared pest species were evaluated (see Table 1 for a summary of the data, see Rosenheim and Harmon, this volume, for a more detailed review). Several studies contained separate experiments on various species of intraguild predators or on various species of intraguild prey. These data were considered independent, resulting in 35 cases (Table 1). To assess the effect of intraguild predation, we used comparisons of the densities or numbers of intraguild prey and pests in presence and in absence of the intraguild predators. When an effect was significant, we scored it as either positive or negative, depending on the sign of the effect, and when an effect was non-significant, it was scored as neutral. Several studies showed variation in the effect (for example with time) and in these cases, we scored just one effect, choosing for the effect that was in agreement with theory. For example, theory predicts negative effects of intraguild predation on populations of the intraguild prey. When the effect on the intraguild prey varied from negative to neutral in a study, we therefore scored this study as showing a negative effect. When effects on the pest varied from positive to neutral, we scored it as positive, which is in agreement with theoretical predictions. When effects on the pest varied from negative to neutral, it was scored as neutral. Hence, the data were skewed to be in agreement with the prediction that intraguild predation has a negative effect on biological control, and our analysis of published data was therefore biased towards finding support for the theory of intraguild predation.

The effect of the presence of intraguild predators on the intraguild prey was often negative, but sometimes no significant effect was detected (Table 1, Fig. 3). Effects of intraguild predation on the dynamics of the pest were much more variable; only 6 out of 35 cases showed a positive effect on the pest, hence, decreased biological control (Table 1, Fig. 3). Of these 6 cases, most did not show a consistent positive effect on the pest (Table 1). Thus, the results of the evaluated studies are often not in agreement with theoretical predictions, despite our bias for finding support.

Table 1. Summary of the studies on the effects of intraguild predation on biological control in plant-inhabiting arthropod food webs.

study	predators	IG prey	target pest	effect on IG–prey[1]	effect on pest[1]	time[2]	set–up[3]	# species
Snyder & Wise, 1999	lycosid spiders and carabids	nonlycosid spiders, nabids, *Geocoris*, scorpion flies	various pests on various crops	–	0/–[4,5]	5	5,4	7 – 10
Snyder & Ives, 2001	*Pterostichus melanarius* (carabid)	*Aphidius ervi* (parasitoid)	*Acyrthosiphon pisum* (aphid) on alfalfa	–/0[5]	+/0[6]	3	5,4	3
Erbilgin *et al.*, 2004	*Anthocoris nemoralis* (predatory bug)	*Psyllaephagus* bliteus (parasitoid)	*Glycapsis brimblecombei* (psyllid) on eucalyptus	–	0	3	3,1	3
Chang, 1996	*Chrysoperla plorabunda* (lacewing), *Coccinella septempunctata*	*Chrysoperla plorabunda* (lacewing), *Coccinella septempunctata*	*Aphis fabae* (aphid) on bean	0	0	1	3,1	3
Rosenheim *et al.*, 1993	*Geocoris* spp. (predatory bug)	*Chrysoperla carnea* (lacewing)	*Aphis gossypii* (aphid) on cotton	–	0	1	2,4	3
Rosenheim *et al.*, 1993	*Nabis* spp., (predatory bug)	*Chrysoperla carnea* (lacewing)	*Aphis gossypii* (aphid) on cotton	–	+/0	1	2,4	3
Rosenheim *et al.*, 1993	*Zelus renardii* (predatory bug)	*Chrysoperla carnea* (lacewing)	*Aphis gossypii* (aphid) on cotton	–	+/0	1	2,4	3

Table 1. continued.

Dinter, 2002	*Erigone atra* (spider)	*Chrysoperla carnea* (lacewing)	*Sitobion avenae* (aphid) on wheat	–	–	2	3,1	3
Dinter, 2002	*Oedothorax apicatus* (spider)	*Chrysoperla carnea* (lacewing)	*Sitobion avenae* (aphid) on wheat	–	–	2	3,1	3
Snyder *et al.*, 2004	*Harmonia axyridis* (coccinellids)	*Aphelinus asychis* (parasitoid)	*Macrosyphum euphorbiae* (aphid) on rose	0	–[4]	4	4.2	3
Denno *et al.*, 2004	*Pardosa littoralis* (spider)	*Grammonota trivittata* (web spider)	*Prokelisia* (planthopper) on Spartina	–	–[4]	5	5,5	>3
Schausberger & Walzer, 2001	*Neoseiulus californicus* (predatory mite)	*Phytoseiulus persimilis* (predatory mite)	*Tetranychus cinnabarinus* (spider mite) on gerbera	–	+	4	4,3	3
Heinz & Nelson, 1996	*Delphastus pusillus* (coccinellid)	*Encarsia formosa*, (parasitoid)	*Bemisia tabaci* (whitefly) on poinsettia	– /0	–[7]	4	5,2	3
Heinz & Nelson, 1996	*D. pusillus* (coccinellid)	*E. pergandiella* (parasitoid)	*Bemisia tabaci* (whitefly) on poinsettia	– /0	–[7]	4	5,2	3
Heinz & Nelson, 1996	*E. pergandiella* (parasitoid)	*Encarsia formosa*, (parasitoid)	*Bemisia tabaci* (whitefly) on poinsettia	–	–	4	5,2	3
Eubanks, 2001	*Solenopsis invicta* (fire ants)	various	various pests on cotton and soybean	–	–[4]	5	5,5	40

Table 1. continued.

study	predators	IG prey	target pest	effect on IG–prey[1]	effect on pest[1]	time[2]	set–up[3]	# species
Snyder & Ives, 2003	various predators	*Aphidius ervi* (parasitoid)	*Acyrthosiphon pisum* (aphid) on alfalfa	0	–	3	5,5	16
Ferguson & Stiling, 1996	*Cycloneda sanguinea* (coccinellids)	*Aphidius floridaensis* (parasitoid)	*Dactynotus* sp. (aphid) on marsh elder	–	0	3	2,4	3
Colfer & Rosenheim, 2001	*Hippodamia convergens* (coccinellid)	*Lysiphlebus testaceipes* (parasitoid)	*Aphis gossypii* (aphid) on cotton	–	–[4]	4	4,5	3
Colfer *et al.*, 2003	*Geocoris* spp. (predatory bug)	*Galendromus occidentalis* (predatory mite)	*Tetranychus urticae* (spider mite) on cotton	–	0[4]	2	1,4	3
Colfer *et al.*, 2003	*Orius tristicolor* (predatory bug)	*Galendromus occidentalis* (predatory mite)	*Tetranychus urticae* (spider mite) on cotton	–	–[4]	2	1,4	0
Sher *et al.*, 2000	*Steinernema* (nematode)	*Diglyphus begini* (parasitoid)	*Liriomyza trifolii* (leafminer) on chrysanthemum	0	–	2	1,1	0
Rosenheim, 2001	*Nabis* spp. (predatory bug)	*Chrysoperla carnea* (lacewing)	*Aphis gossypii* (aphid) on cotton	–/0	0	2	2,4	3
Rosenheim, 2001	*Orius tristicolor* (predatory bug)	*Chrysoperla carnea* (lacewing)	*Aphis gossypii* (aphid) on cotton	0	0	2	2,4	3

Table 1. continued.

Rosenheim, 2001	*Geocoris* spp. (predatory bug)	*Chrysoperla carnea* (lacewing)	*Aphis gossypii* (aphid) on cotton	0	0	2	2,4	3
Rosenheim, 2001	*Misumenops* sp. (spider)	*Chrysoperla carnea* (lacewing)	*Aphis gossypii* (aphid) on cotton	0	0	2	2,4	3
Rosenheim, 2001	*Zelus renardii* (predatory bug)	*Chrysoperla carnea* (lacewing)	*Aphis gossypii* (aphid) on cotton	–	+	0	2,4	3
Rosenheim *et al.*, 2004a	*Nesticodes rufipes* (spider)	*Stethorus siphonulus* (ladybeetle)	*Tetranychus cinnabarinus* (spider mite)	n.r.[8]	+	2	1,5	4
Rosenheim *et al.*, 2004b	*Nesticodes rufipes* (spider)	*Phytoseiulus macropilis* (predatory mite)	*Tetranychus cinnabarinus* (spider mite)	n.r.[8]	0	2	1,5	4
Croft & McRae, 1992b	*Zetzellia mali* (predatory mite)	*Metaseiulus occidentalis*, *Typhlodromus pyri* (predatory mites)	various spider mites on apple	–/0	0/–	5	5,5	6
Croft & McRae, 1992a	*Typhlodromus pyri* (predatory mite)	*M. occidentalis* (predatory mite)	various spider mites on apple	–/0	0/–	5	5,5	5
Parrella *et al.*, 1980	*Orius insidiosus* (predatory bug)	*Leptothrips mali* (predatory thrips)	*Panonychus ulmi* (spider mite) on apple	–	0	1	1,1	3

Table 1. Continued.

study	predators	IG prey	target pest	effect on IG–prey[1]	effect on pest[1]	time[2]	set–up[3]	# species
Lucas & Alomar, 2002	*Dicyphus tamaninii* (predatory bug)	*Macrolophus caliginosus* (predatory bug)	*Trialeurodes vaporariorum* (whitefly)	0	0[7]	1	3,1	3
Mallampalli *et al.*, 2002	*Podisus maculiventris* (predatory bug)	*Coleomegilla maculata* (ladybeetle)	*Leptinotarsa decemlineata* (Colorado potato beetle) on potato	–	0/–[9]	1	3,1	3
Venzon *et al.*, 2001	*Orius laevigatus* (predatory bug)	*Phytoseiulus persimilis* (predatory mite)	*Tetranychus urticae* (spider mite) on cucumber	0	0	4	4,2	3

[1] The results of the experiments that showed variation in the effect were analyzed as follows. Neutral to negative effects (indicated here as 0/-) on the intraguild prey were always scored as negative. Neutral to negative effects on the pest (-/0) were scored as neutral and neutral to positive effects (+/0) were scored as positive. A positive effect on the pest indicates decreased control in presence of intraguild predation.
[2] Time is the duration of the experiment, classified as follows: 1 = < 1 week; 2 = 1 – 2 weeks; 3 = 2 – 4 weeks; 4 = 1 – 2 months; 5 = > 2 months.
[3] Set-up is coded as follows: The first number indicates the spatial complexity of the set-up: 1 = an arena or leaf; 2 = a branch; 3 = a plant; 4 = 2-6 plants; 5 = > 6 plants. The second number gives the spatial scale of the set-up: 1 = laboratory experiment; 2 = cages in a greenhouse; 3= greenhouse; 4 = a closed system in the field; 5 = an open field system.
[4] Control with pest and IG-predator only was not included, so predator may be better competitor.
[5] No effect was found in spring, a negative effect in summer, no effect on plant production.
[6] At high aphid densities on tall plants.
[7] Control by the IG-predator alone was better than that by the IG-prey alone.
[8] Not reported
[9] Depending on the life stage of IG-predator and IG-prey, control by the IG-predator alone was better than by IG-prey alone.

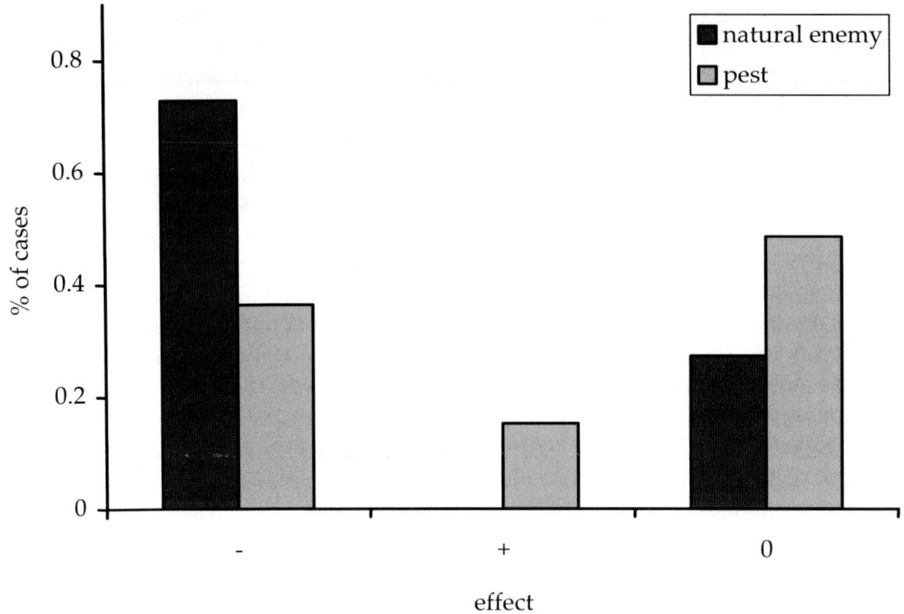

Figure 3. The percentages of cases (35 in total) from the literature (Table 1) that reported a negative, positive or neutral effect of intraguild predation on other natural enemies (i.e. the intraguild prey, dark bars) and pests (light bars). A positive effect on the pest indicates a possible negative effect on biological control. For the natural enemies, effects that were sometimes neutral and sometimes negative were classified as negative. For the shared pest, effects that were sometimes neutral and sometimes negative were classified as neutral; effects that were sometimes neutral and sometimes positive were classified as positive. This skews the data towards positive effects of intraguild predation on pest densities, which is the effect predicted by theory.

Theory predicts a positive effect on the pest when the intraguild predator is the inferior competitor, but a negative effect when the intraguild predator is the superior competitor. There were 4 cases in which the intraguild predator alone reduced the pest to lower densities than the intraguild prey alone, suggesting that the intraguild predator was the superior competitor. Two of these 4 cases showed a negative effect of intraguild predation on pest densities (Table 1). There were also 7 cases without information on the reduction of pest populations with the intraguild predator alone, so the intraguild predator could have been the superior competitor (Table 1). We therefore analysed cases in which the intraguild prey was shown to be the better competitor for the pest, hence, where a positive effect of intraguild predation on the pest was expected. Of the 24 remaining studies, 6 reported a positive effect of intraguild predation on the pest, 13 found a neutral effect and another 5 found a negative effect. Hence, almost as many studies showed a negative effect on pest densities as studies that showed the predicted positive effect, whereas most studies showed no significant effect.

In theory, a cascade of the effect of the intraguild predator down to the pest is expected, so a negative effect of the intraguild predator on the intraguild prey would result in a positive effect of the pest. In practice, some of these effects may not have been significant owing to lack of statistical power of the experiment, resulting in studies where a negative effect on the intraguild prey was found, but no effect on the pest. A result that is clearly in contradiction with theory is when intraguild predation has a negative effect on the intraguild prey as well as the pest, which occurred in 9 of the 33 cases investigated here (Table 1). A possible reason for this is discussed in the next section.

In conclusion, current biological control literature suggests that the effect of intraguild predation on pest densities is often not positive, thus effects on biological control are usually not negative (see also Denno and Finke, this volume).Where does the discrepancy between theory and experiments come from? There are several differences between model assumptions and the reality of experiments: they concern differences in temporal and spatial scale and differences in food web complexity. We attempted to analyse each of these differences separately, even though the number of studies is rather limited.

5. TIME SCALES

Many cropping systems last for short periods, after which the production cycle is started anew. Theory, so far, has addressed equilibrium dynamics and the time to reach such equilibria likely exceeds the production cycle of many crops and certainly exceeds the duration of most of the experiments reviewed here. There is therefore a need to study short-term, transient dynamics in models of systems with intraguild predation. Such dynamics may be distinctly different from long-term dynamics (Briggs and Borer, 2005). For example, if intraguild predators would eventually drive intraguild prey to extinction, the initial presence of intraguild prey might increase the intraguild predator population to above its ultimate density, resulting in a short-term negative effect on pest densities (through apparent competition). Thus, the long-term effect of intraguild predation on the pest can be positive, whereas the short-term effect is negative. We therefore expected that experiments of relatively long duration would show positive effects on pest densities more often than experiments of short duration (Briggs and Borer, 2005).

To evaluate the effects of duration of the experiments on the dynamics, it would be best to scale the experimental period to the generation time of the species involved. This information was not available in most publications, and we therefore classified the experiments according to their length. Experiments were usually short, sometimes much shorter than the growing season of the crop (Table 1). Contrary to what was expected, none of the few long-lasting experiments (3 studies, > 2 months, Table 1) and only 1 out of 9 experiments that lasted for 1 – 2 months showed a positive effect on the pest. There was no clear trend of neutral and negative effects with time.

There is a problem associated with evaluating short-term transient dynamics instead of equilibrium dynamics. Before a system eventually settles at equilibrium, it may go through a transient phase with damped cycles. The period of these cycles is affected by intraguild predation, so systems without intraguild predators cycle out of phase with systems with intraguild predation (Fig. 4). When the amplitudes of the cycles of systems with and without intraguild predation overlap, the pest may temporarily reach higher numbers in systems without intraguild predation than in systems with intraguild predation. This may even occur when the final equilibrium pest densities would show the opposite pattern (Fig. 4). Hence, even when the system would behave exactly as theory predicts, it is possible to find other outcomes in the short term. The effects of such out-of-phase cycling can only be detected when densities are monitored throughout the experimental period. Unfortunately, many studies only give densities at the end of the experimental period. Possibly, some of the studies that reported neutral or negative effects did so only because of such out-of-phase cycles. Clearly, the importance of transient cycles for biological control needs to be evaluated in more detail.

6. SPATIAL SCALES

Another difference between theory and biological control concerns spatial scales. Whereas theory considers well-mixed populations of all species, real populations are often patchy and may actually consist of several subpopulations connected by dispersal. Such dispersal may result in coexistence of predators and prey even when local populations cannot persist (Janssen *et al.*, 1997; Ellner *et al.*, 2001). It is not clear what will be the effect of intraguild predation on pest population levels under such a metapopulation structure. When we assume that the intraguild predator is the inferior competitor for the pest, it is clear that local pest densities will be increased in patches with the intraguild predator compared to patches with the intraguild prey. However, pest densities will be decreased in patches with the intraguild predator compared to patches with the pest only. The overall outcome of these two opposing effects will depend on the dispersal capacity of the pest, the intraguild prey and the intraguild predator.

Populations in small experimental arenas are probably well mixed, in agreement with theoretical assumptions, whereas populations in large arenas or open systems may be viscous. We therefore expected that experiments in small arenas or enclosures would show positive effects of intraguild predation on pest densities more often than experiments in large arenas. Most experiments used enclosures, some of these quite small (Table 1). Contrary to our expectation, the percentage of studies showing any significant effect on the pest (either positive or negative) was slightly higher for open systems (*i.e.* no enclosures) than for closed systems (62.5 % vs 48.1 %). Only 1 of the 13 studies that were done in the laboratory or in the greenhouse, which usually involve less space than field experiments, reported a positive effect of intraguild predation on pest densities. In conclusion, there is no clear trend of space influencing the effect of intraguild predation on pest populations.

7. FOOD WEB COMPLEXITY

Seven of the studies reviewed had more that the three species that are common in models of intraguild predation. The presence of other species can result in other direct and indirect interactions besides intraguild predation, and there is little theory on joint effects of such interactions (McCann et. al, 1998; Briggs and Borer, 2005, see Rosenheim and Harmon, this volume). If anything, we would expect such other interactions to obscure effects of intraguild predation. Five of the 6 cases with a positive effect of intraguild predation on the pest concern studies with 3 species. Thus, 19.2 % of the studies with 3 species showed a positive effect on the pest and 30.8 % of the studies with 3 species showed a negative effect on pest densities, whereas 11 % (1 out of 9) studies with more than 3 species showed a positive effect on the pest and 44 % showed a negative effect (Table 1). This suggests that positive effects of intraguild predation on pest densities perhaps decrease with increasing complexity of the food web, but this clearly needs more theoretical and experimental research.

8. BEHAVIOUR AND INTRAGUILD PREDATION

Besides differences in temporal and spatial scales and differences in food web complexity, the antipredator behaviour of intraguild prey and shared prey may moderate effects of intraguild predation (Heithaus, 2001). It is common knowledge that many prey exhibit antipredator behaviour in response to the presence of predators or of cues associated with predators (Sih, 1980; Lima and Dill, 1990; Lima, 1998; Pallini *et al.*, 1998; Lima and Bednekoff, 1999; Venzon *et al.*, 2000; Janssen *et al.*, 2002; Magalhães *et al.*, 2002; Magalhães *et al.*, 2004). Antipredator behaviour comes in many forms, from escaping and hiding to avoiding areas with predators. Most of these behaviours have one thing in common; spatial complexity is needed for effective antipredator behaviour. Prey that try to escape from predators by running or flying away can only do so if there are other places to go to. Sometimes, the spatial complexity of one single plant is sufficient (Magalhães *et al.*, 2002), but often prey avoid entire plants with predators when searching for food (Pallini *et al.*, 1999; Nomikou *et al.*, 2003) and this behaviour is only manifested if prey have a choice between various plants that differ in presence of predators. Such a choice is often absent in experiments on predator-prey interactions, and this is likely to result in overestimating the interaction strength between prey and predators.

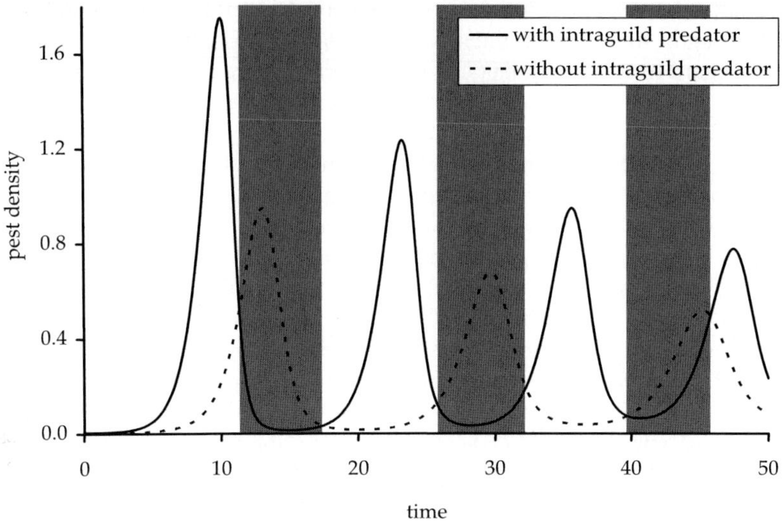

Figure 4. The effect of intraguild predation on transient dynamics according to the model of Holt and Polis (1997). Shown are the dynamics of a pest through time in presence of the intraguild prey and in presence (with IGP, drawn line) or absence (w.o. IGP, broken line) of the intraguild predator. Given enough time, both systems will settle into an equilibrium, and that of the pest in presence of the intraguild prey plus the intraguild predator is higher than that of the pest in presence of the intraguild prey only. Depending on the time at evaluation, the effect of intraguild predation on the pest can be positive (white areas) or negative (gray areas). The equation for the resource is $R' = R(r(1 - R/K) - aN - cP)$; *for the intraguild prey:* $N' = N(abR - m - \alpha P)$; *for the intraguild predator:* $P' = P(dcR + \beta\alpha N - n)$ *with* r = *growth rate of the resource (= 1),* K = *carrying capacity of the resource (= 5),* a = *consumption rate of the resource by the intraguild prey (N) (= 1),* c = *consumption rate of the resource by the intraguild predator (P) (= 0.2),* b = *conversion rate of resource by the intraguild prey (= 1),* m = *mortality of the intraguild prey (= 0.2),* α = *consumption rate of the intraguild prey by the intraguild predator (= 0.6),* d = *conversion rate of resource by the intraguild predator (= 0.5),* β = *conversion rate of intraguild prey by the intraguild predator (= 0.9), and* n = *mortality of the intraguild predator (= 0.5). Initial values were* $R = 0.004$, $N = 1.351$, *and* $P = 0.84$ *(with intraguild predation) or* $P = 0$ *(without intraguild predation).*

The same holds for experiments on intraguild predation; the shared prey, the intraguild prey and intraguild predators are often confined, thus impeding much of the behavioural responses of the prey. We recently found that intraguild prey indeed avoided patches with shared prey and an intraguild predator (Magalhães *et al.*, 2004). The effects of such antipredator behaviour on intraguild predation and its effects on population dynamics remain to be investigated, but it is clear that experiments need to be carried out under conditions of sufficient space and spatial complexity to allow for the full behavioural repertoire of all species in the system. If the intraguild predator would induce antipredator behaviour in the intraguild prey, this might result in a reduction of the strength of intraguild predation in natural systems and complex agricultural systems, whereas the effect of intraguild predation may be strong in simple environments, such as laboratory settings.

With regard to the experiments reviewed here, if there is an effect of antipredator behaviour, one would expect less negative effects on intraguild prey and less positive effects on the pest with increasing spatial complexity. We classified complexity of the experimental system according to the following criteria: the least complex environment consisted of parts of a plant, single leaves, or arenas in the laboratory; entire plants were considered more complex; then groups of 2 – 6 plants; and finally groups of many plants were taken as most complex. For the intraguild prey our expectations were not met; 9 out of 10 studies in systems with high complexity showed a negative effect, whereas this was 15 out of 23 for all other systems together. For the pest, there was no clear trend; of the 6 studies with a positive effect on the pest, 1 was done on a leaf, 3 were done on branches, 1 was done on groups of 2 – 6 plants and 1 was done on many plants (Table 1). This last study (Snyder and Ives, 2001) actually presents evidence for effects of spatial complexity; interactions between intraguild predators, intraguild prey and the pest were weaker when plants were tall (higher complexity) than when plants were short.

9. CONCLUSIONS

Our most important conclusion is that there is only limited evidence for positive effects of intraguild predation on pest densities. Thus, intraguild predation often does not seem to have negative effects on biological control. The use of generalist natural enemies (often intraguild predators) therefore seems less risky for biological control than anticipated so far (van Lenteren *et al.*, 2003). The available studies have mostly looked at the effects of intraguild predation on pest populations, but for biological control, the effect on the crop plants is the ultimate factor to consider. The few studies that did look at effects on plant damage or crop yield found no negative effects of intraguild predation (Snyder and Wise, 1999; Venzon *et al.*, 2001).

What remains to be explained is why intraguild predation often has no negative effect on biological control. Theory predicts that intraguild predators would always have positive effects on the shared prey (pest) as long as they are the inferior competitor. We suggest that the effects of intraguild predation on population dynamics of the intraguild prey and the shared prey may differ from theory because of transient dynamics, increased food web complexity and perhaps because of antipredator behaviour in structured environments.

It should be realised that our analysis is based on a limited data set and is able to show correlations and trends at best. Clearly, experimental studies in which food web complexity and spatial structure are varied systematically should be done to verify the trends gleaned from our review. There is also a dire need for theoretical analyses of the transient effects of intraguild predation. With respect to transient dynamics, there is a more general message: intraguild predation or hyperpredation will often result in a shift of the cycles exhibited before a system reaches a stable state (Fig. 4). Such shifts may result in temporal effects that are different from the long-term effect. Evaluation of the effects of hyperpredation and intraguild predation, not just in biological control systems, but also in natural communities, should therefore be done at various time intervals, and preferably over a period that is long enough for the system to settle in a stable state (if there is one).

10. ACKNOWLEDGEMENTS

Guy Boivin and Jacques Brodeur are thanked for inviting AJ for organizing a wonderful meeting. Martijn Egas is thanked for valuable discussions. While writing this paper, AJ was supported by a scholarship for Visiting Professor (PVE) at the Federal University of Viçosa from CAPES (Brasil), MM was supported by the Dutch Science Foundation (NWO/ALW), and AP was supported by CNPq (Brasil).

11. REFERENCES

Armstrong, R. A. and McGehee, R., 1980. Competitive exclusion. Am. Nat. 115: 151-170.

Borer, E. T., Briggs, C. J., Murdoch, W. W. and Swarbrick, S. L., 2003. Testing intraguild predation theory in a field system: does numerical dominance shift along a gradient of productivity? Ecol. Lett. 6: 929-935.

Briggs, C. J. and Borer, E. T., 2005. Why short-term experiments may not allow long-term predictions about intraguild predation. Ecol. Appl. 15: 1111-1117.

Chang, G. C., 1996. Comparison of single versus multiple species of generalist predators for biological control. Environ. Entomol. 25: 207-212.

Colfer, R. G. and Rosenheim, J. A., 2001. Predation on immature parasitoids and its impact on aphid suppression. Oecologia 126: 292-304.

Colfer, R. G., Rosenheim, J. A., Godfrey, L. D. and Hsu, C. L., 2003. Interactions between the augmentatively released predaceous mite *Galendromus occidentalis* (Acari: Phytoseiidae) and naturally occurring generalist predators. Environ. Entomol. 32: 840-852.

Croft, B. A. and McRae, I. V., 1992a. Biological control of apple mites by mixed populations of *Metaseiulus occidentalis* (Nesbitt) and *Typhlodromus pyri* Scheuten (Acari: Phytoseiidae). Environ. Entomol. 21: 202-209.

Croft, B. A. and McRae, I. V., 1992b. Persistence of *Typhlodromus pyri* and *Metaseiulus occidentalis* (Acari: Phytoseiidae) on apple after inoculative release and competition with *Zetzellia mali* (Acari: Stigmatidae). Environ. Entomol. 21: 1168-1177.

Denno, R. F., Mitter, M. S., Langellotto, G. A., Gratton, C. and Finke, D. L., 2004. Interactions between a hunting spider and a web-builder: consequences of intraguild predation and cannibalism for prey suppression. Ecol. Entomol. 29: 566-577.

Diehl, S. and Feissel, M., 2000. Effects of enrichment on three-level food chains with omnivory. Am. Nat. 155: 200-218.

Diehl, S. and Feissel, M., 2001. Intraguild prey suffer from enrichment of their resources: A microcosm experiment with ciliates. Ecology 82: 2997-2983.

Dinter, A., 2002. Microcosm studies on intraguild predation between female erigonid spiders and lacewing larvae and influence of single versus multiple predators on cereal aphids. J. Appl. Entomol. 126: 249-257.

Ellner, S. P., McCauley, E., Kendall, B. E., Briggs, C. J., Hosseini, P. R., Wood, S. N., Janssen, A., Sabelis, M. W., Turchin, P., Nisbet, R. M. and Murdoch, W. W., 2001. Habitat structure and population persistence in an experimental community. Nature 412: 538-543.

Emmerson, M. and Yearsley, J. M., 2004. Weak interactions, omnivory and emergent food-web properties. Proc. Roy. Soc. B. 271: 397-405.

Erbilgin, N., Dahlsten, D. L. and Chen, P. Y., 2004. Intraguild interactions between generalist predators and an introduced parasitold of *Glycaspis brimblecombei* (Homoptera: Psylloidea). Biol. Contr. 31: 329-337.

Eubanks, M. D., 2001. Estimates of the direct and indirect effects of red imported fire ants on biological control in field crops. Biol. Contr. 21: 35-43.

Ferguson, K. I. and Stiling, P., 1996. Non additive effects of multiple natural enemies on aphid populations. Oecologia 108: 375-379.

Heinz, K. M. and Nelson, J. M., 1996. Interspecific interactions among natural enemies of *Bemisia tabaci* in an inundative biological control program. Biol. Contr. 6: 384-393.

Heithaus, M. R., 2001. Habitat selection by predators and prey in communities with asymmetric intraguild predation. Oikos 92: 542-554.

HilleRisLambers, R. and de Roos, A. M., in prep. Community structure, species coexistence, and the balance between direct and indirect interactions.

HilleRisLambers, R. and Dieckmann, U., 2003. Competition and predation in simple food webs: intermediately strong trade-offs maximize coexistence. Proc. Roy. Soc. B. 270: 2591-2598.

Holt, R. D. and Polis, G. A., 1997. A theoretical framework for intraguild predation. Am. Nat. 149: 745-764.

Holyoak, M. and Sachdev, S., 1998. Omnivory and the stability of simple food webs. Oecologia 117: 413-419.

Huisman, J. and Weissing, F. J., 1999. Biodiversity of plankton by species oscillations and chaos. Nature 402: 407-410.

Janssen, A., van Gool, E., Lingeman, R., Jacas, J. and van de Klashorst, G., 1997. Metapopulation dynamics of a persisting predator-prey system in the laboratory: time series analysis. Exp. Appl. Acarol. 21: 415-430.

Janssen, A., Faraji, F., van der Hammen, T., Magalhães, S. and Sabelis, M. W., 2002. Interspecific infanticide deters predators. Ecol. Lett. 5: 490-494.

Janssen, A., Willemse, E. and van der Hammen, T., 2003. Poor host plant quality causes omnivore to consume predator eggs. J. Anim. Ecol. 72: 478-483.

Kakehashi, N., Suzuki, Y. and Iwasa, Y., 1984. Niche overlap of parasitoids in host-parasitoid systems: its consequence to single versus multiple introduction controversy in biological control. J. Appl. Ecol. 21: 115-131.

Krivan, V., 2000. Optimal intraguild foraging and population stability. Theor. Popul. Biol. 58: 79-94.

Krivan, V. and Diehl, S., 2005. Adaptive omnivory and species coexistence in tri-trophic food webs. Theor. Popul. Biol. 67: 85-99.

Kuijper, L. D. J., Kooi, B. W., Zonneveld, C. and Kooijman, S., 2003. Omnivory and food web dynamics. Ecol. Model. 163: 19-32.

Law, R. and Blackford, J. C., 1992. Self-assembling food webs: a global viewpoint of coesistence of species in Lotka-Volterra communities. Ecology 73: 567-578.

Lawler, S. P. and Morin, P. J., 1993. Food web architecture and population dynamics in laboratory microcosms of protists. Am. Nat. 141: 675-686.

Lima, S. L., 1998. Nonlethal effects in the ecology of predator-prey interactions - What are the ecological effects of anti-predator decision- making? Bioscience 48: 25-34.

Lima, S. L. and Dill, L. M., 1990. Behavioral decisions made under the risk of predation - a review and prospectus. Can. J. Zool. 68: 619-640.

Lima, S. L. and Bednekoff, P. A., 1999. Temporal variation in danger drives antipredator behavior: The predation risk allocation hypothesis. Am. Nat. 153: 649-659.

Lucas, E. and Alomar, O., 2002. Impact of the presence of *Dicyphus tamaninii* Wagner (Heteroptera: Miridae) on whitefly (Homoptera: Aleyrodidae) predation by *Macrolophus caliginosus* (Wagner) (Heteroptera: Miridae). Biol. Control 25: 123-128.

Magalhães, S., Janssen, A., Hanna, R. and Sabelis, M. W., 2002. Flexible antipredator behaviour in herbivorous mites through vertical migration in a plant. Oecologia 132: 143-149.

Magalhães, S., Tudorache, C., Montserrat, M., van Maanen, R., Sabelis, M. W. and Janssen, A., 2004. Diet of intraguild predators affects antipredator behavior in intraguild prey. Behav. Ecol. 16: 364-370.

Mallampalli, N., Castellanos, I. and Barbosa, P., 2002. Evidence for intraguild predation by *Podisus maculiventris* on a ladybeetle, *Coleomegilla maculata*: Implications for biological control of Colorado potato beetle, *Leptinotarsa decemlineata*. Biocontr. 47: 387-398.

May, R. M. and Hassell, M. P., 1981. The dynamics of multiparasitoid-host interactions. Am. Nat. 117: 234-261.

May, R. M., 1973. *Stability and complexity in model ecosystems*. Princeton, New Jersey, USA: Princeton University Press.

McCann, K., Hastings, A. and Huxel, G. R., 1998. Weak trophic interactions and the balance of nature. Nature 395: 794-798.

Morin, P. J., 1999. Productivity, intraguild predation, and population dynamics in experimental food webs. Ecology 80: 752-760.

Mylius, S. D., Klumpers, K., de Roos, A. M. and Persson, L., 2001. Impact of omnivory and stage structure on food web composition along a productivity gradient. Am. Nat. 158: 259-276.

Nomikou, M., Janssen, A. and Sabelis, M. W., 2003. Herbivore host plant selection: whitefly learns to avoid host plants that harbour predators of her offspring. Oecologia 136: 484-488.

Pallini, A., Janssen, A. and Sabelis, M. W., 1998. Predators induce interspecific herbivore competition for food in refuge space. Ecol. Lett. 1: 171-177.

Pallini, A., Janssen, A. and Sabelis, M. W., 1999. Spider mites avoid plants with predators. Exp. Appl. Acarol. 23: 803-815.

Parrella, M. P., McCaffrey, J. P. and Horsburgh, R. L., 1980. Compatibility of *Leptothrips mali* with *Stethorus punctum* and *Orius insidiosus*: predators of *Panonychus ulmi*. Environ. Entomol. 9: 694-696.

Pimm, S. L. and Lawton, J. H., 1978. On feeding on more than one trophic level. Nature 333.

Polis, G. A. and Holt, R. D., 1992. Intraguild predation - the dynamics of complex trophic interactions. Trends Ecol. Evol. 7: 151-154.

Polis, G. A. and Winemiller, K. O., 1996. *Food webs. Integration of patterns and dynamics*. New York, USA: Chapman and Hall.

Polis, G. A., Myers, C. A. and Holt, R. D., 1989. The ecology and evolution of intraguild predation - potential competitors that eat each other. Annu. Rev. Ecol. Syst. 20: 297-330.

Price, J. E. and Morin, P. J., 2004. Colonization history determines alternate community states in a food web of intraguild predators. Ecology 85: 1017-1028.

Rosenheim, J. A., 2001. Source-sink dynamics for a generalist insect predator in habitats with strong higher-order predation. Ecol. Monogr. 71: 93-116.

Rosenheim, J. A., Wilhoit, L. R. and Armer, C. A., 1993. Influence of intraguild predation among generalist insect predators on the suppression of an herbivore population. Oecologia 96: 439-449.

Rosenheim, J. A., Kaya, H. K., Ehler, L. E., Marois, J. J. and Jaffee, B. A., 1995. Intraguild predation among biological control agents - Theory and evidence. Biol. Contr. 5: 303-335.

Rosenheim, J. A., Glik, T. E., Goeriz, R. E. and Rämert, B., 2004a. Linking a predator's foraging behavior with its effects on herbivore population suppression. Ecology 85: 3362-3372.

Rosenheim, J. A., Limburg, D. D., Colfer, R. G., Fournier, V., Hsu, C. L., Leonardo, T. E., *et al.*, 2004b. Herbivore population suppression by an intermediate predator, *Phytoseiulus macropilis*, is insensitive to the presence of an intraguild predator: an advantage of small body size? Oecologia 140: 577-585.

Schausberger, P. and Walzer, A., 2001. Combined versus single species release of predaceous mites: Predator-predator interactions and pest suppression. Biol. Contr. 20: 269-278.

Schröder, A., Persson, L. and de Roos, A. M., 2005. Direct experimental evidence for alternative stable states: a review. Oikos 110: 3-19.

Sher, R. B., Parrella, M. P. and Kaya, H. K., 2000. Biological control of the leafminer *Liriomyza trifolii* (Burgess): Implications for intraguild predation between *Diglyphus begini* Ashmead and *Steinernema carpocapsae* (Weiser). Biol. Contr. 17: 155-163.

Sih, A., 1980. Optimal behavior: can foragers balance two conflicting needs? Science 210: 1041-1043.

Snyder, W. E. and Ives, A. R., 2001. Generalist predators disrupt biological control by a specialist parasitoid. Ecology 82: 705-716.

Snyder, W. E. and Ives, A. R., 2003. Interactions between specialist and generalist natural enemies: Parasitoids, predators, and pea aphid biocontrol. Ecology 84: 91-107.

Snyder, W. E. and Wise, D. H., 1999. Predator interference and the establishment of generalist predator populations for biocontrol. Biol. Contr. 15: 283-292.

Snyder, W. E., Ballard, S. N., Yang, S., Clevenger, G. M., Miller, T. D., Ahn, J. J., Hatten, T. D. and Berryman, A. A., 2004. Complementary biocontrol of aphids by the ladybird beetle *Harmonia axyridis* and the parasitoid *Aphelinus asychis* on greenhouse roses. Biol. Contr. 30: 229-235.

van Lenteren, J. C., Babendreier, D., Bigler, F., Burgio, G., Hokkanen, H. M. T., Kuske, S., Loomans, A. J. M., Menzler-Hokkanen, I., Van Rijn, P. C. J., Thomas, M. B., Tommasini, M. G. and Zeng, Q. Q., 2003. Environmental risk assessment of exotic natural enemies used in inundative biological control. Biocontr. 48: 3-38.

Venzon, M., Janssen, A., Pallini, A. and Sabelis, M. W., 2000. Diet of a polyphagous arthropod predator affects refuge seeking of its thrips prey. Anim. Behav. 60: 369-375.

Venzon, M., Janssen, A. and Sabelis, M. W., 2001. Prey preference, intraguild predation and population dynamics of an arthropod food web on plants. Exp. Appl. Acarol. 25: 785-808.

MULTIPLE PREDATOR INTERACTIONS AND FOOD-WEB CONNECTANCE: IMPLICATIONS FOR BIOLOGICAL CONTROL

Robert F. Denno[1] and Deborah L. Finke[1,2]

[1]Department of Entomology, University of Maryland, College Park, Maryland, 20742; [2]Department of Entomology, Washington State University, Pullman, Washington, 99164

Abstract. The use of single versus multiple natural enemies in biological control remains controversial, largely due to the possibility for antagonistic interactions among predators (e.g., intraguild predation and cannibalism) that can reduce the potential for the top-down control of pest herbivores. Using a natural system, *Spartina* cordgrass and its associated community of arthropods (herbivores, strict predators and intraguild predators), we created 29 different treatment combinations of predators that varied in richness (number of predator species) and trophic composition (proportion of strict to intraguild predators) and measured the ability of each to reduce the density of a key herbivore (the planthopper *Prokelisia dolus*) in the system. We then calculated food-web connectance (the fraction of all possible directed feeding links that are realized in a food web) for each of the experimental food webs. Notably, food-web connectance is enhanced by predator-predator interactions such as intraguild predation and cannibalism. We found a significant negative relationship between food-web connectance and the ability of the predator complex to reduce prey populations. Specifically, well-connected food webs comprised of mostly intraguild predator species were far less effective at suppressing herbivores than webs consisting largely of strict predators. Importantly, trophic composition of the food web was more influential than predator richness in affecting top-down control. We also discovered that a food web comprised of multiple predators was more effective in suppressing herbivores when the structural complexity of the habitat was increased, a result that was attributable to spatial refuges for intraguild prey and relaxed intraguild predation. Thus, in this system, habitat structure has the potential to transform a well-connected food web into a less-connected one by reducing feeding links resulting from intraguild predation and cannibalism. Because of the remarkable similarity of the *Spartina* system to tropical Asian rice, this finding provides encouragement that the effectiveness of the predator complex can be enhanced by management practices that increase the structural complexity of the habitat and thereby dampen intraguild predation. Last, we discuss how food-web analyses might be used to evaluate particular combinations of predators for more effective biological control.

1. INTRODUCTION

A central question in biological control has been how multiple predators interact to collectively suppress populations of pest herbivores (Hochberg, 1996; Riechert and Lawrence, 1997; Rosenheim, 1998; Denoth *et al.*, 2002; Symondson *et al.*, 2002; Wilby and Thomas, 2002; Cardinale *et al.*, 2003; Janssen *et al.* this volume). The issue remains controversial because there is extensive evidence both for and against multiple predators being more effective than single natural-enemy species in reducing herbivore populations. On the positive side, there are numerous studies

J. Brodeur and G. Boivin (eds.), Trophic and Guild Interactions in Biological Control, 45–70.

showing that multiple natural enemies can exert strong collective top-down control on agricultural pests (Heinz and Nelson, 1995; Riechert and Lawrence, 1997; Symondson *et al.*, 2002; Cardinale *et al.*, 2003). However, there is also widespread evidence that employing multiple agents to control pest herbivores disrupts biological control compared to a simple natural-enemy assemblage comprised of one or few predator species (Rosenheim and Wilhoit, 1993; Rosenheim *et al.*, 1993, 1995; Phoofolo and Obrycki, 1998; Snyder and Wise, 1999; Snyder and Ives, 2001; Prasad and Snyder, 2004). The key to understanding when and where a natural-enemy complex promotes or constrains pest suppression likely lies in the sign and strength of interactions among the predators themselves. For example, multiple predators can interact synergistically to enhance pest suppression when one predator forces prey into the foraging ambit of another predator species (Losey and Denno, 1998, 1999). Alternatively, predators may not interact at all in which case their effects on pest populations are simply additive (Chang, 1996; Straub and Snyder, 2006). On the down side, predators may interact antagonistically whereby they consume each other in acts of intraguild predation (IGP) and cannibalism or interfere with each other's searching behavior and capture success, thus relaxing top-down controls on herbivores (Rosenheim *et al.*, 1995; Hodge, 1999; Finke and Denno, 2003; Prasad and Snyder, 2004).

Overall, however, the widespread occurrence of IGP appears to be a major factor contributing to the ineffectiveness of multiple predators in biological control (Rosenheim *et al.*, 1995; Rosenheim, 1998; Hodge, 1999; Snyder and Wise, 1999; Prasad and Snyder, 2004), although its strength admittedly varies seasonally (Walde *et al.*, 1997), in the presence of alternative prey (Lucas *et al.*, 1998), and with the size, stage class, feeding niche, mobility, foraging style, and aggressiveness of the shared prey, intraguild prey and predators in the complex (Polis *et al.*, 1989; Lucas *et al.*, 1998; Wilby and Thomas, 2002; Matsumura *et al.*, 2004, Rosenheim *et al.*, 2004). Notably, IGP can be moderated in complex-structured habitats where intraguild prey find refuge from predation (Finke and Denno, 2002; Denno *et al.*, 2005 a, 2005 b; Finke, 2005). In such complex habitats, enemy effects on shared herbivorous prey are intensified (Denno *et al.*, 2002; Finke and Denno, 2002; Finke, 2005), providing encouragement to pest managers that the effectiveness of the predator complex in cropping systems might be enhanced with appropriate structural manipulations of the habitat. Nonetheless, IGP is a common phenomenon in many simple-structured agricultural systems (Rosenheim *et al.*, 1995; Hawkins *et al.*, 1999; Eubanks, 2001; Snyder and Ives, 2001) as well as in a diversity of natural systems where predator effects on subtending trophic levels often attenuate (Polis *et al.*, 1989; Sih *et al.*, 1998; Finke and Denno, 2002, 2003, 2004; Arim and Marquet, 2004). Thus, in a biological control context it becomes essential to critically assess the nature of interactions among predators (antagonistic, synergistic or non-interactive), the frequency and strength of such interactions in the food web (particularly IGP), how such interactions affect pest suppression, and how habitat and landscape structure might temper predator-predator interactions and enhance top-down controls.

In addition to their importance in biological control, multi-predator interactions are known to have significant consequences for food-web dynamics (Polis and Strong, 1996; Fagan, 1997; McCann *et al.*, 1998; Closs *et al.*, 1999), the occurrence of trophic cascades (Finke and Denno, 2004, 2005), and ecosystem functioning (Montoya *et al.*, 2003). Multi-predator interactions including IGP often arise from and are promoted by increased species diversity at higher trophic levels (Finke and Denno, 2004, 2005). Such interactions, whereby generalist consumers feed across trophic levels (omnivores including intraguild predators) were historically thought to destabilize food-web dynamics (May, 1973). For example, generalist predators whose population dynamics are not closely coupled with that of their "preferred prey" can drive such prey to extinction because their abundance can be sustained on alternative prey (see Closs *et al.*, 1999). Using similar reasoning, earlier modelling efforts focused on the destabilizing effects of omnivory on food-web dynamics that result in strong density fluctuations and the eventual loss of species (see Polis, 1998). This way of thinking led to the view that complexity in food webs, food webs rich in species and with high degrees of omnivory, was destabilizing, a notion that was fundamentally counterintuitive to most ecologists (see Polis, 1998). However, there was a growing body of empirical evidence at the time suggesting that omnivory, including IGP, was a main source of complexity in food webs and that such factors could be strongly stabilizing forces in food-web dynamics (Strauss, 1991; Polis and Strong, 1996; Fagan, 1997). Thus, a paradox emerged such that empirical evidence was at odds with theory and modelling efforts (see Fagan, 1997).

Notably, recent modelling approaches (e.g., McCann *et al.*, 1998) provide a mathematical and theoretical basis for a paradigm shift that is now overtaking ecology, namely that complexity does promote food-web stability (Polis, 1998). New modelling efforts focus on non-equilibrium assumptions, non-linear terms that incorporate well-substantiated behavior for feeding interactions such that consumers (e.g., predators) cannot maintain high feeding efficiencies on many prey species at the same time, and many weak linkages that dampen oscillations between resources and consumers (McCann *et al.*, 1997, 1998; Polis, 1998; McCann, 2000). Even more recently there has been a wealth of new empirical data suggesting that complex-structured food webs with many reticulate interactions (omnivory and IGP) diminish top-down control on herbivores (Rosenheim, 1998; Snyder and Ives, 2001; Finke and Denno, 2003), dampen trophic cascades (Hart, 2002; Finke and Denno, 2004, 2005), and lend stability to food-web dynamics and ecosystem services (Closs *et al.*, 1999; Montoya *et al.*, 2003).

Thus, current theory, modelling efforts, and experimental evidence from natural systems (Fagan, 1997; Finke and Denno, 2004) are converging on the view in managed systems that food webs comprised of multiple predators, and in particular ones dominated by intraguild predators, can disrupt biological control (Rosenheim *et al.*, 1995; Snyder *et al.*, 2005). The relevant emerging message is that increased species richness at higher trophic levels, a high incidence of connectance among predators (e.g., predator-predator links created by IGP), and weak to intermediate interaction strengths among component predators can promote food-web stability and weaken top-down effects on shared prey (Fagan, 1997; McCann *et al.*, 1998; Finke and Denno, 2004). Nonetheless, the debate continues as to the components of species richness that contribute to prey suppression, food-web stability, and ultimately ecosystem function (Naeem and Li, 1997; Downing and Liebold, 2002; Wilby and Thomas, 2002; Chalcraft and Resetarits, 2003; Finke and Denno, 2004). Specifically, with an increase in species richness, is it trophic composition (e.g. proportion of IG predators in the enemy complex), species complementarity (total niche space occupied), or species composition (species-specific effects) that dampen top-down effects? We have argued that increasing predator richness *per se* in arthropod-based food-webs is not the primary contributor to relaxed predator effects on herbivores and dampened trophic cascades (Finke and Denno, 2004, 2005). Rather, we suggest that diversifying the trophic composition of the predator assemblage (proportion of intraguild to strict predators in the food web) determines the impact of increasing predator species richness on herbivore control and the occurrence of trophic cascades (Finke and Denno, 2005). Thus, the consequences of increasing predator richness on prey suppression should depend on the trophic composition of the predator assemblage, specifically the proportion of intraguild predators.

Three testable predictions emerge that are important in the context of assessing multiple versus single predator effects for enhanced biological control. First, increasing food-web connectance (e.g., due to linkages that arise from omnivory and IGP) should diminish prey suppression. Second, enhancing strict predator (predators that feed only on herbivores) richness should increase the suppression of shared prey, but not beyond the additive expectation because such predators are largely non-interactive. Last, increasing the richness of intraguild predators should diminish prey suppression due to relaxed predation on shared prey, again with predator density controlled.

We have conducted experiments that allow for the testing of these hypotheses (Finke and Denno, 2005), but the data have never been analyzed in a food-web connectance context. The system we used for our experiments was the assemblage of arthropods (phloem-feeding herbivores, strict predators, and intraguild predators) associated with *Spartina* cordgrass, the dominant vegetation cover of inter-tidal salt marshes that fringe the Atlantic coast of North America (Denno *et al.*, 1996). The design of our experiment allowed us to separate the effects of predator species richness from trophic composition on herbivore suppression. For this experiment we manipulated both predator richness (number of species) and the trophic composition (proportion of strict to intraguild predators) and measured the effect of the resulting food-webs on planthopper density (*Prokelisia dolus*) (Homoptera: Delphacidae), one of the major herbivores in the system. For this chapter, we now calculate food-web connectance (the fraction of all possible feeding links that are realized in a food web, Dunne *et al.*, 2002) for each of the respective experimental food webs and measure the relationship between connectance and herbivore suppression. Moreover, we have also conducted similar experiments in both simple and complex-structured habitats (Finke, 2005) allowing for an assessment of how habitat structure might influence food-web connectance and prey suppression by altering the incidence of IGP.

There is extensive variation in the structure of natural-enemy complexes in agroecosystems, where both the number of predator species as well as the prevalence of intraguild predators differ considerably (Rosenheim *et al.*, 1995; Symondson *et al.*, 2002). Such complex trophic variation clearly leads to differences in the effectiveness of enemy complexes in suppressing pests as evidenced by the controversy surrounding the issues of multiple-predator use and effectiveness in biological control (Rosenheim *et al.*, 1993, 1995; Heinz and Nelson, 1995; Snyder and Wise, 1999; Snyder and Ives, 2001; Symondson *et al.*, 2002; Cardinale *et al.*, 2003; Snyder *et al.*, 2005). Our experimental approach was to mimic such predator diversity and provide a new analysis of the role of predator diversity and IGP in pest suppression. Toward this end, we assess the relationship between prey suppression and food-web connectance, a synthetic metric that captures the complex species linkages arising from omnivory and IGP. We conclude by discussing how food-web metrics might be useful in determining the structure of specific natural-enemy assemblages for successful biological control and how ineffective predator combinations can be transformed into more effective ones by habitat management practices that minimize food-web connectance and IGP.

2. FOOD WEB OF PLAYERS IN THE SALT MARSH SYSTEM

The vegetation of mid-Atlantic salt marshes is dominated by the perennial cordgrass *Spartina alterniflora*, where it grows in extensive pure stands within the intertidal zone (Gallagher *et al.*, 1988; Ornes and Kaplan, 1989; Bertness, 1991; Denno *et al.*, 1996; Bertness and Pennings, 2000). Within this zone, though, the structure of *Spartina* varies considerably with elevation due to differences in tidal flooding, nutrient subsidy, and leaf litter (thatch) removal and decay (Gallagher *et al.*, 1988; Denno *et al.*, 1996). In tidally-subsidized low-marsh habitats, *Spartina* plants accumulate a large standing crop, grow vigorously tall, are very nutritious, but are relatively devoid of leaf litter because it is carried away by tidewaters (Denno and Grissell, 1979; Denno, 1983; Ornes and Kaplan, 1989; Denno *et al.*, 2002). By contrast, in high marsh meadows where nutrient subsidy and tidal flushing are much less, standing crop is low, plants are dwarfed, less nutritious, but they are frequently surrounded by a dense entanglement of thatch (Ornes and Kaplan, 1989; Denno *et al.*, 2002). Thus, moving up the elevational gradient from low-marsh habitats (tidal creek banks) to high-marsh plateaus (meadows), *Spartina* plants generally decrease in biomass and nitrogen content, but increase in the amount of associated thatch (Denno and Grissell, 1979; Denno, 1983; Ornes and Kaplan, 1989). However, even on the high marsh subtle differences in soil porosity, standing water, and salinity result in the differential decay of dead leaf material and combine to produce a mosaic of habitats that vary in the amount of thatch (Gallagher *et al.*, 1988; Denno *et al.*, 1996). Notably, *Spartina* growing in meadows accumulates much more thatch than plants occupying mud flats (Redfield, 1972; Denno *et al.*, 2002).

Spartina serves as the only host plant for a diversity of insect herbivores, mostly sap-feeders such as planthoppers, leafhoppers and phytophagous mirid bugs (Denno *et al.*, 2003). Chewing herbivores, mainly stem-boring caterpillars and beetles are not nearly as common. Of the sap-feeding herbivores, *Prokelisia* planthoppers (*P. dolus* and *P. marginata*) are by far the most abundant, with adult densities often exceeding several thousand individuals m^{-2} and nymphal densities frequently exceeding 10,000 m^{-2} (Denno *et al.*, 2000). Both *Prokelisia* species are extremely abundant in a variety of marsh habitats, however, *Prokelisia dolus* is the predominant planthopper found in high-marsh meadow habitats, a habitat where predators abound (Denno *et al.*, 1996). By contrast, *P. marginata*, due to its greater mobility, is able to colonize and exploit low-marsh habitats (Denno *et al.*, 1996). Host plant nutrition plays an important and direct role in determining the potential for population growth in *Prokelisia* planthoppers. For example, in the absence of predators, planthopper populations grow rapidly and frequently outbreak on nitrogen-rich *Spartina* (Denno, 1983; Cook and Denno, 1994; Denno and Peterson, 2000). Population increases result from a combination of enhanced colonization, oviposition, development rate, survival, and fecundity on plants rich in amino nitrogen (Cook and Denno, 1994; Denno *et al.*, 2002). The most striking planthopper outbreaks occur annually in low-marsh habitats where they are associated with nitrogen-rich *Spartina* and low densities of natural enemies (Denno and Peterson, 2000; Denno *et al.*, 2002, 2005b).

Numerous natural enemies, both invertebrate predators and parasitoids, attack *Prokelisia* planthoppers, but predators are a far more important source of mortality on the mid-Atlantic marshes of North America (Döbel and Denno, 1994). The most abundant predators consuming planthoppers are the hunting spiders *Pardosa littoralis* (Araneae: Lycosidae), *Clubiona saltitans* (Araneae: Clubionidae), and *Marpissa pikei* (Araneae: Salticidae), the web-building spider *Grammonota trivittata* (Araneae: Linyphiidae), the lady beetle *Naemia seriata* (Coleoptera: Coccinellidae), and the mirid bug *Tytthus vagus* (Hemiptera: Miridae) (Döbel and Denno, 1994; Denno *et al.*, 2002, 2003, 2004; Finke and Denno, 2002, 2003). All of these predators attack planthopper adults and nymphs, with the exception of *Tytthus* which is a specialist on planthopper eggs that have been oviposited within the leaf blades of *Spartina* (Döbel and Denno, 1994; Finke and Denno, 2002). This predator complex includes predators that have been identified as intraguild predators (the hunting spiders) as well as strict predators (the web-building spider, the coccinellid, and the mirid bug) (Finke and Denno, 2002, 2003, 2005; Denno *et al.*, 2002, 2003, 2004; Finke, 2005). Therefore, the opportunity exists for a diversity of interactions among predators.

3. METHODS

3.1. Experimental design

We investigated the importance of food-web connectance in mediating predator diversity effects on planthopper (*Prokelisia dolus*) population suppression by factorially manipulating predator species richness (1, 2, or 3 species) and predator trophic composition (strict predators only, intraguild predators only, or a mixture of both strict and intraguild predators) in mesocosms and then assessing the impact of these predator diversity treatments on planthopper density (see Finke and Denno, 2005). For this experiment, 20 field-collected planthopper adults were released into each mesocosm and predator diversity treatments were established shortly thereafter. Using a species pool of 6 predator species, 3 strict predators (*Tytthus, Grammonota,* and *Naemia*) and 3 intraguild predators (*Pardosa, Clubiona,* and *Marpissa*) (Finke, 2005), predator diversity treatments were created by crossing predator species richness (1, 2 or 3 species) with trophic composition (strict predators only, intraguild predators only, or a mixture of both strict and intraguild predators) (Finke and Denno, 2005). In this way, we created 29 different treatment combinations of predators that varied in their level of food-web connectance. Each treatment combination was replicated 6 times. In addition to the 29 predator treatments, there was also a predator-free control where only planthoppers were present. See Finke and Denno (2005) for the real-world densities of planthoppers and predators used in the experiment. An additive treatment design was used so that predator diversity was not confounded with changes in the abundance of individual predator species (Jolliffe, 2000). As a result, departures from the null hypothesis of independent effects among predators were attributable to changes in interspecific interactions among predators such as IGP, rather than intraspecific effects (Sih *et al.*, 1998; Jolliffe, 2000).

After two planthopper generations, the effects of the predator diversity treatments on planthopper density were assessed. Final planthopper densities were used to calculate the mean "per capita effect size" for each of the 29 predator treatment combinations on planthopper density, where per capita effect size = ln(prey density in predator-free control/prey density in predator treatment) ÷ initial combined density of all predators in treatment (see Denno *et al.*, 2003). Thus, predator combinations resulting in large per capita effect sizes better suppress planthopper populations than treatments with small effect sizes. Effect sizes were calculated on a per capita basis to account for differences in the initial abundance of predators across treatments due to the additive treatment design (see Finke and Denno, 2005).

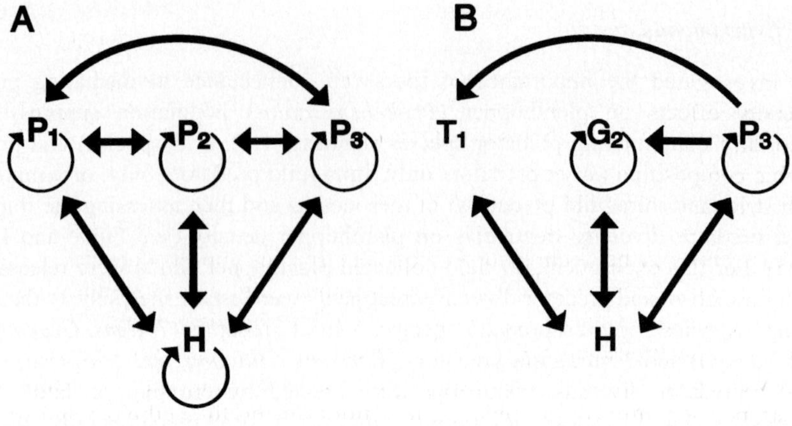

Figure 1. (A) Hypothetical food web with four species (S = 4) including one herbivore (H) and three predators (P1, P2, and P3). Directed feeding links (L) shown by arrows are maximized at 16 (S2) assuming that: the herbivore shows strong intraspecific effects (competition or facilitation indicated by the return arrow), the herbivore is fed on extensively by all 3 predators and populations of the 3 predators increase as a result of feeding on the herbivore (bidirectional arrows), all three predators exhibit reciprocal IGP (bidirectional arrows), and that all predators are cannibalistic (return arrows). For this hypothetical food web, connectance (C = L/S2) is maximized at 1.0. (B) Directed feeding links (arrows) in a real-world food web comprised of the herbivore Prokelisia dolus *(H) and the three predators* Tytthus *(T1),* Grammonota *(G2), and* Pardosa *(P3).* P. dolus *is very insensitive to intraspecific density effects (no return arrow). All predators suppress populations of the herbivore, and the predators exhibit strong numerical responses (aggregative or reproductive) to herbivore density (bidirectional arrows).* Pardosa *consumes both* Tytthus *and* Grammonota, *but the reverse does not occur, nor does* Pardosa *respond numerically to densities of either* Tytthus *or* Grammonota *(unidirectional arrows). Both* Grammonota *and* Pardosa *are cannibalistic (return arrows). For this example food web, L = 10 and connectance = 0.63.*

3.2. Food-web connectance

Food-web connectance values were also calculated for each of the 29 predator-treatment webs in our experiment (Finke and Denno, 2005), and were ultimately used to establish the relationship between food-web connectance and per capita effect size (effectiveness of the various predator treatments in reducing planthopper population size). We calculated food-web connectance as $C = L/S^2$ (the fraction of all possible feeding links that are realized in a food web), where S is the number of species (nodes) in the food web, L is the total number of directed feeding links in the web, L/S is the linkage density (average number of feeding links per species), and S^2 is the maximum number of feeding links including both cannibalism and IGP (Morin, 1999; Dunne *et al.*, 2002; Montoya *et al.*, 2003; Melián and Bascompte,

2004). For example, considering a simple food web with one herbivore and three predators (S = 4), the maximum number of feeding links (S^2) is 16 if one assumes that: the herbivore shows strong intraspecific effects (competition or facilitation), the herbivore is fed on extensively by all 3 predators, populations of the 3 predators increase as a result of feeding on the herbivore, all three predators exhibit reciprocal IGP, and that all predators are cannibalistic (Fig. 1A). For such a hypothetical food web, connectance ($C = L/S^2$) is maximized at 1.0.

Directed feeding links (L) represent the net effect of each species in a food web on the densities of each of the other species, rather than merely specifying the presence or absence of interaction between two species (undirected links). Directed feeding links may be unidirectional, when one species impacts densities of the other (e.g. through consumption or by forming a numerical response) but the reverse does not occur, or they may be bidirectional, when both species impact densities of the other. In order to establish the number of directed feeding links in each of the 29 food webs stemming from our experiment, we used published information on the occurrence and strength of feeding interactions among the herbivore and predators in the system (e.g. consumption rates and incidence of IGP and cannibalism for particular combinations of species [Döbel and Denno, 1994; Denno *et al.* 2000, 2002, 2003, 2004; Finke and Denno 2002, 2003, 2005; Langellotto 2002; Finke, 2005]). To score the presence or absence of a feeding link between two species, we used threshold criteria and ranked links as strong, moderate or weak depending on consumption rates or the fraction that one species comprised of the other's diet (see Winemiller, 1990). If links were either strong or moderate we scored them present; if links were non-existent or weak, we scored the link as absent. As an example, consider the food web comprised of the herbivore *Prokelisia* and the three predators *Tytthus*, *Grammonota*, and *Pardosa* (Fig. 1B). *Prokelisia dolus* is very insensitive to conspecific density effects (Denno *et al.*, 2000) and we scored this intraspecific link as "weak" and did not include it as a link in the food web. All predators suppress populations of the herbivore, and the predators exhibit strong numerical responses (aggregative or reproductive) to herbivore density (Döbel and Denno, 1994; Denno *et al.*, 2002, 2004; Finke and Denno, 2002, 2003, 2005). Thus, we scored all links between predators and the herbivore as present and bidirectional. *Pardosa* consumes both *Tytthus* and *Grammonota*, but the reverse does not occur (Finke and Denno, 2002, 2005; Denno *et al.*, 2004), nor does *Pardosa* respond numerically to densities of either *Tytthus* or *Grammonota* in the field (Lewis and Denno, unpublished data) so we scored these two links as unidirectional. Both *Grammonota* and *Pardosa* are cannibalistic, but *Tytthus* is not (Finke and Denno, 2002, 2005; Langellotto, 2002; Denno *et al.*, 2004), adding two more links. Thus, for this example food web, the total number of directed links (L) = 10 and connectance (C) = 0.63.

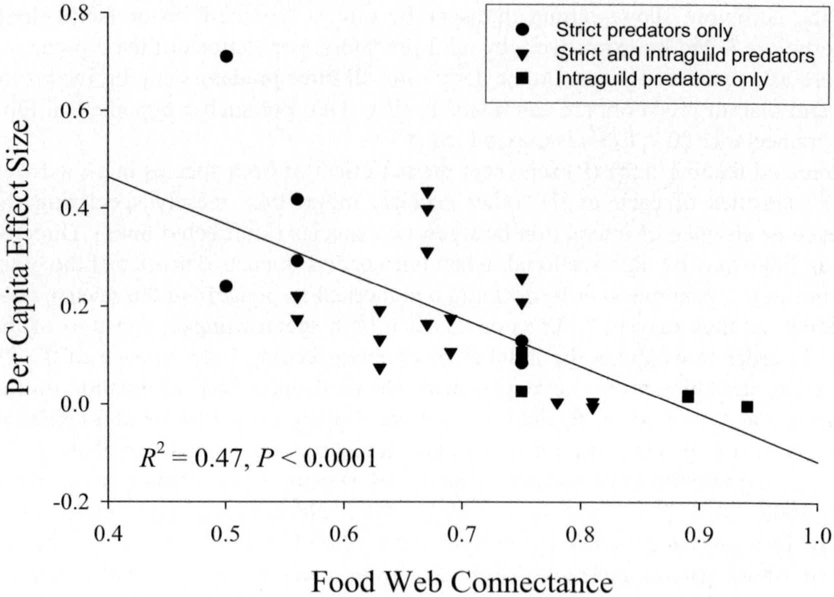

Figure 2. Relationship between food-web connectance (C) and the ability of the predator combination to reduce herbivore density (per capita effect size) for 29 different treatment combinations of strict and intraguild predators. Food webs with low connectance (those dominated by strict predators) had a greater adverse effect on herbivore populations than food webs with high connectance (those dominated by intraguild predators).

Hypotheses were tested by regressing "per capita effect size" against (1) overall connectance (C) for all food webs, (2) C for food webs containing only strict predators, (3) C for food webs containing only IGP predators, (4) overall species richness of the food web (S) for all food webs, (5) S for webs containing only strict predators, and (6) S for webs containing only IGP predators (SAS Institute, 2000). ANCOVA, with S as the covariate, was used to assess the difference in relationship between effect size and species richness for food-webs comprised of only strict or only intraguild predators. The overall relationship between species richness and connectance was also determined for all food webs, and then separately for those food webs containing only strict or only intraguild predators.

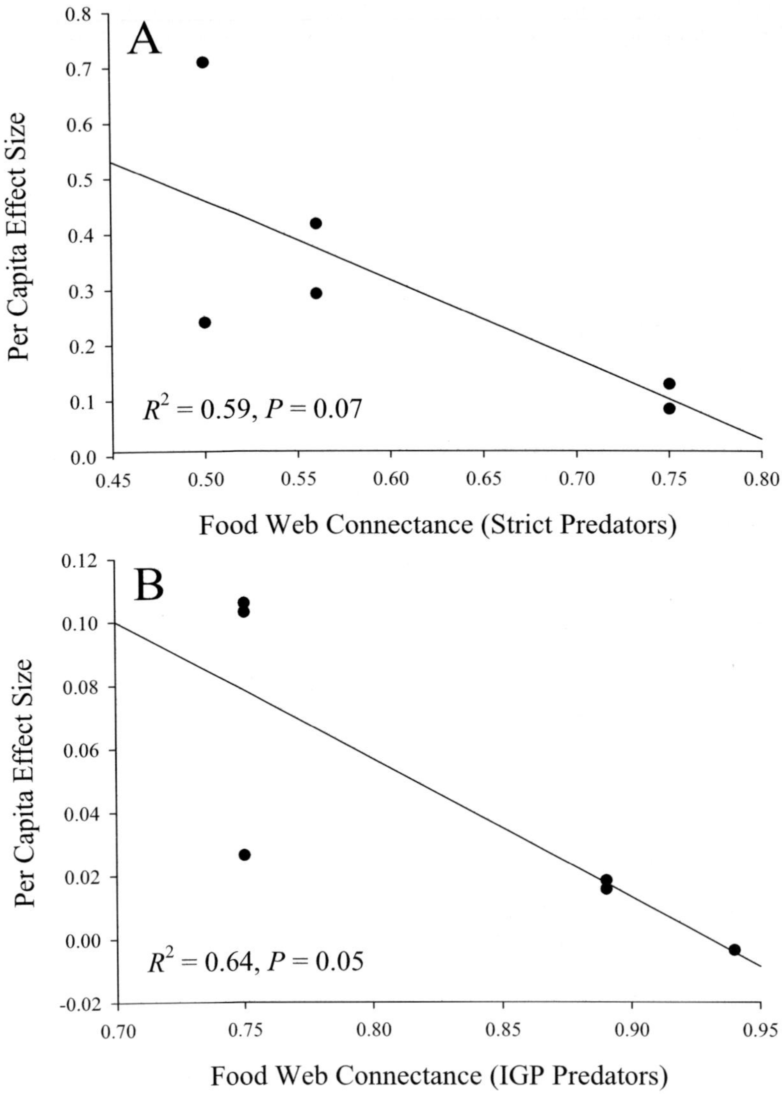

Figure 3. Relationship between food-web connectance (C) and the ability of the predator combination to reduce herbivore density (per capita effect size) for food webs comprised of (A) only strict predators, and (B) only intraguild predators.

Table 1. Predator species identity, food-web species richness (S), food-web connectance (C), and per capita effect size of predators on the herbivore population for each of 29 food-web configurations.

Predator species identity	S	C	Per capita effect size
G. trivittata	2	0.75	0.08
N. seriata	2	0.75	0.13
T. vagus	2	0.50	0.71
C. saltitans	2	0.75	0.10
M. pikei	2	0.75	0.03
P. littoralis	2	0.75	0.11
G. trivittata + *T. vagus*	3	0.56	0.29
N. seriata + *T. vagus*	3	0.56	0.42
N. seriata + *P. littoralis*	3	0.67	0.31
G. trivittata + *P. littoralis*	3	0.78	0.004
M. pikei + *T. vagus*	3	0.67	0.43
P. littoralis + *T. vagus*	3	0.67	0.40
C. saltitans + *T. vagus*	3	0.67	0.16
C. saltitans + *P. littoralis*	3	0.89	0.02
M. pikei + *P. littoralis*	3	0.89	0.02
G. trivittata + *N. seriata* + *T. vagus*	4	0.50	0.24
G. trivittata + *T. vagus* + *P. littoralis*	4	0.63	0.07
G. trivittata + *T. vagus* + *C. saltitans*	4	0.63	0.19
G. trivittata + *T. vagus* + *M. pikei*	4	0.63	0.13
T. vagus + *N. seriata* + *P. littoralis*	4	0.56	0.17
T. vagus + *N. seriata* + *C. saltitans*	4	0.56	0.22
T. vagus + *N. seriata* + *M. pikei*	4	0.56	0.22
C. saltitans + *P. littoralis* + *G. trivittata*	4	0.81	0.005
C. saltitans + *P. littoralis* + *N. seriata*	4	0.69	0.11
C. saltitans + *P. littoralis* + *T. vagus*	4	0.75	0.21
P. littoralis + *M. pikei* + *G. trivittata*	4	0.81	-0.006
P. littoralis + *M. pikei* + *N. seriata*	4	0.69	0.17
P. littoralis + *M. pikei* + *T. vagus*	4	0.75	0.26
C. saltitans + *M. pikei* + *P. littoralis*	4	0.94	-0.004

3.3. Habitat structure and food-web connectance

IGP can be reduced in complex-structured habitats where intraguild prey find refuge from top predators, a situation that often enhances top-down impacts on shared prey (Finke and Denno, 2002). Using published data from an experiment where a diverse predator assemblage (mix of strict and IGP predators) was crossed with habitat complexity (Finke, 2005), we calculated food-web connectance in both simple- and complex-structured habitats and its association with the ability of the predator complex to suppress populations of shared planthopper prey (per capita effect size). The experiment was conducted in laboratory mesocosms using a subset of the dominant arthropods that inhabit inter-tidal wetlands (Finke, 2005). After establishing herbivore populations (*Prokelisia dolus*) on *Spartina* transplants, two predator treatments were assigned: no predators or a mix of two strict predators (the egg predator *Tytthus* and the sheet-web building spider *Grammonota*) and one intraguild predator (*Pardosa*). Both predator treatments were crossed with a habitat complexity treatment that consisted of either adding or withholding leaf litter (thatch) from mesocosms. After two months (two planthopper generations), the per capita effect size of the predator treatments were calculated from the remaining number of planthoppers in treatment mesocosms (for design details see Finke, 2005). Food-web connectance values and directed links were determined as above for the species complex in both in the presence and absence of leaf litter, with the expectation that IGP, and thus food-web connectance, would be reduced in complex-structured habitats.

4. RESULTS

4.1. Relationships between food-web connectance, species richness and effect size

Predator species richness, food-web connectance values, and per capita effect sizes of predators on the herbivore population for each of the 29 food webs used in this assessment are provided in Table 1. There was a significant negative relationship between food-web connectance and the effect of the predator complex on suppressing planthopper populations ($R^2 = 0.47$, $P < 0.0001$; Fig. 2). Overall, food webs with low connectance (those dominated by strict predators such as *Tytthus*, *Naemia*, and *Grammonota*) had a greater adverse effect on herbivore populations than those with high connectance (those dominated by intraguild predators like *Pardosa* and *Clubiona*). Notably, herbivore suppression declined with an increase in food-web connectance for webs containing only strict predators ($R^2 = 0.59$, $P = 0.07$; Fig. 3A) or exclusively intraguild predators ($R^2 = 0.64$, $P = 0.05$; Fig. 3B). Linkage density (L/S), the mean number of directed feeding links per species, averaged 2.25 ± 0.12 across the 29 predator treatment combinations.

Table 2. ANCOVA results for the effects of predator species richness (covariate) and predator functional group (food webs comprised of either strict or intraguild predators) on the ability of the predator functional group to suppress the herbivore population (per capita effect size).

Source of variation	NDF	DDF	*F*	*P*
Species richness	1	11	0.007	0.67
Predator functional group	1	11	6.53	0.03
Species richness x functional group	1	11	0.04	0.85

Notably, there was no clear overall effect of increasing predator richness of the food web on the ability of the predator complex to suppress herbivore populations ($R^2 = 0.03$, $P = 0.39$; Fig. 4). Moreover, per capita effect size did not increase with species richness for food webs comprised of only strict predators ($R^2 = 0.004$, $P = 0.91$; Fig. 5A). In contrast, there was a negative relationship between per capita effect size and species richness for webs consisting exclusively of intraguild predators ($R^2 = 0.60$, $P = 0.07$; Fig. 5 B). ANCOVA, with predator richness as the covariate, found a significant difference in per capita effect size between webs comprised of intraguild predators (per capita effect size = 0.04 ± 0.07) and strict predators (per capita effect size = 0.31 ± 0.07) with webs consisting of intraguild predators exerting much less top-down control ($F_{1,11}$ = 0.20, P = 0.03; Table 2). However, the interaction between species richness and predator functional category was not significant ($F_{1,11}$ = 0.04, P = 0.85). Small sample size (n = 6 for webs including strict and intraguild predators) likely precluded a more robust result. Finally, there was no overall relationship between connectance (C) and species richness (S) (R^2 = 0.01, P = 0.56; Fig. 6). Likewise, connectance was not influenced by an increase in species richness when only strict predators were present (R^2 = 0.38, P = 0.19). However, there was a positive relationship between species richness and connectance for those food webs containing only IGP predators (R^2 = 0.94, P = 0.002). Overall, our results suggest that it is food-web connectance (e.g. as influenced by the proportion of intraguild to strict predators in the food web) rather than species richness *per se* that affects the ability of the predator complex to suppress herbivores.

4.2. Habitat Structure and Food-web Connectance

In a simple-structured habitat without thatch there were 4 species (*Prokelisia*, *Tytthus*, *Grammonota*, and *Pardosa*), 10 feeding links and a connectance value of 0.63. The high number of feeding links and large connectance value resulted from the frequent IGP of *Tytthus* and *Grammonota* by *Pardosa* and moderate levels of cannibalism in *Grammonota* and *Pardosa* (Finke and Denno, 2002, 2003; Langellotto, 2002; Denno *et al.*, 2004; Fig. 7A). Adding thatch to experimental arenas effectively reduced the number of feeding links to 6 and resulted in lower food-web connectance (0.38) due to the virtual elimination of IGP and cannibalism (Finke, 2002; Langellotto, 2002; Fig. 7B). The lower level of food web connectance in the presence of thatch was associated with an enhanced ability of the predator complex to suppress planthoppers (per capita effect size = 0.10 ± 0.005) compared to when thatch is absent (0.05 ± 0.007). Thus, adding thatch in this system was tantamount to shifting the composition of the predator complex from one dominated by IGP predators to one comprised largely of strict predators, a composition that resulted in a reduced number of feeding links, lowered food-web connectance, and enhanced herbivore suppression. Overall, our results suggest that habitat structure moderates intraguild predation by providing spatial refuges for intraguild prey, thus reducing food-web connectance and increasing the effectiveness of the predator complex in suppressing shared prey.

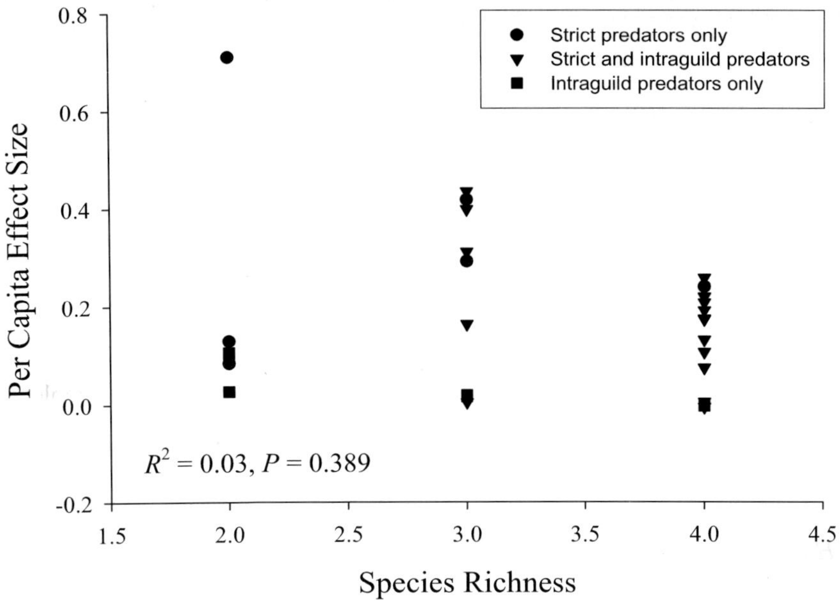

Figure 4. Relationship between species richness of the food web (achieved by varying the number of predator species) and the ability of the predator combination to reduce herbivore density (per capita effect size) for 29 different treatment combinations of strict and intraguild predators. There was no significant relationship.

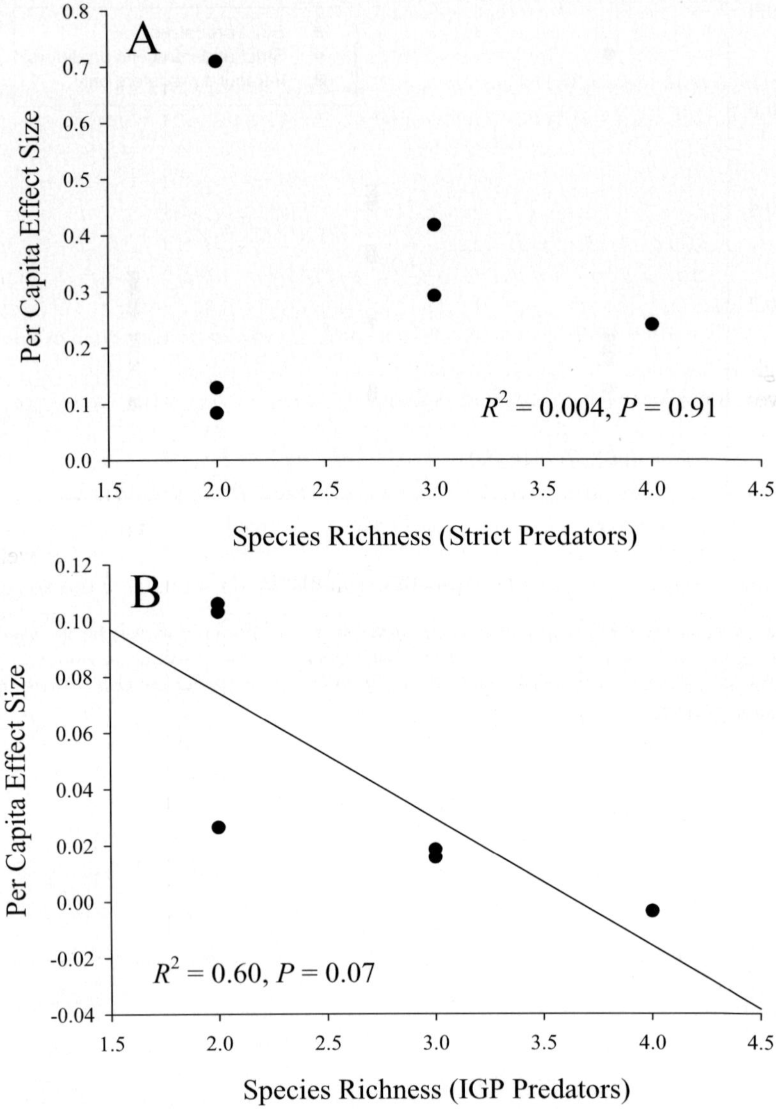

Figure 5. Relationship between species richness of the food web (achieved by varying the number of predator species) and the ability of the predator treatment combination to reduce herbivore density (per capita effect size) for food webs comprised of (A) only strict predators, and (B) only intraguild predators.

5. DISCUSSION

We found a significant negative relationship between food-web connectance and the effectiveness of the predator complex in reducing the population size of shared herbivore prey (Fig. 2). A negative relationship between parasitism rate and food-web complexity has also been shown for parasitoid-dominated communities, whereby parasitoids were better able to suppress pest herbivores in simple-structured enemy complexes characterized by low connectance due to few hyperparasitoid species (Montoya *et al.*, 2003). High levels of connectance in our constructed food-webs resulted from extensive IGP and cannibalism among predators (see also Finke and Denno, 2002, 2003, 2005; Langellotto, 2002; Denno *et al.*, 2004). In general, hunting spiders such as *Pardosa*, *Clubiona*, and *Marpissa* were voracious predators of each other as well as most of the strict predators in the system (*Tytthus* and *Grammonota*) (Finke 2005), which resulted in an increased number of feeding links and high levels of connectance for food webs containing these intraguild predators. However, the role of enhanced food-web connectance in dampening predator effects on prey was not solely due to the occurrence of intraguild predation, since the negative relationship between connectance and prey suppression existed for food webs containing only strict predators (Fig. 3A), as well as those containing only IGP predators (Fig. 3B).

Overall, simply increasing the total number of predator species in the web did not result in enhanced per capita prey suppression (Fig. 4). Thus, it is not surprising that food-web connectance was not related to species richness across all experimental treatments (Fig. 6). However, the ability of predator complexes to suppress herbivore populations in food webs dominated by strict or IGP predators showed different responses to increasing species richness. For food webs comprised of only strict predators, where there was no relationship between species richness and connectance, per capita prey suppression was unaffected by in increase in species richness (Fig. 5A). On the contrary, for food webs comprised of only IGP predators, where species richness was positively related to connectance, per capita prey suppression was diminished as species richness was enhanced (Fig. 5B). Overall, we conclude that it is primarily the composition of the predator complex (proportion of intraguild to strict predators) and not species richness that affects the ability of the predator complex to suppress herbivores through changes in food-web connectance.

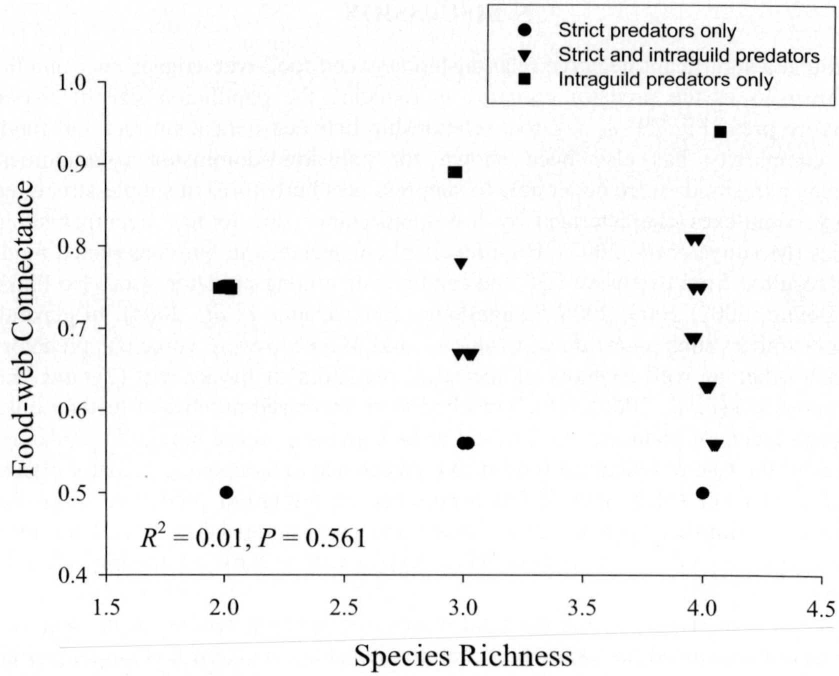

Figure 6. Relationship between food-web connectance (C) and food-web species richness (achieved by varying the number of predator species) for 29 different treatment combinations of strict and intraguild predators. There was no significant relationship

We have argued previously, and here as well, that increasing predator diversity raises the probability for including intraguild predators in the complex and thus diminishes prey suppression and dampens any top-down cascade to basal resources (Finke and Denno, 2004, 2005). Importantly, the goal of biological control programs in agricultural systems is to initiate trophic cascades by manipulating predator complexes that result in enhanced crop yield (Snyder *et al.*, 2005). Toward this end, our results suggest that the extent to which multiple natural enemies enhance pest suppression depends more on the composition of predator complex than on the number of component species. Notably, the food webs we created for our experiments captured the range in taxonomic composition (e.g. spiders, heteropterans, and coccinellid beetles), trophic composition (proportion of strict and intraguild predators) and richness (number of important natural enemies) of predator complexes present in many agroecosystems (e.g., Hochberg, 1996; Riechert and Lawrence, 1997; Rosenheim, 1998; Denoth *et al.*, 2002; Symondson *et al.*, 2002; Wilby and Thomas, 2002; Cardinale *et al.*, 2003). Moreover, the mean linkage density (L/S) across all of our experimental food webs was 2.25, a value very similar to that for food-webs at large (2.0; Pimm *et al.*, 1991) and for insect-dominated food webs (2.2; Schoenly *et al.*, 1991). Thus, the message drawn from our experiment is likely general and can be extended legitimately to other systems.

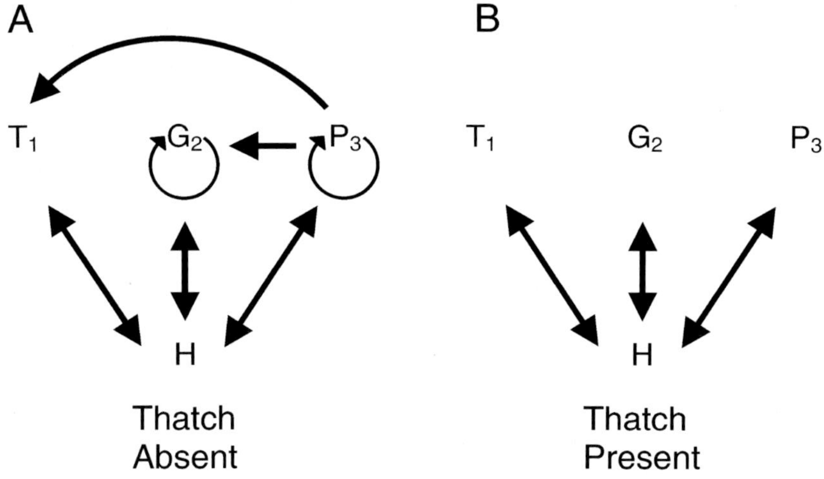

Figure 7. Effect of habitat complexity on the number of feeding links (L) and food-web connectance in an experimentally constructed food web. (A) Directed feeding links (arrows) in a food web comprised of the herbivore Prokelisia dolus *(H) and the three predators* Tytthus *(T1),* Grammonota *(G2), and* Pardosa *(P3) in a simple-structured habitat without thatch.* P. dolus *is very insensitive to intraspecific density effects (no return arrow). All predators suppress populations of the herbivore, and the predators exhibit strong numerical responses (aggregative or reproductive) to herbivore density (bidirectional arrows).* Pardosa *consumes both* Tytthus *and* Grammonota, *but the reverse does not occur, nor does* Pardosa *respond numerically to densities of either* Tytthus *or* Grammonota *(unidirectional arrows). Both* Grammonota *and* Pardosa *are cannibalistic (return arrows). For this example food web, L = 10 and C = 0.63. (B) Adding thatch to experimental arenas effectively reduced the number of feeding links in this web to 6 and resulted in lower food-web connectance (0.38) due to the virtual elimination of IGP by* Pardosa *(P3) on* Tytthus *(T1) and* Grammonota *(G2) (no unidirectional arrows) and cannibalism (no return arrows).*

Our results suggest that increasing predator diversity by using multiple natural enemies to control arthropod pests has different effects depending on the trophic composition of the predator complex. Most natural-enemy assemblages in cropping systems contain a large number of species, however only a few are sufficiently abundant to exert significant control on pests. For example, in Philippine rice more than 90 predator species and 30 parasitoid taxa have been recorded (Schoenly *et al.*, 1998). Of these, however, only a few predator species stand out as being persistent and abundant (Heong and Schoenly, 1998; Schoenly *et al.*, 1998). These include the heteropteran predators *Microvelia douglasi* (Hemiptera: Veliidae), *Mesovelia vittigera* (Hemiptera: Mesoveliidae), and the planthopper egg predator *Cyrtorhinus lividipennis* (Hemiptera: Miridae), and the spiders *Atypena formosana* (Araneae: Linyphiidae), *Tetragnatha javana* (Araneae: Tetragnathidae), and *Lycosa pseudoannulata* (Araneae: Lycosidae) (Heong and Schoenly, 1998; Schoenly *et al.*, 1998). Other cropping systems, cotton for example, support a very different complex of important predator species including the lacewing *Chrysopa carnea* (Neuroptera: Chrysopidae) and the heteropterans *Geocoris* sp. (Hemiptera: Lygaeidae), *Nabis* spp. (Hemiptera: Nabidae), *Zelis* spp. (Hemiptera: Reduviidae) and *Orius tristicolor* (Hemiptera: Anthocoridae) (Rosenheim *et al.*, 1993; Rosenheim and Wilhoit, 1993). Yet other systems such as potato support an even different predator assemblage comprised of the lady beetle *Coleomegilla maculata* (Coleoptera: Coccinellidae), the lacewing *Chrysoperla rufilabris* (Neuroptera: Chrysopidae) and the midge *Aphidoletes aphidimyza* (Diptera: Cecidomyiidae) (Lucas *et al.*, 1998). Thus, for successful biological control and management of the enemy complex it becomes critical to identify the important predators in the system and elucidate the nature and strength of interactions among them.

Because our experiments included many of the above-mentioned predators, one can ask if there are insights from our *Spartina* system that might help target particular combinations of predator taxa to encourage or avoid when managing multiple predators for biocontrol? The least effective predator combinations (effect sizes < 0.05) were the two-species combinations of *Pardosa* and *Marpissa*, *Pardosa* and *Clubiona*, and *Pardosa* and *Grammonota*, and the three-species combinations of *Pardosa*, *Marpissa* and *Grammonota*, *Pardosa*, *Marpissa* and *Clubiona*, and *Pardosa*, *Clubiona* and *Grammonota,* all sub-webs with high connectance (> 0.8). Thus, predator combinations dominated by generalist hunting spiders tended to be least effective in suppressing prey.

This conclusion appears on the surface to be at odds with several studies suggesting that assemblages of generalist spiders can lead to effective biological control (Riechert and Bishop, 1990; Riechert and Lawrence, 1997; Hodge, 1999). Variation in the structural complexity of the habitat may underlie discrepancies in the literature concerning the effectiveness of spider complexes and other generalist predator assemblages in suppressing prey. We have shown that by increasing the structural complexity of the habitat by including leaf litter, IGP was significantly reduced, food-web connectance dropped from 0.63 to 0.38 as a result, and the effectiveness of the predator complex (*Pardosa*, *Tytthus* and *Grammonota*) in reducing planthopper populations more than doubled (per capita effect size = 0.10) when compared to litter-free conditions (per capita effect size = 0.05). Both *Tytthus* and *Grammonota* find a spatial refuge from *Pardosa* predation, largely because intraguild prey locate hiding sites from the visually orienting *Pardosa* (Finke and Denno, 2002; Finke, 2005). In the field, the combination of reduced IGP and *Pardosa* aggregation leads to effective planthopper suppression in litter-rich habitats (Döbel and Denno, 1994; Denno *et al.*, 2005 a, 2005 b). In fact, many invertebrate predator species aggregate in complex-structured habitats and are more effective in suppressing prey there (reviewed in Langellotto and Denno, 2004). Although numerous mechanisms have been proposed to explain predator aggregation in complex habitats, one that surfaces more often than not is reduced IGP and cannibalism (Langellotto, 2002; Langellotto and Denno, 2004; Denno *et al.*, 2005 a). Altogether, our data suggest that increasing the structural complexity of the habitat reduces IGP and enhances the effectiveness of a natural-enemy complex dominated by generalist predators.

This finding provides encouragement that agricultural habitats can be managed in ways that enhance the success of multiple predators for biological control. A particularly relevant comparison can be made between the *Spartina* system and Asian rice because the two systems are remarkably similar with regard to taxonomic composition, diversity and trophic structure. Both systems have grass monocultures as the resource base, planthoppers and leafhoppers are the major herbivores, and both strict predators (heteropteran egg predators) and intraguild predators (spiders) abound (Kenmore *et al.*, 1984; Döbel and Denno, 1994; Heong and Schoenly, 1998; Schoenly *et al.*, 1998; Denno and Peterson, 2000; Denno *et al.*, 2003). Moreover, intraguild predation is well documented in both systems, particularly between hunting spiders and mirid egg predators (Döbel and Denno, 1994; Fagan *et al.*, 1998; Finke and Denno, 2002, 2003). Thus, opportunities abound for manipulating habitat structure in ways (e.g., use of architecturally complex varieties, altering harvest practices to enhance litter, and managing the vegetation on levees) that might improve pest management. In fact, habitat manipulations have been used previously in tropical rice to increase the efficacy of generalist natural enemies (Settle *et al.*, 1996).

With some knowledge of the important predators in a system and their interactions with each other and habitat complexity, we suggest that food-web connectance techniques can be useful in targeting particular predator combinations to encourage or discourage biological control. Estimating directed feeding links among the component species in a food web, however, is not a simple task. Experiments can be conducted to determine consumption rates from which interaction strengths within and between species can be calculated in a diversity of ways (Berlow *et al.*, 1999, 2004; Montoya *et al.*, 2003; Melián and Bascompte, 2004). A possible short cut involves estimating the proportion of a predator's diet each prey species comprises, ranking such links from weak to strong, and then using threshold criteria to include or exclude the feeding link in the food web analysis (Winemiller, 1990). However, such techniques provide estimates for only direct consumptive effects (e.g. predation, IGP, or cannibalism), they do not allow for a measure of indirect non-consumptive effects (e.g., trait-mediated effects, exploitative competition, or interference phenomena) (Ohgushi *et al.*, in press). For example, hunting spiders can reduce the fitness of grasshoppers by forcing them to switch to poor-quality host plants (Schmitz *et al.*, 1997; Schmitz, 1998), predators can promote the immigration of insect herbivores from optimal resources (Cronin *et al.*, 2004), and interference between predator species (anti-predator behavior and altered foraging) can result in lowered fitness (Magnhagen, 1991; Lima, 1998; Parsons *et al.*, 2002). Although such risk-avoiding behaviors may deter predation or IGP, they certainly lead to fitness reductions and thus form a legitimate link between two species in a food web, albeit indirect. As a consequence of such indirect effects among and within consumers, link density and connectance among players in a food web will most certainly increase and in some cases may even double (see Ohgushi *et al.*, in press).

While at first estimating all the direct and indirect interactions among the important consumers in a food web appears daunting, the answers to predictive food-web dynamics and biological control likely lie with such an approach. Toward this end, we feel that our findings here take an initial first step. Despite its somewhat preliminary nature, our analysis has produced a very robust result, namely that food-web connectance, as it is enhanced by predator-predator interactions, is inversely related to efficacy of the predator complex in reducing herbivore density. Thus, our results support the view that IGP, more often than not, can disrupt biocontrol (Rosenheim *et al.*, 1993, 1995; Snyder *et al.*, 2005). However, it is most encouraging that increasing the structural complexity of crop habitats might reduce IGP by providing spatial refuges for intraguild prey and thus provide the opportunity for using multiple predators for successful biological control.

6. ACKNOWLEDGEMENTS

This article extends from a presentation at a symposium on “Trophic and Guild Interactions in Biological Control” at the joint meeting of the IOBC-NRS/Biocontrol Network held in Magog-Orford, Quebec, Canada on 8-11 May 2005. We sincerely thank both Jaques Brodeur and Guy Boivin for inviting RFD to participate in the symposium and to Lucie Levesque for her superb organization of the conference. We also gratefully acknowledge the many sponsors including NSERC for supporting this stimulating conference. Two anonymous reviewers provided insightful comments on an earlier draft of this chapter. This research was supported by National Science Foundation grants to RFD (DEB-0313903) and DLF (DEB-0410572).

7. REFERENCES

Arim, M., and Marquet, P.A., 2004, Intraguild predation: A widespread interaction related to species biology, Ecol. Lett. 7: 557-564.

Berlow, E.L., Navarrete, S.A., Briggs, C.J., Power, M.E., and Menge, B.A., 1999, Quantifying variation in the strengths of species interactions. Ecology 80: 2206-2224.

Berlow, E.L., Neutel, A-M, Cohen, J.E., de Ruiter, P.C., Benjamin, B., Emmerson, M., Fox, J.W., Jansen, V.A.A., Jones, J.I., Kokkoris, G.D. , Logofet, D.O. , McKane, A.J. , Montoya, J.M., and Petchey. O., 2004, Interaction strengths in food webs: Issues and opportunities, J. Ecol. 73: 585-598.

Bertness, M.D., 1991, Zonation of *Spartina patens* and *Spartina alterniflora* in a New England salt marsh, Ecology 72: 138-148.

Bertness, M.D., and Pennings, S.C., 2000, Spatial variation in process and pattern in salt marsh plant communities in eastern North America, in: Concepts and Controversies in Tidal Marsh Ecology, M.P. Weinstein and D.A. Kreeger, eds., Kluwar Academic Publishers, Boston, pp. 39-57.

Cardinale, B.J., Harvey, C.T., Gross, K., and Ives, A.R., 2003, Biodiversity and biocontrol: Emergent impacts of a multi-enemy assemblage on pest suppression and crop yield in an agroecosystem, Ecol. Lett. 6: 857-865.

Chalcraft, D.R., and Resetarits. W.J.J., 2003, Predator identity and ecological impacts: Functional redundancy or functional diversity? Ecology 84: 2407-2418.

Chang, G.C., 1996, Comparison of single versus multiple species of generalist predators for biological control, Environ. Entomol. 25: 207-212.

Closs, G.P., Balcombe, S.R., and Shirley, M.J., 1999, Generalist predators, interaction strength and food-web stability, Adv. Ecol. Res. 28: 93-126.

Cook, A., and Denno, R.F., 1994, Planthopper-plant interactions: Feeding behavior, plant nutrition, plant defense, and host plant specialization, in: Planthoppers: Their Ecology and Management, R..F. Denno and T.J. Perfect, eds., Chapman and Hall, New York, pp. 114-139.

Cronin, J.T., Haynes, K.J., and Dillemuth, F., 2004, Spider effects on planthopper mortality, dispersal, and spatial population dynamics, Ecology 85: 2134-2143.

Denno, R.F., 1983, Tracking variable host plants in space and time, in: Variable Plants and Herbivores in Natural and Managed Systems, R.F. Denno and M.S. McClure, eds., Academic Press, New York, pp. 291-341.

Denno, R.F., Finke, D.L., and Langellotto, G.A., 2005 a, Direct and indirect effects of vegetation structure and habitat complexity on predator-prey and predator-predator interactions, in Ecology of Predator-Prey Interactions, P. Barbosa and I. Castellanos, eds., Oxford University Press, London, pp. 211-239.

Denno, R.F., Gratton, C., Döbel, H., and Finke, D.L., 2003, Predation risk affects relative strength of top-down and bottom-up impacts on insect herbivores, Ecology 84: 1032-1044.

Denno, R.F., Gratton, C., Peterson, M.A., Langellotto, G.A., Finke, D.L., and Huberty, A.F., 2002, Bottom-up forces mediate natural-enemy impact in a phytophagous insect community, Ecology 83: 1443-1458.

Denno, R.F., and Grissell, E.E., 1979, The adaptiveness of wing- dimorphism in the salt marsh-inhabiting planthopper, *Prokelisia marginata* (Homoptera: Delphacidae), Ecology 60: 221-36.

Denno, R.F., Lewis, D., and Gratton, C., 2005 b, Spatial variation in the relative strength of top-down and bottom-up forces: causes and consequences for phytophagous insect populations, Annal. Zool. Fenn. 42: 295-311.

Denno, R.F., Mitter, M.S., Langellotto, G.A., Gratton, C., and Finke, D.L., 2004, Interactions between a hunting spider and a web-builder: Consequences of intraguild predation and cannibalism for prey suppression, Ecol. Entomol. 29: 566-577.

Denno, R.F., and Peterson, M.A., 2000, Caught between the devil and the deep blue sea, mobile planthoppers elude natural enemies and deteriorating host plants, Am. Entomol. 46: 95-109.

Denno, R.F., Peterson, M.A., Gratton, C., Cheng, J., Langellotto, G.A., Huberty, A.F., and Finke, D.L., 2000, Feeding-induced changes in plant quality mediate interspecific competition between sap-feeding herbivores, Ecology 81: 1814-1827.

Denno, R.F., Roderick, G.K., Peterson, M.A., Huberty, A.F., Döbel, H.G., Eubanks, M.D., Losey, J.E., and Langellotto, G.A., 1996, Habitat persistence underlies the intraspecific dispersal strategies of planthoppers, Ecol. Mono. 66: 389-408.

Denoth, M., Frid, L., and Myers, J.H., 2002, Multiple agents in biological control: improving the odds? Biol. Cont. 24: 20-30.

Döbel, H.G., and Denno, R.F., 1994, Predator-planthopper interactions, in: Planthoppers: Their Ecology and Management, R.F. Denno and T.J. Perfect, eds., Chapman and Hall, New York, pp. 325-399.

Downing, A.L., and Leibold, M.A., 2002, Ecosystem consequences of species richness and composition in pond food webs, Nature 416: 837-841.

Dunne, J.A., Williams, R.J., and Martinez, N.D., 2002, Food web structure and network theory: role of connectance and size, Proc. Nat. Acad. Sci. USA, 99: 12917-12922.

Eubanks, M.D., 2001, Estimates of the direct and indirect effects of red imported fire ants on biological control in field crops, Biol. Cont. 21: 35-43.

Fagan, W.F., 1997, Omnivory as a stabilizing feature of natural communities, Am. Nat. 150: 554-567.

Fagan, W.F., Hakim, A.L., Ariawan, H., and Yuliyantiningsih, S., 1998, Interactions between biological control efforts and insecticide applications in tropical rice agroecosystems: The potential role of intraguild predation, Biol. Cont. 13: 121-126.

Finke, D.L., 2005, Predator diversity, habitat complexity and the strength of terrestrial trophic cascades, Doctoral dissertation, Department of Entomology, University of Maryland, College Park.

Finke, D.L., and Denno, R.F., 2002, Intraguild predation diminished in complex-structured vegetation: Implications for prey suppression, Ecology 83: 643-652.

Finke, D. L., and Denno, R. F., 2003, Intra-guild predation relaxes natural enemy impacts on herbivore populations, Ecol. Entomol. 28: 67-73.

Finke, D.L., and Denno, R.F., 2004, Predator diversity dampens trophic cascades, Nature 429: 407-410.

Finke, D.L., and Denno, R.F., 2005, Predator diversity and the functioning of ecosystems: The role of intraguild predation in dampening trophic cascades, Ecol. Lett. 8: 1299-1306.

Gallagher, J.L., Somers, G.F., Grant, D.M., and Seliskar, D.M., 1988, Persistent differences in two forms of *Spartina alterniflora*: a common garden experiment. Ecology 69: 1005-1008.

Hart, D.R. 2002. Intraguild predation, invertebrate predators, and trophic cascades in lake food webs, J. Theor. Biol. 218: 111-128.

Hawkins, B.A., Mills, N.J., Jervis, M.A., and Price, P.W., 1999, Is the biological control of insects a natural phenomenon? Oikos 86: 493-506.

Heinz, K.M., and Nelson, J.M., 1996, Interspecific interactions among natural enemies of *Bemisia* in an inundative biological control program, Biol. Cont. 6: 384-393.

Heong, K.L., and Schoenly, K.G., 1998, Impact of insecticides on herbivore-natural enemy communities in tropical rice ecosystems, in: Ecotoxicology: Pesticides and Beneficial Organisms, P.T. Haskell and P. McEwen, eds, Chapman and Hall, London, pp. 381-403.

Hochberg, M.E., 1996, Consequences for host population levels of increasing natural enemy species richness in classical biological control, Am. Nat. 147: 307-318.

Hodge, M.A., 1999, The implications of intraguild predation for the role of spiders in biological control, J. Arach. 27: 351-362.

Huston, M.A., 1997, Hidden treatments in ecological experiments: re-evaluating the ecosystem function of biodiversity, Oecologia 110: 449-460.

Jolliffe, P.A., 2000, The replacement series, J. Ecol. 88: 371-385.

Kenmore, P.E., Cariño, F.O., Perez, C.A., Dyck, V.A., and Gutierrez, A.P., 1984, Population regulation of the rice brown planthopper (*Nilaparvata lugens*) within rice fields in the Philippines, J. Plant Prot. Trop. 1: 19-37.

Langellotto, G., 2002, The aggregation of invertebrate predators in complex habitats: Ecological mechanisms and practical applications, Doctoral Dissertation, Department of Entomology, University of Maryland, College Park, MD.

Langellotto, G.A., and Denno, R.F., 2004, Responses of invertebrate natural enemies to complex-structured habitats: A meta-analytical synthesis, Oecologia 139: 1-10.

Lima, S.L., 1998, Stress and decision making under the risk of predation: recent developments from behavioral, reproductive, and ecological perspectives, Adv. Stud. Behav. 27: 215-290.

Losey, J.E., and Denno, R.F., 1998, Positive predator-predator interactions: Enhanced predation rates and synergistic suppression of aphid populations, Ecology 79: 2143-2152.

Losey, J.E., and Denno, R.F., 1999, Factors facilitating synergistic predation: The central role of synchrony, Ecol. Appl. 9: 378-386.

Lucas, E., Coderre, D., and Brodeur, J., 1998, Intraguild predation among aphid predators: Characterization and influence of extraguild prey, Ecology 79: 1084-1092.

Magnhagen, C., 1991, Predation risk as a cost of reproduction, Trends Ecol. Evol. 6: 183-186.

Matsumura, M., Trafelet-Smith, G.M., Gratton, C., Finke, D.L., Fagan, W.F., and Denno, R.F., 2004, Does intraguild predation enhance predator performance? A stoichiometric perspective, Ecology 89: 2601-2615.

May, R.M., 1973, Stability and Complexity in Model Ecosystems, Princeton University Press, Princeton, New Jersey..

McCann, K.S., 2000, The diversity-stability debate, Nature 405: 228-233.

McCann, K.S., and Hastings, A., 1997, Re-valuating the omnivory-stability relationship in food webs, Proc. Roy. Soc. Lond. B 264: 1249-1254.

McCann, K S., Hastings, A., and Huxel, G.R., 1998, Weak trophic interactions and the balance of nature, Nature 395: 794-798.

Melián, C.J. and Bascompte, J., 2004, Food web cohesion, Ecology 85: 352-358.

Montoya, J.M., Rodríguez, M.A., and Hawkins, B.A., 2003, Food web complexity and higher-level ecosystem services, Ecol. Lett. 6: 587-593.

Morin, P.J., 1999, Community Ecology, Blackwell Science, Inc., London, UK.

Naeem, S., and Li, S., 1997, Biodiversity enhances ecosystem reliability, Nature 390: 507-509.

Ohgushi, T., Craig, A., and Price, P.W., eds, in press, Indirect Interaction Webs: Nontrophic Linkages Through Induced Plant Traits, Cambridge University Press, London, UK.

Ornes, W.H., and Kaplan, D.I., 1989, Macronutrient status of tall and short forms of *Spartina alterniflora* in a South Carolina salt marsh, Mar. Ecol. Prog. Ser. 55: 63-72.

Parsons, M.H., Walker, S.E., and Rypstra, A.L., 2002, Fitness costs and benefits of antipredator behaviour mediated by chemotactical cues in the wolf spider *Pardosa milvina* (Araneae: Lycosidae), Beh. Ecol. 13: 386-392.

Pimm, S.L., Lawton, J.H., and Cohen, J.E., 1991, Food web patterns and their consequences, Nature 350: 669-674.

Phoofolo, M.W., and Obrycki, J.J., 1998, Potential for intraguild predation and competition among predatory Coccinellidae and Chrysopidae, Entomol. Exp. Appl. 89: 47-55.

Polis, G., 1998, Stability is woven by complex webs, Nature 395: 744-745.

Polis, G., Myers, C., and Holt, R., 1989, The ecology and evolution of intraguild predation: Potential competitors that eat each other, Ann. Rev. Ecol. Syst. 20: 297-330.

Polis, G., and Strong, D.R., 1996, Food web complexity and community dynamics, Am. Nat. 147: 813-846.

Prasad, R.P., and Snyder, W.E., 2004, Predator interference limits fly egg biological control by a guild of ground-active beetles, Biol. Cont. 31: 428-437.

Redfield, A. C., 1972, Development of a New England salt marsh, Ecol. Mono. 42: 201-237.

Riechert, S.E. and Bishop, L., 1990, Prey control by an assemblage of generalist predators: Spiders in a garden test system, Ecology 71: 1441-1450.

Riechert, S.E. and Lawrence, K., 1997, Test for predation effects of single versus multiple species of generalist predators: spiders and their insect prey, Entomol. Exp. Appl. 84: 147-155.

Rosenheim J.A., 1998, Higher order predators and the regulation of insect herbivore populations, Ann. Rev. Entomol. 43: 421-447.

Rosenheim, J.A., Glik, T.E., Goeriz, R.E., and Rämert, B., 2004, Linking a predator's foraging behavior with its effects on herbivore population suppression, Ecology 85: 3362-3372.

Rosenheim, J.A., Kaya, H.K., Ehler, L.E., Marois, J.J., and Jaffee, B.A., 1995, Intraguild predation among biological-control agents: theory and evidence, Biol. Cont. 5: 303-335.

Rosenheim J.A., and Wilhoit, L.R., 1993, Predators that eat other predators disrupt cotton aphid control, Cal. Ag. 47: 7-9.

Rosenheim J.A., Wilhoit, L.R., and Armer, C.A., 1993, Influence of intraguild predation among generalist insect predators on the suppression of an herbivore population, Oecologia 96: 439-449.

SAS Institute, 2000, JMP Statistics and Graphics Guide, SAS Institute, Cary, North Carolina.

Schmitz, O.J., 1998, Direct and indirect effects of predation and predation risk in old-field interaction webs, Am. Nat., 151: 327-342.

Schmitz, O.J., Beckerman, A.P., and O'Brien, K.M., 1997, Behaviorally mediated trophic cascades: Effects of predation risk on food web interactions, Ecology 78: 1388-1399.

Schoenly, K., Beaver, R.A., and Heumier, T.A., 1991, On the trophic relations of insects: A food-web approach, Am. Nat. 137: 597-638.

Schoenly, K.G., Justo, H.D., Barrion, A.T., Harris, M.K., and Bottrell, D.G., 1998, Analysis of invertebrate biodiversity in a Philippine farmer's irrigated rice field, Environ. Entomol. 27: 1125-1136.

Settle, W. H., Ariawan, H., Astuti, E. T., Cahyana, W., Hakim, A. L., Hindayana, D., Lestari, A. S., and Sartanto, P., 1996, Managing tropical rice pests through conservation of generalist natural enemies and alternative prey, Ecology 77: 1975-1988.

Sih, A., Englund, G., and Wooster, D., 1998, Emergent impacts of multiple predators on prey, Trends Ecol. Evol. 13: 350-355.

Snyder, W.E., Chang, G.C., and Prasad, R.P., 2005, Conservation biological control: Biodiversity influences the effectiveness of predators, in: Ecology of Predator-Prey Interactions, P. Barbosa and I. Castellanos, eds., Oxford University Press, London, pp. 324-343.

Snyder, W.E., and Ives, A.R., 2001, Generalist predators disrupt biological control by a specialist parasitoid, Ecology 82: 1571-1583.

Snyder, W.E., and Wise, D.H., 1999, Predator interference and the establishment of generalist predators populations for biocontrol, Biol. Cont. 15: 283-292.

Straub, C.S., and Snyder, W.E., 2006, Species identity dominates the relationship between predator biodiversity and herbivore suppression, Ecology.

Strauss, S. Y., 1991, Indirect effects in community ecology: Their definition, study and importance, Trends Ecol. Evol. 6: 206-210.

Symondson, W.O.C, Sunderland, K.D., and Greenstone, M.H., 2002, Can generalist predators be effective biocontrol agents? Ann. Rev. Entomol. 47: 561-594.

Walde, S.J., Hardman, M., and Magagula, C.N., 1997, Direct and indirect species interactions influencing within-season dynamics of apple rust mite, *Aculus schlechtendali* (Acari: Eriophyidae), Exp. Appl. Acar. 21: 587-614.

Wilby, A., and Thomas, M.B., 2002, Natural enemy diversity and pest control: patterns of pest emergence with agricultural intensification, Ecol. Lett. 5: 353-360.

Winemiller, K.O., 1990, Spatial variation in tropical fish networks, Ecol. Mono. 60: 331-367.

INTER-GUILD INFLUENCES ON INTRA-GUILD PREDATION IN PLANT-FEEDING OMNIVORES

D.R. Gillespie[1,2] and B.D. Roitberg[2]

[1]Pacific Agri-Food Research Centre, Agriculture and Agri-Food Canada, POB 1000, Agassiz, British Columbia, Canada, V0M 1A0; [2]Department of Biology, Simon Fraser University, Burnaby, British Columbia, Canada, V5A 1S6

Abstract: Omnivorous predators that also feed on plants (omnivores) are important natural enemies of pest herbivores in many biological control programs. Bottom-up influences of the plant on the biology of these omnivores have been shown to have considerable influence on their interactions with target prey species. Likewise, plants should affect how these omnivores interact with other members of the natural enemy guild in biological control settings. Using community models, we show that intra-guild predation (IGP) by omnivores, in contrast to "pure predators" that do not feed on plants, could be affected by plant quality, and plant toxins. These models suggest that plants should affect IGP by omnivores in two ways: in the short-term by affecting the tendency of omnivores to engage in IGP, and in the long-term, by affecting both herbivore and omnivore population dynamics. We review the available literature effects of plant traits (plant quality, plant toxins, plant hairs, and plant kairomones) on IGP in omnivorous and pure predator species. Although plant traits do affect the nature and intensity of IGP by omnivores, we conclude that the effects are idiosyncratic, and are specific to the species involved.

1. INTRODUCTION

Omnivorous arthropods that feed on both plants and herbivores are used in biological control programs in many parts of the world, and are components of conservation biocontrol programs and of augmentation and inundation release programs in greenhouse and field crop systems. Some of these species are true bugs (Hemiptera: Heteroptera) and their biology has been reviewed in Alomar and Wiedenmann (1996) and Coll and Ruberson (1998). Some predatory mites are omnivorous, and feed on pollen and plant sap as well as on arthropod prey (McMurtry and Croft, 1997; Magalhães and Bakker, 2002; Nomikou, Janssen, and Sabelis, 2003; van Rijn, van Houten, and Sabelis, 2002; van Rijn and Sabelis, 1990). Western flower thrips, which is a serious pest in many horticultural crops, also feeds on eggs of spider mites and on eggs of predaceous mites and can be important in the population dynamics of spider mites on some crops (Agrawal, Kobayashi and Thaler, 1999; Janssen, Willemse, and van der Hammen, 2003; Magalhães, Janssen, Montserrat, and Sabelis, 2005). Some Carabidae, for example *Harpalus* spp. (Coleoptera) are omnivorous, and feed on seeds as well as arthropod prey (eg Menalled, Lee, and Landis, 2001, Poole, Gormalley, and Skeffington, 2003, Hough-Goldstein, Vangessel, and Wilson, 2004). Similarly, some Orthoptera feed

J. Brodeur and G. Boivin (eds.), Trophic and Guild Interactions in Biological Control, 71–100.

omnivorously and may be important in conservation biocontrol programs (Bright, Bernays, and Moran, 1994, Hough-Goldstein, *et al.*, 2004). These examples are not intended to be exhaustive summaries of the extent of omnivorous feeding in biocontrol and natural communities but rather to point to the diversity of omnivorous biocontrol agents. Concomitant with taxonomic diversity are great differences in foraging ecology of omnivorous arthropods. For example, a foliage-dwelling true bug and a ground beetle on the soil surface face different concentrations and availabilities of plant and animal resources.

The omnivorous natural enemies that are used as biological control agents feed on both the animal and plant kingdoms, and are often generalist feeders in the animal kingdom. This fits one definition of omnivores, as animals that feed from both the plant and animal kingdoms (e.g. Lincoln, Boxshall, and Clark, 1998). A broader definition by Pimm and Lawton (1978) defined omnivores as animals that feed at more than one trophic level. This includes a great many predatory interactions, including intra-guild predation (IGP) at secondary consumer trophic levels and above (Polis, Myers, and Holt, 1989; Fox, 1975; Agrawal, 2003). IGP is thought to have a very important influence on the natural enemy guild in biological control programs, and can affect, sometimes dramatically, the effects of natural enemies on target prey populations (Rosenheim, Kaya, Ehler, Marois, and Jaffe, 1995; Rosenheim, 2001).

How should presence of plant-feeding omnivores (hereafter "omnivores") in biological control systems affect IGP, and ultimately, the impacts of these omnivores on the target pests? Because omnivores can feed on both the host plant and on animal prey, their tendency to IGP is of importance in biological control for two reasons. Firstly, under conditions of prey scarcity, omnivores can switch to, or increase their degree of plant feeding (Coll and Guershon 2002). The tendency of the omnivore to forage for other animal prey, such as Intra-Guild prey (IGprey) rather than feed on plants, is therefore important because otherwise the beneficial omnivore might cause plant damage and thus act as a pest. Secondly, IGP has been found to reduce overall mortality in target pest populations (Polis *et al.*, 1989; Rosenheim; 2001). If an omnivorous predator concentrates on plants for food and reduces attacks on IGprey then omnivorous predators could have considerably less impact on IGprey populations than "pure predators", defined here as predators that feed exclusively within the animal kingdom. Complexity of IGP in food webs containing plant-feeding omnivores will depend on the trophic level at which the omnivore occurs (Fig. 1). At the secondary consumer trophic level, there is a third source of food for the omnivore, which can contribute in some way to reproduction, and a third feeding activity for the omnivore to engage in, which alters the way the omnivore spends its time. At the primary consumer trophic level, IGP involving an omnivore is similar in structure to food webs containing IGP among pure predators at the secondary trophic level.

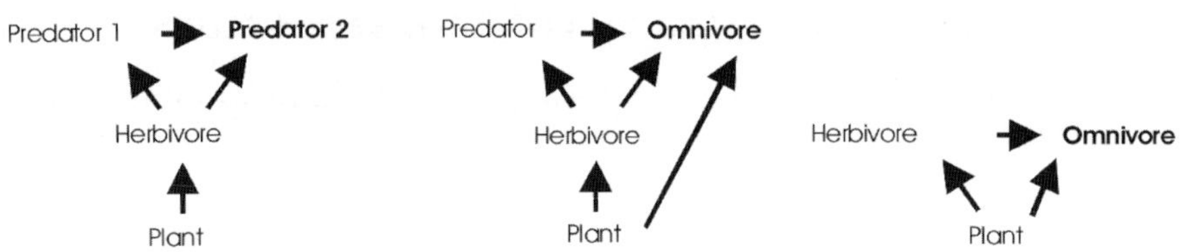

Figure 1. Hypothetical food webs containing either an intra-guild predator (left, bold) or an intra-guild omnivore.

Omnivorous species are frequently involved in IGP in nature. Arim and Marquet (2004) examined the frequency of IGP in food webs, and noted that among the potential IGpredators and IGprey, IGP among intermediate omnivores, defined as organisms that fed on both plants and animals, occurred much more frequently than would be expected under their null model of IGP interactions between species pairs purely by chance.

If omnivores respond to IGP opportunities differently from "pure predators" then these differences are likely to be mediated through the influence of plant resources on the omnivore. Therefore, an examination of the bottom-up effects of plants on intraguild relationships in pure predators and omnivorous predators might suggest how plants affect IGP by plant-feeding omnivores.

We developed a tritrophic community model which allows for omnivory and IGP at the third trophic level, and used this to explore the potential effects of omnivorous feeding on population dynamics. We reviewed the literature on the effects of plant characteristics on tritrophic food webs. We used this literature review and the results from our model to compare descriptive models of the responses of omnivorous predators and pure predators to plant characteristics. We particularly wanted to consider how recruitment to plant patches and the per capita tendency to IGP (PCIGP, i.e. the per capita instantaneous tendency to IGP) might be influenced by attributes of the plant that differently affect pure predators and omnivores. We then developed some predictions of how plant characteristics might affect PCIGP in a predatory omnivore. Our purpose in this chapter was not to provide an exhaustive review of the effects of plants on tritrophic relationships. We particularly wanted to examine how omnivores might interact with other natural enemies differently from pure predators, and to consider plants as a context for IGP and other IG interactions. The bottom-up effects of attributes of individual plants and of plant communities on tri-trophic interactions has been considered in a number of studies in biological control and community ecology, but there appears to be a dearth of general theory or experimental study on bottom-up influences on IGP. Studies on IGP in omnivorous species used for biological control (e.g. McGregor and Gillespie, 2005; Cloutier and Johnson, 1993; Lucas and Alomar 2001; Gillespie and Quiring 1992) have focussed on behaviour and the outcomes of the intra-guild interactions, and not on the role of omnivorous feeding behaviour *per se* on the tendency of predators to engage in IGP.

2. PLANTS AS CONTEXT FOR GUILD INTERACTIONS

Plants can interact directly and indirectly with the natural enemies of herbivores. Plant quality, toxins, plant hairs, plant architecture, herbivore-induced plant volatiles, and plant nectaries are among the traits that have been shown to affect IGP (see below). Plants influence the nature and intensity of interactions between the member species of communities and thereby determine the structure of those communities (Kagata and Ohgushi, 2006). These plant-driven inter-guild interactions are fundamental processes in community ecology. Top-down interactions of top consumers in plant communities can change the abundance of individual plant species, and the strength and outcome of these top-down effects of natural enemies of herbivores can be moderated by interactions between the natural enemies, or intra-guild interactions (Schmitz and Suttle, 2001).

3. COMMUNITY MODELS

To illustrate the potential for plants to mediate IGP, we begin with a simple predator-prey model that comprises several trophic levels. The model is a modified version of Coll and Izraylevich (1997) and Lalonde, McGregor, Roitberg and Gillespie (1999); it is not meant to represent any particular biological system but rather provide a starting point for discussion. Our model is based upon discrete time steps within generations. In this form, the model is very intuitive, it allows for within-generation changes in behaviour and generates similar dynamics to the analogous continuous time models at least for the parameter values chosen for this exercise (Lalonde *et al.*, 1999). Here, we assume that an herbivore or pest species (V) feeds exclusively on plants that contribute directly to herbivore reproduction. Second, our hypothetical system comprises a strict carnivore species (P) that feeds exclusively upon the aforementioned herbivore. Third, there is an omnivore (O) that feeds upon plants and on two kinds of prey, i.e. both herbivores and predators.

The dynamics of this system are determined as follows: First, the intrinsic herbivore birth rate b_V is modified by intraspecific and interspecific competition from herbivores and omnivores, respectively as:

$$1-\left(\frac{\left(V_t+O_t f\left(V_t,P_t,\mu\right)\right)}{K_v}\right) \tag{1}$$

where: V_t = number of herbivores at time t, O_t = number of omnivores at time t, $f(V_t,P_t,\mu)$ is the plant feeding function for the omnivore that considers the presence of both herbivores and predators as potential prey and K_V is the carrying capacity of the environment from the herbivore's perspective. We assume that plant density is independent of herbivore and omnivore feeding i.e., donor controlled. The realized herbivore death rate is a product of the intrinsic herbivore death rate from predation

d_V, and the number of predators (P_t) and omnivores. For the omnivores, we assume that prey and plant feeding are mutually exclusive such that the proportion of time devoted to plant and prey feeding is a function of prey availability. This plant feeding function f(Vt, Pt , μ) is:

$$1 - ((V_T + P_T)/V_{max}))^{\mu}\ (0,1) \tag{2}$$

This function indicates that omnivores determine their degree of plant feeding by the summed density of herbivores and predators in the environment. When prey are at their perceived maximum, P_{max}, arbitrarily set at the herbivore's carrying capacity, all feeding is on prey but as they become vanishingly rare, the $((Vt + Pt)/\ V_{max})^{\mu}$ term approaches zero and all of the omnivore feeding is on plants. The values for plant feeding by omnivores are constrained to (0,1) i.e. from no plant feeding to complete herbivory. Thus, in the absence of herbivores (and the herbivores' carnivorous enemies), omnivores would feed as exclusive herbivores and experience population growth to K_0 (see Eq. 5). The mu term is a shape parameter that describes the sensitivity of omnivore's plant feeding response to prey density. For the purposes of this exercise, we set μ to unity. Herbivore population dynamics are defined by:

$$V_{t+1} = V_t + b_V V_t \left(1 - \frac{V_t + O_t f(V_t, P_t, \mu)}{K_V}\right) - d_v V_t \left(P_t + O_t \left(1 - f(V_t, P_t, \mu)\right)\right) \tag{3}$$

Second, predator population dynamics are determined by intrinsic predator birth rates b_P and intrinsic predator death rates, d_P, further modified by the number of prey and omnivores, respectively:

$$P_{t+1} = P_t + b_P V_t P_t - d_P P_t O_t \left(1 - f(V_t, P_t, \mu)\right) \tag{4}$$

Finally, omnivores recruit via birth from feeding on prey and predators (b_O) and plants (b'_O). We assume that omnivores could not be sustained indefinitely by feeding on plants alone. Thus, the omnivore dynamics are:

$$O_{t+1} = O_t + b_O(V_t + P_t)O_t\left(1 - f(V_t, P_t, \mu)\right) + b'_O\left(1 - \frac{V_t + O_t f(V_t, P_t, \mu)}{K_O}\right)O_t\left(f(V_t, P_t, \mu)\right) \tag{5}$$

Notice that there are a number of common terms in equations 3 through 5. Thus, the dynamics of any one player in this simple food web has the potential to impact the other two. To elucidate these impacts, particularly with regard to IGP, we solved the model numerically on a computer using mostly the same values as Lalonde *et al.* (1999) (the predator values, by definition, are new) such that: b_V (2.8, 3.4), $d_V = 0.1$,

$b_P = 0.0001$, $K_V = 3000$, $P_{max} = 3000$, $d_P = 0.1$, $b_O = (0.001)$, $b'_O = (1.2, 1.45)$, $K_O = 3000$, $d_O = 1.5$.

We considered 2 scenarios wherein plants could have significant mediator effects on population dynamics and IGP, via plant quality and plant toxins. The former refers to nutrient quality of plants from the herbivore's (or omnivore's) perspective and could result from quantity of available nitrogen (e.g. Kyto, Neimla, and Larssen, 1996) which is known have impact on survivorship and reproduction of a large number of plant feeding insects (White, 1984). To investigate the impact of plant quality, we varied intrinsic birth rates b_V, and b'_O from 2.8 to 3.4 and from 1.2 to 1.45, for herbivores and omnivores, respectively. We further subdivided the plant quality scenarios into 2 sub-scenarios, the first being the case when plant quality only affects herbivore birth rates (i.e. $b'_O = 1.2$) and the second being the case where both herbivore and omnivore birth rates are impacted by plant quality. In the plant toxin scenario, we assumed that toxins affect only the death rates of herbivores and omnivores though we acknowledge that, in the real world, birth rates could be affected as well (Ashouri, Michaud, and Cloutier, 2001). Our goal here is to contrast plant mediation in birth versus death processes. Death rates from toxins, d_t, varied from 0 to 50% reduction in intrinsic predator-independent survivorship such that:

$$V_{t+1} = V_t + b_V V_t\left(1 - \frac{V_t + O_t f(V_t, P_t, \mu)}{K_V}\right) - d_v V_t\left(P_t + O_t\left(1 - f(V_t, P_t, \mu)\right)\right) - d_t V_t \quad (6)$$

We illustrate the effects of plant mediation in several ways. First, we generated bifurcation plots for each of the aforementioned scenarios. We accomplished this by numerically solving Eq. 3-5 simultaneously for 1000 generations then plotting prey densities for the succeeding 25 generations. Visually, the plots can be interpreted as follows: if a single line is observed across a range of plant trait values then the population has been shown to continuously track the same population levels, generation after generation. By contrast, a single bifurcation (i.e. 2 points at a single value for plant quality) would indicate oscillations between two values, 4 points would indicate more complex oscillations and so on until chaos is indicated by movement to different population values every generation. To evaluate the impact of plant traits (quality and toxin level) on IGP, we plot total IGP, variation in IGP (indexed by coefficient of variation) and per capita IGP (PCIGP) as a function of plant trait values.

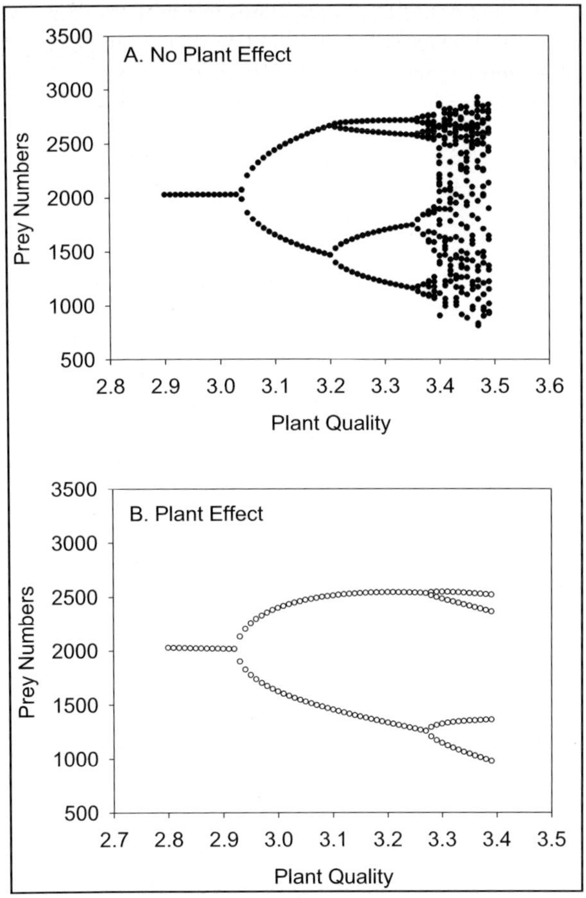

Figure 2. Bifurcation plots showing prey populations where plant quality affects herbivore only (A) or herbivore and omnivore (B).

Plant quality affects population stability in a context-dependent manner. Classic population dynamics theory predicts that population stability will decrease with increasing prey birth rates (Gotelli, 2001). This clearly is the case here; the system exhibits a single equilibrium for low plant quality (i.e. low prey birth rates), cyclic dynamics for higher plant quality and chaos for high plant quality (Figs. 2 A, 2B). Obviously, the dynamics are more complicated when only herbivores benefit improved plant quality (Fig 2A) than when omnivores profit as well (Fig 2B). When omnivores also benefit from increased plant quality, their numbers are robust and thus prevent prey from temporarily escaping and setting in motion chaotic dynamics. In fact, there are regions in parameter space where stable oscillations are generated when omnivores benefit directly from plant quality whereas system extinction occurs (not shown) when benefits to omnivores are indirect (Fig. 2A, 2B). How do changes in plant quality impact IGP? First, notice that mean per capita IGP (PCIGP) is relatively stable across plant quality values when omnivores do not benefit directly from plants whereas PCIGP declines when omnivores benefit directly (Fig 3).

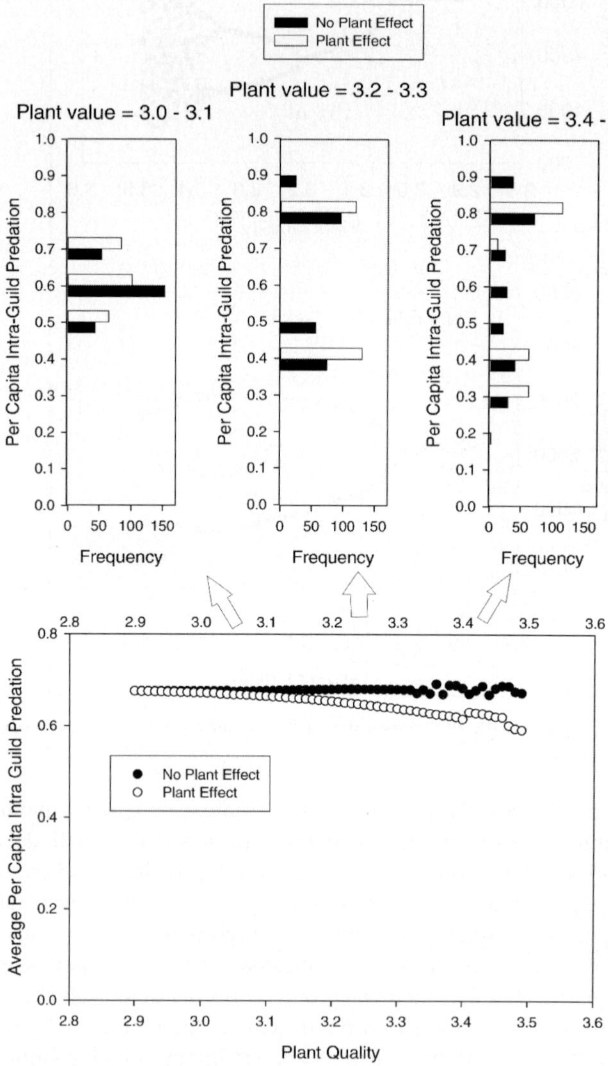

Figure 3. Effect of plant quality on Per Capita Intra Guild Predation when plant quality either does not, or does affect an omnivorous IGpredator.

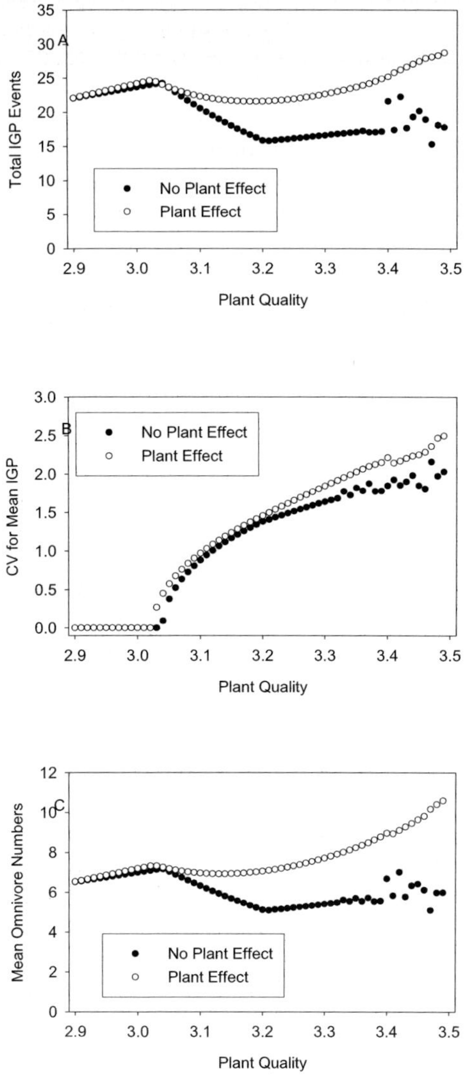

Figure 4. Effects of plant quality on number of IGP events (A), CV of IGP events (B) and number of omnivores (C) when plant quality either affects the herbivore alone, or affects both the herbivore and the omnivore.

Although this change in average PCIGP is small, this is not the whole story. The variance in PCIGP is larger when plant quality affects omnivores than when it only affects the herbivores, especially for high levels of plant quality (Fig 3). One pattern is clear: as plant quality increases, so does the average number of PCIGP's, though not in a linear fashion. These patterns are driven solely by the plant feeding function (Eq.2), i.e. they are driven strictly by prey numbers. If one contrasts population level IGP events (Fig. 4A) with PCIGP, the former is more closely allied with omnivore numbers (4C) than with the IGP tendencies (Fig. 3).

Finally, in terms of predicting and managing IGP as part of a biocontrol strategy, variance in IGP clearly increases with plant quality, even when weighted by the mean (Fig 4B). This is an expected phenomenon that biocontrol practitioners should consider, both in terms of their ability to manage systems as well as promoting biocontrol strategies; growers are generally variance adverse. Paradoxically, growers generally maximize crop yield through plant quality (e.g nitrogen fertilizers), and our models would imply that this practice increases variance in IGP.

Now, consider plant toxins. Here, we find that prey dynamics differ significantly from those caused by manipulating plant quality. We considered the effects of toxins at 2 plant quality levels, low ($b_V = 2.8$) and high ($b_V = 3.4$) (Fig 5 A, B).

Once again, the patterns are more complex at the high plant nutrition quality level, where, bifurcations are quite common, whereas single stable prey values are generated at low plant nutrition levels. In the latter, as plant toxicity values increase so do prey populations as they benefit from decreasing impact from omnivores and reduced interspecific and intraspecific competition. At high toxicity values however, high toxin-induced prey death rates overwhelm the aforementioned benefits. In contrast to the prey dynamics, PCIGP'S follow prey population trends and generate parabolic effects (i.e. low PCIGP at high and low toxin values) at both plant quality levels (Fig 6).

In this simple model, omnivores simply track prey density and adjust prey and plant feeding in kind. It is important to look at distributions and not just PCIGP means. Here, the distribution of PCIGP is nearly unimodal at low plant quality, but at high quality levels that PCIGP diversity *declines* with toxin level (Fig. 7A). Again, contrast, PCIGP with total IGP and once again these appear to be disconnected. Population IGP peaks at lowest toxin levels and drops to near zero at high toxin levels. At high toxin levels, omnivores suffer a double whammy (Fig 7C). First, there is a shortage of prey, for both herbivores and pure predators, which causes omnivores to feed more on plants, which of course, are toxic, and this further reduces omnivore numbers. IGP is rare under such circumstances because both omnivores and pure predators are rare, thus greatly negating the effects of increased PCIGP. These sorts of patterns are rather obvious in retrospect but may not be *a priori*. Finally, variance in IGP follows the population trends with an exponential increase in the variance of IGP at very high toxicity levels. However, this may not be biologically important because IGP levels are so low under those conditions (Fig. 7C).

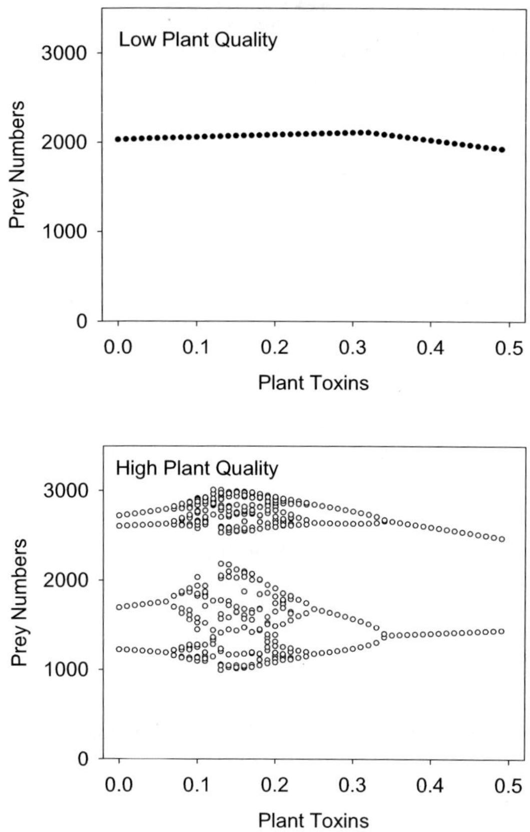

Figure 5. Bifurcation plots showing prey populations where plant toxins affect both herbivores and omnivores and plant nutritional quality is low (A) or high (B).

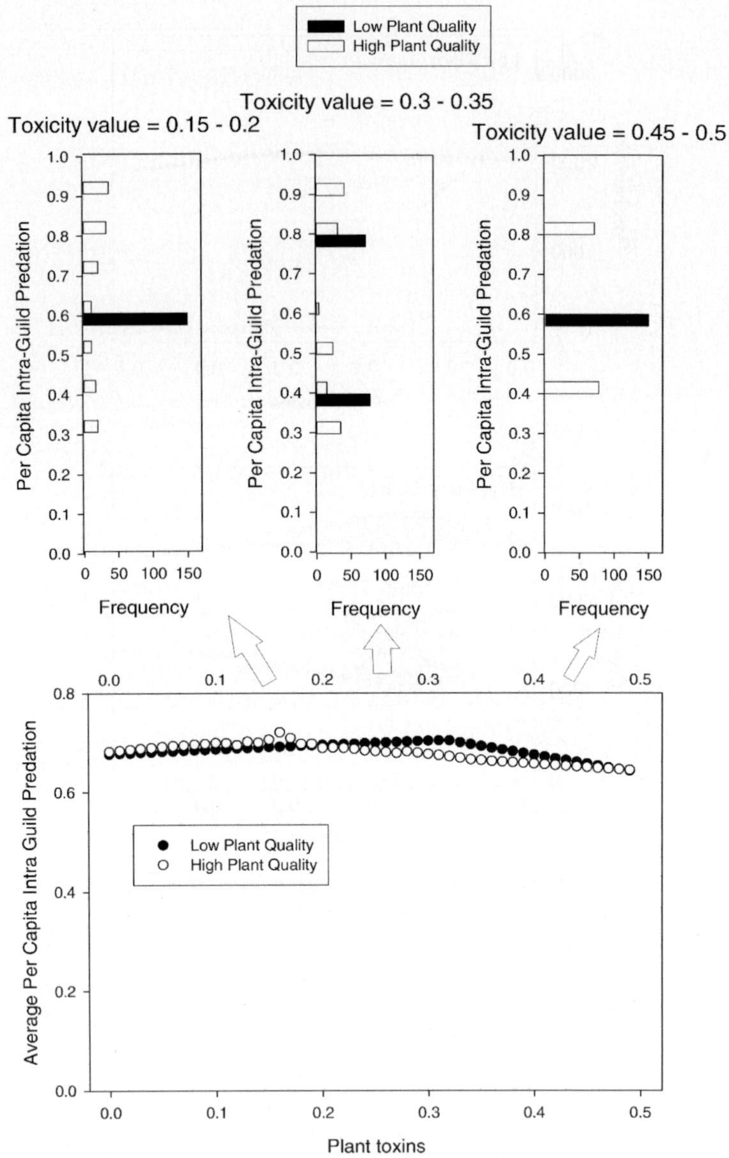

Figure 6. Effects of plant toxins on Per Capita Intra-guild predation when plant quality is either low or high.

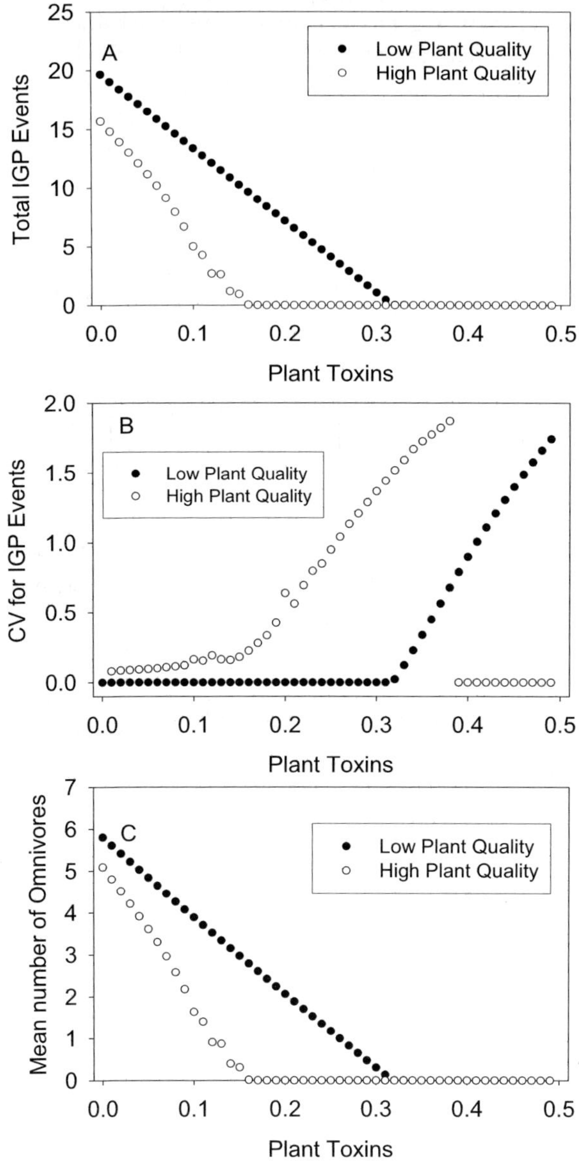

Figure 7. Effect of plant toxins on number of IGP events (A), CV of IGP events (B) and number of omnivores (C) when plant quality is either low or high.

In summary, we asked the simple question: Does the plant have an impact on IGP when omnivores are part of the food web? The answer is clearly yes. The two plant traits that we considered, plant nutrient quality and plant toxicity can affect IGP in several ways, including the mediation of prey densities and effects on per capita IGP, effects on omnivore growth rates and thus affecting total IGP and in more complex ways by causing omnivores to shift feeding rates and thus impacting encounters with and attacks on IGprey. How do our findings match up with what we know of the "real world"? In the next section, we consider a number of mechanisms by which plants are known to mediate IGP in biocontrolled systems.

4. A SEGUE TO THE "REAL WORLD"

Based on the results of our models, it is quite reasonable to predict that omnivores might engage in IGP differently on different host plants. To determine that this might be so in the real world, we performed an experiment with an omnivorous mirid bug, *Dicyphus hesperus* Knight (Hemiptera: Miridae). This insect is used for biological control of greenhouse whitefly and other pests on greenhouse tomato crops (McGregor, Gillespie, Quiring and Foisy, 1999; Sanchez, Gillespie, and McGregor, 2003). It is omnivorous, and must feed on plants and on prey to develop and reproduce (Gillespie and McGregor, 2000), and is known to engage in IGP by feeding on other natural enemy species (McGregor and Gillespie, 2005). Gillespie and McGregor (2000) demonstrated that feeding on tomato plants and prey by *D. hesperus* provided an approximately 10% decrease in developmental time relative to a diet of water and prey. Sanchez, Gillespie, and McGregor (2004) demonstrated that development and reproduction in *D. hesperus* differs on different plant species. Prey species also affected developmental time, with spider mites being less favourable prey than greenhouse whitefly or moth eggs (McGregor *et al.*, 1999). Finally, the predatory mite *Phytoseiulus persimilis* (Acari: Phytoseiidae) is an IGprey of *D. hesperus* (Gillespie unpub obs.). We hypothesized that plant species might affect the intensity of IG-predation by *D. hesperus* on *P. persimilis* in the presence of two-spotted spider mites.

The effect of plant species on IGP was evaluated on three plant species, *Chrysanthemum coronarium* L. (Asteraceae), *Capsicum anuum* L. CV Enza 444 (Solanaceae), and *Lycopersicon esculentum*, CV Rhapsodie (Solanaceae). The petiole of a leaflet of one of the species was placed through a hole in a plastic 59.2 ml cup (Solo Cup Co. Illinois) which was placed inside a 250 ml Styrofoam cup (Solo Cup Corp), so that the smaller cup rested tightly against the wall of the larger, and the petiole of the leaf was immersed in a reservoir of water in the bottom of the larger cup. A plastic lid was used to seal the outer cup. Twenty cups were constructed with each plant species. Each cup received approximately 40 adult spider mite herbivores, and a single adult female of *P. persimilis*. A single female *D. hesperus*, < 7 d old, was placed into 10 of the cups of each plant species, and 10 cups were left as Dicyphus-free controls. After 5 days, the cups were inspected to determine survival of *P. persimilis*. Two non-exclusive measures of IGP were counted: the female being alive; and the presence of immature stages. The data were analyzed by a nominal logistic model in JMP (SAS Institute, 2003).

The proportion of *P. persimilis* producing eggs and nymphs, either in the presence or absence of *D. hesperus* was not affected by plant species (Fig. 8, $\chi^2_{2,5} = 0.44$, $p = 0.8$). The overall proportion of *P. persimilis* producing immatures, was strongly affected by the presence of *D. hesperus* ($\chi^2_{1,5} = 22.34$, $P < 0.0001$).

This simple experiment suggests that IGP by *D. hesperus* on *P. persimilis* was not affected by the plant host. The IGprey (*P. persmilis*) produced offspring less frequently than expected in the presence of the IGpredator (*D. hesperus*). The three plant species were chosen for their ability to support spider mite populations and *P. persimilis*. These plants differed in their value as hosts for nymphs of *D. hesperus*, but not in their effects on adult longevity (Sanchez *et al.*, 2004), so the adult bugs may not have perceived a difference in plant substrate. The spatial and temporal scale of this experiment may also have prevented detection of plant effects on IGP. Our models above suggest that the impact of host plants on IGP of omnivores would be expressed over relatively longer time intervals.

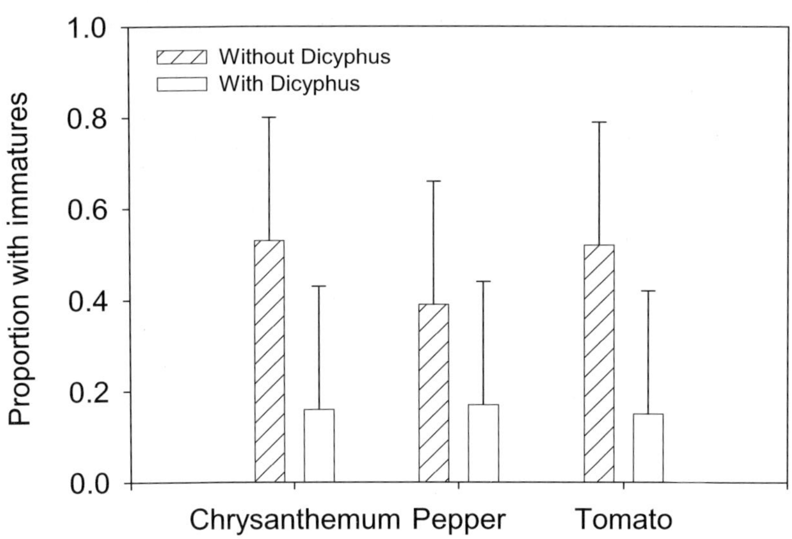

Figure 8. Effects of plant species on IGP by D. hesperus *on the proportion of* P. persimilis *that successfully produced offspring in small cages. Error bars represent the experiment-wide SE.*

There are some obvious complications to investigating the plant context of IGP in omnivores. Firstly, plant species should actually provide a demonstrably different context for the omnivore. In our experiment, three plant species were chosen that would support the common prey species and the IGprey, but the plants may not have contrasted sufficiently to affect the behaviour of the IGpredator. The space in which the experiment was conducted did not allow a full suite of behaviours to be expressed. Patch abandonment by either the IGprey or the IGpredator, or the effects of plant architecture on search behaviour might be particularly important (Roitberg, van Lenteren, van Alphen, Galais and Prokopy, 1982; Van Lenteren and Bakker, 1978). Alternatively, behaviour with respect to plant substrate may be relatively unimportant in the context of IGP in omnivores, and the effects of the plant substrate on reproduction and longevity of the IGpredator may be far more significant. The effects of plants on omnivores could be mediated through plant nutrition, toxins, physical attributes such as leaf surface architecture, or plant kairomones, to suggest a few. In the following sections, we use a combination of verbal models and literature review to ask how these aspects of the plant substrate might provide a context for IGP interactions involving omnivores in the real world.

5. PLANT QUALITY

Plant quality is often discussed in relation to herbivore and omnivore feeding, and the term can be used as a synonym for nutrition, toxicity, or various physical attributes such as hairiness or architecture. Here, as in the community model section, we define plant quality to mean the nutrition available from feeding on plants. As plant quality increases, and as a consequence, the fitness returns from plant feeding increase for both prey and omnivores, recruitment of both pure predators and omnivores should increase and PCIGP should decrease (Fig 9), but for different reasons.

On an intra-generation time-scale, recruitment of pure predators could be expected to increase as plant quality increases, because of the effects of plant nutrition on prey quality and quantity that would then recruit more natural enemies to patches via increased attraction or retention (Fig 9a). Similarly, the presence of more nutritious prey might in the short term, divert foraging effort away from potential IGprey. Apparent competition between the pure predator and the IGprey would, of course, complicate the interactions, and, ultimately, the resultant dynamics are not obvious. Recruitment of omnivores should increase with increasing plant quality because of the inherent value of the plant, but omnivores might or might not be oblivious to the indirect effects of plant quality on prey quality. Increased plant quality should reduce the tendency to per capita intra-guild predation if nutritional needs of the omnivores can be increasingly met by plant feeding.

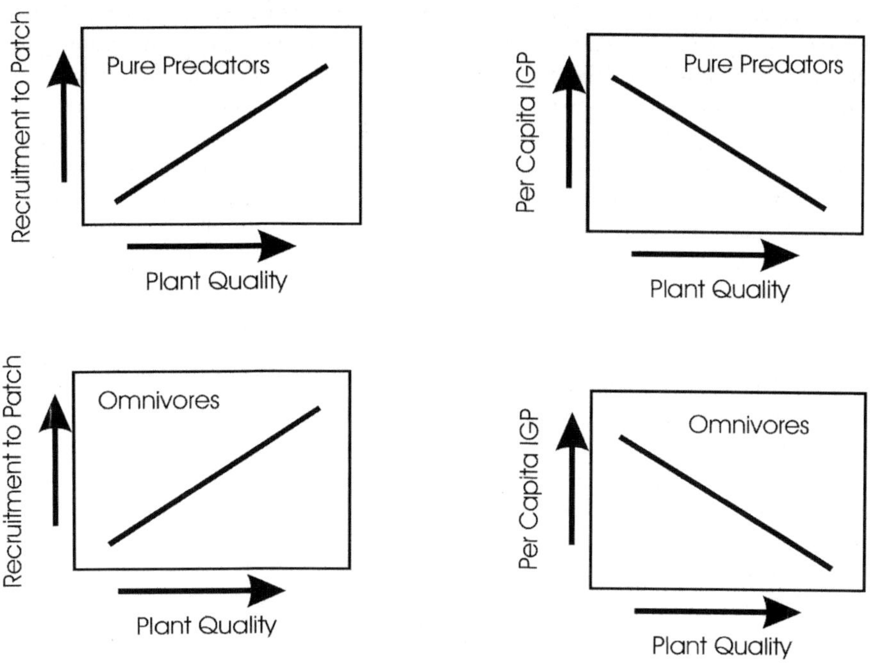

Figure 9. A theoretical depiction of the effects of plant quality on patch recruitment and per capita IGP in pure predators and omnivores. Through direct feeding on the plant, recruitment of omnivores should increase with increasing quality of the plant and availability of prey; and as plant feeding becomes increasingly rewarding, individual omnivores should feed decreasingly on IGprey. Recruitment and PCIGP of pure predators should be dependent on effects of plant quality on prey only, but the relationships should be of the same shape, but of different magnitudes.

Plant quality varies both within and between plant species. Different plant species and plant parts have different values for omnivorous feeders (e.g. Sanchez *et al.*, 2004; Eubanks and Denno, 1999). Similarly, N fertilization is positively correlated with recruitment and numerical increases in many herbivores (Dixon, 1985; Brodek, Stavisky, Funderburk, Andersen and Olson, 2001; Schuh, Redak and Bethke, 1998; Jansson, Leibee, Sanchez and Lecrone, 1991; Moon and Stiling; 2002). In general, recruitment of herbivores increases with moderate to high increases in plant N. Increases in prey numbers should tend to decrease PCIGP, because IGP events have been shown to be inversely correlated with prey numbers (Polis *et al.*, 1989). Prey supply is positively correlated with predator immigration and reproduction. According to our preliminary analysis, as natural enemy numbers increase in response to increased prey quality and numbers, the actual number of IGP events would tend to increase and the tendency to PCIGP would ultimately increase as prey populations decrease in response to predation.

Nitrogen limitation has been proposed as a major factor involved in the promotion of omnivory (Denno and Fagan, 2003). In general, fertilization with N increases recruitment to the enemy guild, but effects of plant quality have been found to be greater for herbivores than for parasitoids (Teder and Tammaru, 2002). Given that N is much more available in animal than plant foods (Denno and Fagan, 2003), it stands to reason that small increases in N in plants would have proportionately greater effects on herbivores than on predators. Parasitism of diamond back moths, aphids and gall midges increased on N fertilized plants (Jansson *et al.* 1991; Wurst and Jones, 2003; Moon and Stiling, 2002). A greater rate of host feeding by the parasitoid *Encarsia formosa* (Hymenoptera: Aphelinidae) on nymphs of *Bemesia tabaci* was observed on fertilized, as opposed to unfertilized poinsettia (Bentz, Reeves, Barbosa and Francis, 1996). Parasitism rates of aphids were negatively correlated with salinity, which decreases plant N (Moon and Stiling, 2002). The effect of predatory mites on pest mite numbers on apple was reduced at high N, and the per capita effects of one *Typhlodromus pyri* (Acarina: Phytoseiidae), were highest at low levels of N fertilization (Walde, 1995). Because *T. pyri* includes pollen in its diet (McMurty and Croft, 1997, Croft McCrae, and Curran, 1992), low inputs from plants may have driven the increased predation. Denno *et al.* (2002) showed that the numbers of wolf spiders, *Pardosa littoralis* (Araneae: Lycosidae), and a specialist predatory mirid, *Tytthus vagus*, both increased in numbers with increases in N fertilization. The numbers of the mirid however decreased with increasing spider numbers and the effects of the spider were higher in high than in low N regimes (Denno *et al.* 2002). Reductions in plant quality result in the omnivorous western flower thrips, *F. occidentalis* increasing feeding on spider mites (Agrawal *et al.*, 1999), and on eggs of predatory mites (Janssen *et al.*, 2003). Populations of the omnivorous bug *Geocoris punctipes* (Hemiptera: Geocoridae) can be maintained on soybean in the presence low density or poor quality prey, if high quality plant resources (pods) are present (Eubanks and Denno, 1999). In this system, the population dynamics of *G. punctipes* were driven by plant, rather than prey, resources.

Recruitment of herbivores, plant feeding omnivores and pure predators generally increases with increasing plant N in agreement with our initial predictions (Fig. 9), but per capita effects of plant-feeding omnivores on prey and IG-prey decrease with increasing plant N. It appears that the prediction that per capita IGP by omnivores should decrease with increasing plant N holds, and is consistent with our model predictions. Denno et al (2002), however, showed that plant N facilitated production of high quality prey, resulting in higher reproduction by a spider, which in turn reduced numbers of the specialist mirid predator. The effects of N on IGP were therefore mediated not by behaviour but by population dynamics. If plant N has proportionately greater effects on herbivores than on secondary consumers, IGP in food webs containing omnivores should be driven by long-term effects of plant N on recruitment of herbivores and omnivores.

6. PLANT TOXINS

Plant toxins (produced via induced resistance and constitutive defense) reduce recruitment and reproduction in the herbivore guild (Schoonhoven, Jermy and van Loon, 1998). This effect should also include the plant-feeding omnivores.

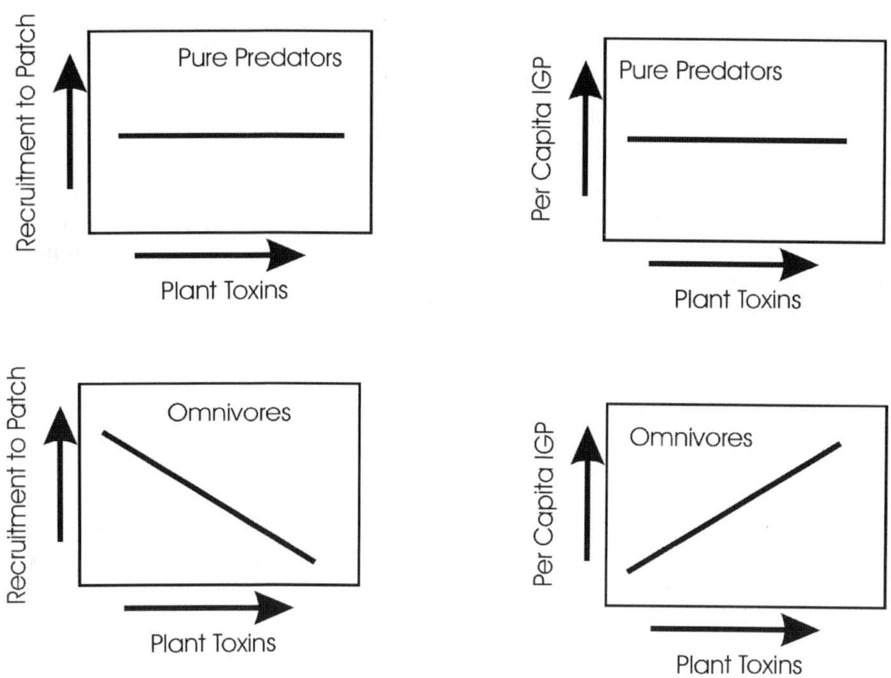

Figure 10. A theoretical depiction of the effects of plant quality on patch recruitment and per capita IGP in pure predators and omnivores. Pure predators should not be directly affected by plant toxins, because they do not feed on the plant. Through direct feeding on the plant, recruitment of omnivores should decrease with increasing toxicity of the plant; and as plant feeding becomes increasingly costly, individual omnivores should feed increasingly on IGprey.

Pure predators should not be directly affected by increasing levels of plant toxins, because they do not feed on the plants (Fig. 10). Numbers of predators on patches would be an indirect function of effects of toxins on prey. Similarly, PCIGP should also be a function of predator numbers and per capita, both of which are independent of plant toxins. Omnivores, on the other hand, feed on the plant and would be directly affected by plant toxins through plant feeding, and would also be indirectly affected by plant toxins through effects of toxins on prey. Recruitment would decrease with increasing toxin levels because toxins could cause omnivores to abandon plants, or would affect reproduction of the omnivore. PCIGP would increase in patches if omnivores reduce their rate of feeding on plants and increase their rate of feeding on prey.

Toxins expressed through transgenic plants provide a way to examine the effects of toxins on predators and omnivores, since there are clearly no co-evolved relationships in a tritrophic system involving toxins in transgenic plants. The presence of *Bacillus thuringiensis* (*Bt*) toxins from transgenic plants in artificial diets (direct exposure) does not influence survival of *Chrysoperla carnea* (Neuroptera: Chrysopidae) or *Crytorhinus lividipennis* (Bernal, Aguda and Cohen, 2002; Romeis, Dutton, and Bigler, 2004). Lacewing larvae, in particular, are pure predators that do not feed omnivorously on plants, and *C. lividipennis* feeds on plants to a very limited extent when prey are absent (Wheeler, 2001). In contrast, Hilbeck Moar, Pusztai-Carey, Fillipini and Bigler (1999) showed that *C. carnea* larvae that were fed caterpillars that were reared on a diet with Bt-toxin suffered increased mortality and delayed development. The difference between the former and the latter was that the Bt toxin was fed directly to predators in the former study and indirectly through prey in the latter. Romeis *et al.* (2004) argue that *C. carnea* is not susceptible to Bt toxins and that the results in Hilbeck *et al.* (1999) were a result of the effects of poor prey quality on predators. Bouchard, Michaud, and Cloutier (2003) found that the presence of a rice-derived proteinase inhibitor implanted in potato did not affect the predator *Perillus bioculatus* (Hemiptera: Pentatomidae), despite the toxin being present in the prey, the Colorado potato beetle. These latter authors also demonstrated that survivorship of *P. bioculatus* was not affected when constrained to feed on the transgenic potato foliage, compared to non-transformed foliage. The omnivorous predator, *Orius majusculus* (Heteroptera: Anthocoridae) was not adversely affected when feeding on a Bt-immune prey species, (*Anaphothrips obscurus*) (Thysanoptera: Thripidae) on transgenic *Bt* corn (Zwahlen, Nentwig, Bigler, and Hilbeck, 2000). In contrast, Ponsard, Gutierrez and Mills, (2002) demonstrated that feeding on Bt cotton plus a Bt-tolerant prey species adversely affected *O. tristicolor* and *Geocoris punctipes*, but not a *Nabis* sp. and *Zelus renardii*. The former two species are known for their omnivorous feeding habits, but neither Nabidae nor Reduviidae are known to feed extensively on plants (Ambrose, 2000; Braman, 2000).

It is premature to draw conclusions from studies on the effects of transgenic toxins in plants on tritrophic systems. These systems do, however, mimic the situation where a generalist predator is naïve, in an evolutionary sense to toxins expressed in plants and thus provide a useful starting point. Our assumption that pure predators will not be directly affected by plant-based toxins is borne out by the results, but pure predators might be affected indirectly by prey quality. Furthermore, it appears that omnivorous predators feeding on either plants with toxins or on prey that are tolerant of toxins will be adversely affected by those toxins.

In non-transformed plants, the results are somewhat different. The presence of a strong bitter flavour, curcubitacin, in cucumber was correlated with the production of volatile chemicals known to be involved in recruitment of predatory mites

(Agrawal, Janssen, Bruin, Posthumus and Sabelis, 2002), but despite this, plants that produced curcubitacins were less attractive to predators than those that did not. Toxins that were induced by herbivory in cotton reduced numbers of *Orius tristicolor* but not of *Geocoris punctipes* or *F. occidentalis* (Agrawal, Karban and Colfer, 2000), but Agrawal and Klein (2000) showed that induced resistance in cotton reduced thrips numbers through both direct and indirect effects, as would be predicted for an omnivorous species. Omnivorous predatory mites are killed by systemic pesticides in plants, whereas predator mites that function as pure predators are not (Nomikou *et al.* 2003). A specialist caterpillar *Milionia basalis* Druce (Geometridae) feeds on leaves of *Podocarpus macrophyllus* and converts constituent precursors into a highly toxic glucoside. A generalist pentatomid predator, *Eocanthecona furcellata* dies when feeding on larvae of the caterpillar, but is nonetheless recruited to the podocarp trees as a predator (Yasui, 2001). Dyer, Dodson, Beihoffer and Letourneau (2001) showed that amides were induced by herbivory on *Piper cenocladum* in the absence of mutualistic ants. These amides deterred omnivorous ants as well as generalist herbivores. Contrary to our assumptions, toxins in plants can reduce recruitment of pure predators as well as omnivores. The effects of plant-based toxins on survival seem different for pure predators and omnivores, however, with omnivores being more severely affected than pure predators, except in systems where specialist herbivores can appropriate the toxins for their own defence.

7. PLANT HAIRS

Plant hairs are a trait that we did not address in our earlier models; however, this trait does affect natural enemy foraging (e.g. Treacy, Benedict, Lopez and Morisson, 1987; Sutterlin and Van Lenteren, 1997). If, as plant hairiness increases, plant substrates become less hospitable for both pure predators and omnivores then we assume that recruitment to patches should decrease with increasing hairiness (Fig. 11).

As a simple starting point, PCIGP might decrease for both pure predators and omnivores because movement on plants would become increasingly difficult and encounters between guild members would therefore decrease. Plant hairs are highly varied, and range from simple, erect hairs to glandular trichomes that may contain toxins, adhesives and repellents of various sorts (Rodriguez, Healy, and Mehta, 1984; Wagner, 1991). The relative effects of these hairs on specialist and omnivorous predators in a tritrophic system should depend on the size of the species, on the relative densities and effects of the hairs on them, and on the degree of co-adaptation among the members of the system. We predict however that the decrease in IGP with increasing hairiness should occur at a higher rate for omnivores, all other things being equal. This would be because, in a simple system, the omnivore could increasingly rely on plant feeding to satisfy its nutritional requirements, whereas a predator is constrained to find prey.

Glandular hairs on tomato and potato stems and leaves contain a sticky substance and the density of these hairs was correlated with resistance to spider mites (van Haren, Steenhuis, Sabelis and de Ponti, 1987) and aphids (Obrycki, Tauber, and Tingey, 1983). Glandular hairs on tomato reduce the movement and survival of the spider mite predator *Phytoseiulus persimilis* (Van Haren *et al.*, 1987). Tomato glandular hairs however, did not affect development time or survival in another predatory mite, *Neoseiulus californicus* (Cedola and Sanchez, 2003). The density of glandular hairs on potato did not affect parasitoids or large-bodied predators (Obrycki *et al.*, 1983). The density of Type IV glandular hairs on *Lycopersicon* spp. was positively correlated with cannibalism and negatively correlated with prey consumption in a lacewing, *Mallada signata* (Neuroptera: Chrysopidae) (Simmons and Gurr, 2004).

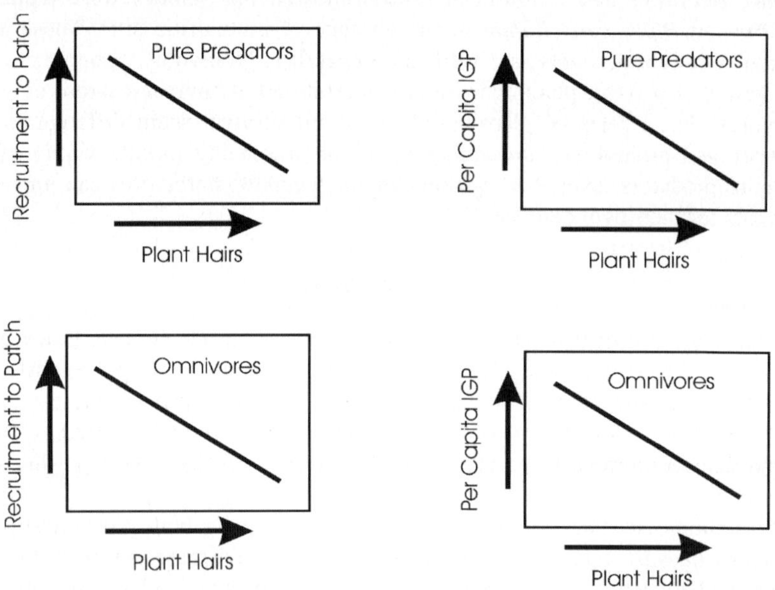

Figure 11 . A theoretical depiction of the effects of plant hairs on patch recruitment and per capita IGP in pure predators and omnivores. As plant hairs increase, recruitment of both predators and omnivores should decrease through declining prey availability. As foraging becomes more difficult, encounters with guild-members should also decrease, and consequently, PCIGP should decrease.

On gerbera, the efficiency of *P. persimilis* searching for *T. urticae* was negatively correlated with density of non-glandular leaf hairs (Krips, Kleijn, Willems, Gols and Dicke, 1999), but these effects interacted with prey density; predation efficiency was not affected at high prey densities. Trichome density was negatively correlated with prey capture and movement of *Chrysoperla rufilabris* on cotton (Treacy *et al.*, 1987) and the effects of hair density were greatest for smaller individuals. Walking speed and predation in one coccinellid predator of mealybugs, *Cryptolaemus montrouzei* (Coleoptera: Coccinellidae) was not affected by leaf pubescence on various plant species, but was affected in a second species, *Nephus reunioni* (Heideri, 1999). Size of the predators played a role, with the larger, *C. montrouzei* not being affected whereas the smaller *N. reunioni* was (Heideri, 1999).

Among omnivorous species, predation of corn earworm eggs by *Geocoris punctipes* was not affected by leaf pubescence on soybean (Powell and Lambert, 1993). Moreover, *G. punctipes* prefers oviposition sites on hairy leaves and tends to oviposit at the base of hairs. IG-predation by the western flower thrips on the predatory mite *Typhlodromus pyri* was reduced on pubescent apple varieties (Roda, Nyrop, Dicke, and English-Loeb, 2000). In this case, pubescence provided the IG prey with a refuge, and complex leaf topography might in general function to provide IGprey with refuges. The predator mite *Iphiseius degenerans* (Phytoseiidae) tends to lay its eggs at the tips of leaf hairs, which provides it with a refuge from IG-predation (Faraji, Janssen and Sabelis, 2002). Agrawal et al (2000) showed that the presence of artificial leaf domatia on cotton enhanced populations of both pure and omnivorous predators. *Coleomegilla maculata* (DeGeer) (Coleoptera: Coccinellidae) suffers greater mortality from cannibalism on plants without glandular hairs as opposed to those with glandular hairs, and preferentially lays eggs on plants with glandular hairs (Cottrell and Yeargan, 1998; Staley and Yeargan, 2005).

Some species may be particularly adapted to forage among leaf hairs. Tarsal modifications in the bryocorine mirids (Hemiptera: Miridae: Bryocorinae) allow omnivorous and herbivorous species to move efficiently on plants with dense pubescence and glandular hairs (Wheeler, 2002). The mirid predators *Macrolophus caliginosus*, *Dichyphus hesperus* and *D. tamaninii* used in augmentative and conservation biocontrol (Alomar and Albajes, 1996; McGregor *et al.* 1999; Gabarra, Castane, Bordas and Albajes, 1988) on greenhouse tomato crops belong to this subfamily, show such tarsal modification, and move about easily on pubescent plants. Laycock, Camm, VanLaerhoven and Gillespie (in press) showed that cannibalism by *D. hesperus* was lower on smooth than on hairy leaves but in their study it was not clear if this was due to reduced hunting capacity by the cannibal or better escape ability by the victims.

Plant hairs in general may reduce foraging efficiency of predators, and provide both herbivores and IG prey with refuges from predation. If hairs reduce efficiency of foraging through interference or refuges, herbivores and IG prey should both be affected and the per capita IG predation should decline with increasing density of plant hairs, for both pure and omnivorous predators. In one case, plant hairs resulted in an increase in IG-predation (cannibalism) in a pure predator, apparently as a result of reduced dispersal and increased encounters between conspecifics. In other systems, plant hairs provide refuges for IG-prey and made them less vulnerable IG-predation. The response of omnivorous and pure predators to plant hairs appears to be situation-specific, and dependent on the size of hairs and the biology and behaviour of the species involved.

8. PLANT KAIROMONES

Plant kairomones (herbivore-induced plant volatiles) could increase recruitment by immigration of both omnivores and pure predators to patches (Fig 12), assuming there is no competition between omnivores and herbivores for plant resources.

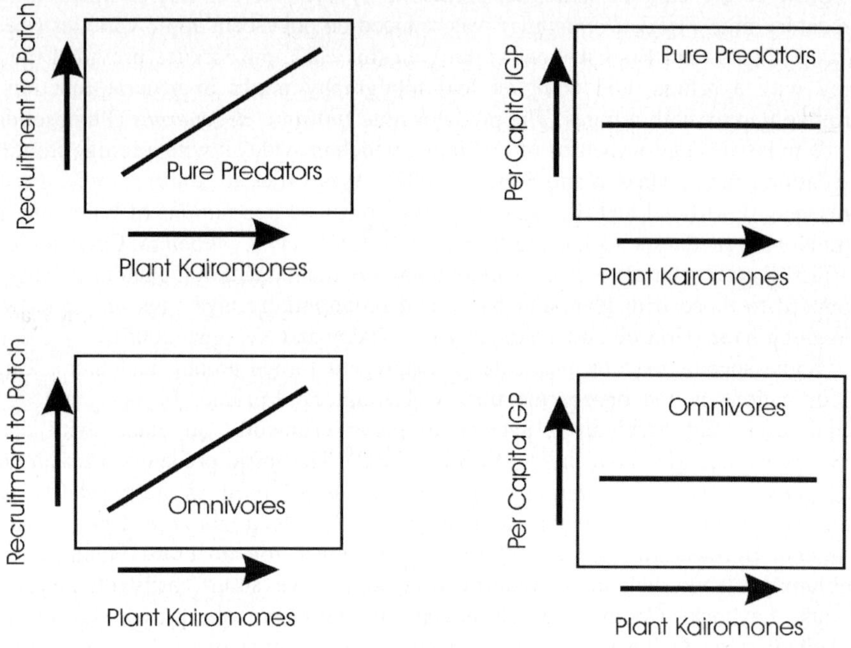

Figure 12. A theoretical depiction of the effects of plant kairomones on patch recruitment and per capita IGP in pure predators and omnivores. As the tendency to emit plant kairomones increases, recruitment of both omnivores and predators might increase. The effects of plant kairomones are not expressed post-recruitment, and therefore PCIGP should be unaffected by levels of kairomone emission.

Recruitment by immigration might however, tend to level off, or at least increase at a decreasing rate as the pool of available predators is depleted. The per capita tendency to IGP should not change with tendency of the plant to emit kairomones because IGP is a post-arrival behaviour.

It is generally accepted that plants produce volatile compounds (kairomones) in response to herbivore feeding and that these compounds will attract natural enemies (Dicke and Vet, 1999). If feeding on a plant by omnivores such as western flower thrips and some true bugs induces such compounds, then recruitment of other natural enemies will result, increasing the opportunities for IGP. In contrast to our initial predictions, the presence of herbivores might reduce the recruitment of some omnivores to patches if competition among herbivores is a constraint. Whether or not the omnivore becomes an IGpredator or an IGprey will depend on the species recruited to the patch. Kessler and Baldwin (2001, 2004) showed that herbivory by a mirid, *Tupiocoris notatus* induced a plant defensive response that protected the plant against more damaging defoliators by attracting a generalist (and omnivorous) predator, *Geocoris pallens*, which also preyed on the mirid. Although *T. notatus* is not omnivorous, the mechanisms illustrate the principle.

Competition among omnivores for plant resources may affect recruitment to plants through herbivore-induced plant volatiles. For example, western flower thrips recruitment to patches of cotton with spider mites was lower when conspecifics were present due to WFT-specific odours (Agrawal and Colfer, 2000).

9. CONCLUSIONS

At the outset, we asked a simple question: "Do plants provide a context that helps to explain IGP in omnivores?" Our theoretical exploration and literature review lead us to conclude that IGP by omnivores is a special form of IGP and that plants can mediate this intra-guild interaction. Differences between IGP by omnivores and pure predators occur in both behaviour of individuals and in long-term consequences expressed in prey population growth. Interestingly, behavioural differences do not necessarily translate into community level effects. Although the plant context adds another level of complexity, we feel this complexity is important to a better understanding of an inherently intricate process. By understanding how plants may influence IGP and the outcomes of biological control processes, it may be possible to manipulate plant traits to improve efficacy and predictability of omnivorous natural enemies applied for biological control. Plant resistance, and plant architecture are just a few of the plant traits that are commonly manipulated in modern farming systems. All of these agricultural tactics may provide avenues for optimizing pest suppression through the interactions among natural enemy guild members.

10. ACKNOWLEDGEMENTS

We thank the editors and organizers, J. Brodeur and G. Boivin, for inviting us to participate in the symposium, Jackson Chu for assistance conducting the experiment on plant substrate and IGP and A. Janssen and M. Sabelis for helpful comments on an earlier version of this manuscript. This work was supported by funding from NSERC, AAFC, and the NSERC Biocontrol Network.

11. REFERENCES

Agrawal, A. A., 2003, Why Omnivory? Ecology 84: 2521.

Agrawal, A. A., and Colfer, R. G., 2000, Consequences of thrips-infested plants for attraction of conspecifics and parasitoids. Ecological Entomology 25: 493-496.

Agrawal, A. A., Janssen, A., Bruin, J., Posthumus, M. A., and Sabelis, M. W., 2002, An ecological cost of plant defense: attractiveness of bitter cucumber plants to natural enemies of herbivores. Ecology Letters 5: 377-285.

Agrawal, A.A., Karban, R., and Colfer, R.G., 2000, How leaf domatia and induced plant resistance affect herbivores, natural enemies and plant performance. Oikos 89: 70-80

Agrawal, A. A., Kobayashi, C. and Thaler, J. S., 1999, Influence of prey availability and induced host-plant resistance on omnivory by western flower thrips. Ecology 80: 518-523.

Agrawal, A. A., and Klein. C. N., 2000, What omnivores eat: direct effects of induced plant resistence on herbivores and indirect consequences for diet selection by omnivores. Journal of Animal Ecology 69: 525-535.

Alomar, O., and Albajes, R., 1996, Greenhouse whitefly (Homoptera: Aleyrodidae) predation and tomato fruit injury by the zoophytophagous predator *Dicyphus tamaninii* (Heteroptera: Miridae). In: O. Alomar, and R. N. Wiedenmann, eds., Zoophytophagous Heteroptera: Implications for Life History and IPM, pp. 155-177. Washington, DC: Entomological Society of America. Proceedings: Thomas Say Publications in Entomology.

Alomar, O., and Wiedenmann, R. N. eds., 1996, Zoophytophagous Heteroptera: Implications for Life History and IPM.. Washington, DC, Entomological Society of America. Proceedings: Thomas Say Publications in Entomology.

Ambrose, D. P., 2000, Assassin Bugs (Reduviidae excluding Triatominae). In: C. W. Schaefer and A. R. Panizi, eds., Heteroptera of Economic Importance, pp. 695-712, CRC Press, New York.

Arim, M., and Marquet P.A., 2004, Intraguild predation: a widespread interaction related to species biology. Ecology Letters 7: 557-564.

Ashouri A., Michaud, D., and Cloutier, C., 2001, Unexpected effects of different potato resistance factors to the Colorado potato beetle (Coleoptera: Chrysomelidae) on the potato aphid (Homoptera: Aphididae). Environmental Entomology 30: 524-532.

Bentz, J., Reeves, J.I., Barbosa, P., and Francis, B., 1996, The effect of nitrogen fertilizer applied to *Euphorbia pulcherrima* on the parasitization of *Bemesia argentifolii* by the parasitoid *Encarsia formosa*. Entomologia Experimentalis et Applicata 78: 105-110.

Bernal, C. C., Aguda, R. M., and Cohen, M. B., 2002, Effect of rice lines transformed with *Bacillus thuringiensis* toxin genes on the brown planthopper and its predator *Cyrtorhinus lividipennis*. Entomologia Experimentalis et Applicata 102: 21-28.

Bouchard, E., Michaud, D., and Cloutier, C., 2003, Molecular interactions between an insect predator and its herbivore prey on transgenic potato expressing a cysteine proteinase inhibitor from rice. Molecular Ecology 122: 429-2437.

Braman, S. K., 2000, Damsel Bugs (Nabidae), In: C. W. Schaefer and A. R. Panizi, eds., Heteroptera of Economic Importance, pp. 695-712, CRC Press, New York.

Bright, K. L., Bernays, E. A., and Moran, V. C., 1994, Foraging patterns and dietary mixing in the field by the generalist grasshopper *Brachystola magna* (Orthoptera: Acrididae). Journal of Insect Behavior 7: 779-793.

Brodek, B. V., Stavisky, J., Funderburk, F. E., Andersen, P. C., and Olson, S. M., 2001, Flower nitrogen status and populations of *Frankliniella occidentalis* feeding on *Lycopersicon esculentum*. Entomologia Experimentalis et Applicata 99: 165-172.

Cedola, C.V., and Sanchez, N.E., 2003, Effect of tomato pubescence on development, survival and fecundity of *Tetranychus urticae* Koch and *Neoseiulus californicus* (McGregor) (Acari: Tetranychidae: Phytoseiidae). Acarologia 43: 255-260.

Cloutier, C., and Johnsen, S., 1993, Predation by *Orius tristicolor* (Hemiptera: Anthocoridae) on *Phytoseiulus persimilis* (Acarina: Phytoseiidae): Testing for compatibility between biocontrol agents. Environmental Entomology 22: 477-482.

Cohen, A. C., 1996, Plant feeding by predatory Hemiptera: Evolutionary and adaptational aspects of trophic switching. In: R. Wiedenmann, and O. Alomar, eds., Zoophytophagous Heteroptera: Implications for Life History and Integrated Pest Management, pp. 1-17, Thomas Say Publications in Entomology, Entomol. Soc. Am., Lanham.

Coll, M, and Guershon, M., 2002, Omnivory in Terrestrial Arthropods: Mixing Plant and Prey Diets. Annual Review of Entomology 47: 267-297.

Coll, M., and Izraylevich, S., 1997, When predators also feed on plants: Effects of competition and plant quality on omnivore-prey population dynamics. Annals of the Entomological Society of America 90: 155-161.

Coll, M., and Ruberson, J. R., eds., 1998, Predatory Heteroptera: Their Ecology and Use in Biological Control. Lanham, Maryland, Entomological Society of America. Thomas Say Publications in Entomology: Proceedings.

Cottrell, T.E., and Yeargan, K.V., 1998, Influence of a native weed, *Acalypha ostryaefolia* (Euphorbiaceae), on *Coleomegilla maculata* (Coleoptera: Coccinellidae) population density, predation, and cannibalism in sweet corn. Environmental Entomology 27: 1375-1385.

Croft, B. A., MacRae, I. V., and Currans, K. G., 1992, Factors affecting biological control of apple mites by mixed populations of *Metaseiulus occidentalis* and *Typhlodromus pyri*. Experimental and Applied Acarology 14: 343-355.

Denno, R. F., and Fagan, W. F., 2003, Might nitrogen limitation promote omnivory among carnivorous arthropods? Ecology 84: 2522-2531.

Denno, R. F., Gratton, C., Peterson, M. A., Langellotto, G. A., Finke, D. L., and Huberty, A. F., 2002, Bottom-up forces mediate natural-enemy impact in a phytophagous insect community. Ecology 83: 1443-1458.

Dicke, M., and Vet, L. E. M., 1999, Plant-carnivore interactions: evolutionary and ecological consequences for plant, herbivore and carnivore. In: H. Olff, V. K. Brown and R. H. Drent, eds., Herbivores: Between Plants and Predators, pp. 483-520, Blackwell Science, Oxford.

Diehl, S., 2003, The evolution and maintenance of omnivory: dynamic constraints and the role of food quality. Ecology 84: 2557-2567.

Dixon, A.F.G., 1985, Aphid ecology. Blackie, Glasgow and London, United Kingdom.

Dyer, L. A., Dodson C. D., Beihoffer, J., and Letourneau, D. K., 2001, Trade-offs in anthiherbivore defenses in *Piper cenocladum*: ant mutualists versus plant secondary metabolites. Journal of Chemical Ecology 27: 581-592.

Eubanks, M. D., and Denno R. F., 1999, The ecological consequences of variation in plants and prey for an omnivorous insect. Ecology 80: 1253-1266.

Faraji, F., Janssen, A., and Sabelis, M.W., 2002, Oviposition patterns in a predatory mite reduce the risk of egg predation caused by prey. Ecological Entomology 27: 660-664.

Fox, L. R., 1975, Factors influencing cannibalism, a mechanism of population limitation in the predator *Notonecta hoffmanii*. Ecology 56: 933-941.

Gabarra, R., Castane, C., Bordas, E., and Albajes, R., 1988, *Dicyphus tamaninii* as a beneficial insect and pest in tomato crops in Catalonia, Spain. Entomophaga 33: 219-228.

Gillespie, D. R., and Quiring, D. J. M., 1992, Competition between *Orius tristicolor* (White) (Hemiptera: Anthocoridae) and *Amblyseius cucumeris* (Oudemans) (Acari: Phytoseiidae) feeding on *Frankliniella occidentalis* (Pergande) (Thysanoptera: Thripidae). The Canadian Entomologist 12: 1123-1128.

Gillespie, D.R., and McGregor, R.R., 2000, The functions of plant feeding in the omnivorous predator *Dicyphus hesperus*: water places limits on predation. Ecological Entomology 25: 380-386.

Gotelli, N., 2001, A Primer Of Ecology. Sinauer Press, Sunderland, MA.

Heidari, M., 1999, Influence of host-plant physical defenses on the host searching behaviour and efficacy of two coccinellid predators of the obscure mealybug, *Pseudococcus viburni* (Signoret). Entomologica 33: 397-402.

Hilbeck, A., Moar, W. J., Pusztai-Carey, M., Filippini, A., and Bigler, F., 1999, Prey-mediated effects of Cry1Ab toxin and protoxin and Cry2A protoxin on the predator *Chrysoperla carnea*. Entomologia Experimentalis et Applicata 91: 305-316.

Hough-Goldstein, J. A., Vangessel, M. J., and Wilson, A. P., 2004, Manipulation of weed communities to enhance ground-dwelling arthropod populations in herbicide-resistant field corn. Environmental Entomology 33: 577-586.

Janssen, A., Willemse, E., and van der Hammen, T., 2003, Poor host quality causes omnivore to consume predator eggs. Journal of Animal Ecology 72: 478-483.

Jansson, R. K., Leibee, G. L., Sanchez, C. A., and Lecrone, S. H., 1991, Effects of nitrogen and foliar biomass on population parameters of cabbage insects. Entomologia Experimentalis et Applicata 61: 7-16.

Kagata, H., and Ohgushi, T., 2006, Bottom-up trophic cascades and material transfer in terrestrial food webs. Ecological Research 21: 26-34.

Kessler, A., and Baldwin, I. T., 2001, Defensive function of herbivore-induced plant volatile emission in nature. Science 291: 2141-2144.

Kessler, A., and Baldwin, I. T., 2004, Herbivore induced plant vaccination. Part 1. The orchestration of plant defenses in nature and their fitness consequences in the wild tobacco *Nicotiana attenuata*. The Plant Journal 38: 639-649.

Krips, O. E., Kleijn, P. W., Willems, P. E. L., Gols, G. J. Z., and Dicke, M., 1999, Leaf hairs influence searching efficiency and predation rate of the predatory mite *Phytoseiulus persimilis* (Acari: Phytoseiidae). Experimental and Applied Acarology 23: 119-131.

Kyto, M., Neimela, P., and Larsson, S., 1996, Insects on trees: Population and individual response to fertilization. Oikos 75: 148-159.

Lalonde, R. G., McGregor, R. R., Roitberg, B.D., and Gillespie, D. R., 1999, Plant-feeding by arthropod predators contributes to the stability of predator-prey population dynamics. Oikos 87: 603-609.

Laycock, A., Camm, E., VanLaerhoven, S., and Gillespie, D., in press, Cannibalism in a zoophytophagous omnivore is mediated by prey availability and plant substrate. Journal of Insect Behavior.

Lincoln, R., Boxshall, G., and Clark, P., 1998, Dictionary of Ecology, Evolution, and Systematic, 2nd edition. Cambridge University Press, Cambridge.

Lucas, E. and Alomar, O., 2001, *Macrolophus caliginosus* (Wagner) as an intraguild prey for the zoophytophagous *Dicyphus tamaninii* Wagner (Heteroptera: Miridae). Biological Control 20: 147-152.

Magalhães, S. and Bakker R. M., 2002, Plant feeding by a predatory mite inhabiting cassava. Experimental and Applied Acarology 27: 27-37.

Magalhães, S., Janssen, A., Montserrat, M., and Sabelis, M. W., 2005, Host-plant species modifies the diet of an omnivore feeding on three trophic levels. Oikos 111: 47-56.

McGregor, R.R., and Gillespie, D.R., 2005, Intraguild predation by the generalist predator *Dicyphus hesperus* on the parasitoid *Encarsia formosa*. Biocontrol Science and Technology 15: 219-227.

McGregor, R., Gillespie, D., Quiring, D., and Foisy, M., 1999, Potential use of *Dicyphus hesperus* Knight (Heteroptera: Miridae) for Biological Control of Pests of Greenhouse Tomatoes. Biological Control 16: 104-110.

McMurtry, J. A., and Croft, B. A., 1997, Lifestyles of phytoseiid mites and their roles in biological control. Annual Review of Entomology 42: 291-321.

Menalled, F. D., Lee, J. C., and Landis, D. A., 2001, Herbaceous filter strips in agroecosystems: Implications for ground beetle (Coleoptera: Carabidae) conservation and invertebrate weed seed predation. The Great Lakes Entomologist 34: 77-91.

Moon, D. C., and Stiling, P., 2002, The effects of salinity and nutrients on a tritrophic salt-marsh system. Ecology 83: 2465-2476.

Nomikou, M., Janssen, A., and Sabelis, M. W., 2003, Phytoseiid predator of whitefly feeds on plant tissue. Experimental and Applied Acarology 31: 27-36.

Obrycki, J.J., Tauber, M.J., and Tingey, W.M., 1983, Predator and parasitoid interaction with aphid-resistant potatoes to reduce aphid densities: a two-year field study. Journal of Economic Entomology 76: 456-462.

Pimm, S.L., and Lawton, J.H., 1978, On feeding on more than one trophic level. Nature 268: 329-331.

Polis, G. A., Myers, C. A., and Holt, R. D., 1989, The Ecology and Evolution of Intraguild Predation: Potential Competitors That Eat Each Other. Annual Review of Ecology and Systematics 20: 297-330.

Ponsard, S., Gutierrez, A. P., and Mills, N. J., 2002, Effect of Bt-toxin (Cry1Ac) in transgenic cotton on the adult longevity of four heteropteran predators. Environmental Entomology 31: 1197-1205.

Poole, A., Gormally, M., and Skeffington, M. S., 2003, The flora and carabid beetle fauna of a mature and regenerating semi-natural oak woodland in south-east Ireland. Forest Ecology and Management 177: 207-220.

Powell, J.E., and Lambert, L., 1993, Soybean genotype effects on bigeyed bug feeding on corn earworm in the laboratory. Crop Science 33: 556-559.

Roda, A., Nyrop, J., Dicke, M., and English-Loeb, G., 2000, Trichomes and spider-mite webbing protect predatory mite eggs from intraguild predation. Oecologia 125: 428-435.

Rodriguez, E., Healy, P. L., and Mehta, I., eds, 1984, Biology and Chemistry of Plant Trichomes, Plenum Press, New York.

Roitberg, B.D., Van Lenteren, J.C., Van Alphen, J.M.J., Galis, F., and Prokopy, R.J., 1982, Foraging behaviour of *Rhagoletis pomonella*, a parasite of hawthorn (*Crataegus viridis*), in nature. Journal of Animal Ecology 51: 307-325.

Romeis, J., Dutton, A., and Bigler, F., 2004, *Bacillus thuringiensis* toxin (Cry1Ab) has no direct effect on larvae of the green lacewing *Chrysoperla carnea* (Stephens) (Neuroptera: Chrysopidae). Journal of Insect Physiology 50: 175-183.

Rosenheim, J., 2001, Source-sink dynamics for a generalist insect predator in habitats with strong higher-order predation. Ecological monographs 71: 93-116.

Rosenheim, J.A., Kaya, H.K., Ehler, L.E., Marois, J.J., and Jaffee, B.A., 1995, Intraguild predation among biological-control agents: theory and evidence. Biological Control 5: 303-335.

Sanchez, J. A., Gillespie, D. R., and McGregor, R. R., 2003, The effects of mullein plants (*Verbascum thapsus*) on the population dynamics of *Dicyphus hesperus* (Heteroptera: Miridae) in tomato greenhouses. Biological Control 28: 313-319.

Sanchez, J. A., Gillespie, D. R., and McGregor, R. R., 2004, Plant preference in relation to life history traits in the zoophytophagous predator *Dicyphus hesperus*. Entomologia Experimentalis et Applicata 112: 7-19.

SAS Institute, 2003, JMP Statistical Discovery Software. SAS Institute Inc.

Schmitz, O. J., and Suttle, K. B., 2001, Effects of top predator species on direct and indirect interactions in a food web. Ecology 82: 2072-2081.

Schoonhoven, L. M., Jermy, T., and. Van Loon, J. J. A., 1998, Insect–Plant Biology. Chapman and Hall, London.

Schuch, U. K., Redak, R. A., and Bethke, J. A., 1998, Cultivar, fertilizer, and irrgation affect vegetative growth and susceptiblity of *Chrysanthemum* to western flower thrips. Journal of the American Society for Horticultural Science 123: 727-733.

Simmons, A. T., and Gurr, G. M., 2004, Trichome-based host plant resistance of *Lycopersicon* species and the biocontrol agent *Mallada signata*: are they compatible. Entomologia Experimentalis et Applicata 113: 95-101.

Staley, A.C., and Yeargan K.V., 2005, Oviposition behaviour of *Coleomegilla maculata* (Coleoptera: Coccinellidae): Diel periodicity and choice of host plants. Environmental Entomology 32: 440-445.

Sutterlin, S., and van Lenteren, J.C., 1997, Influence of hairiness of *Gerbera jamesonii* leaves on the searching efficiency of the parasitoid *Encarsia formosa*. Biological Control 9: 157-165.

Teder, T., and Tammaru, T., 2002, Cascading effects of variation in plant vigour on the relative performance of insect herbivores and their parasitoids. Ecological Entomology 27: 94-104.

Treacy, M. F., Benedict, J. H., Lopez, J. D., and Morrison, R. K., 1987, Functional response of a predator (Neuroptera: Chrysopidae) to bollworm (Lepidoptera: Noctuidae) eggs on smoothleaf, hirsute, and pilose cottons. Journal of Economic Entomology 80: 376-379.

Van Haren, R. J. F., Steenhuis, M. M., Sabelis, M. W., and de Ponti, O. M. B., 1987, Tomato stem trichomes and dispersal success of *Phytoseiulus persimilis* relative to its prey *Tetranychus urticae*. Experimental and Applied Acarology 3: 115-121.

Van Lenteren, J, and Bakker, K., 1978, Behavioral aspects of the functional responses of a parasite *Pseudeucoila bochei* to its host *Drosophila melanogaster*. Netherlands Journal of Zoology 28: 213-233

van Rijn, P., and Sabelis, M. W., 1990, Pollen as an alternative food source for predatory mites and its effect on the biological control of thrips in greenhouses. Proceedings of the Experimental and Applied Entomology Section of the Netherlands Entomological Society Meeting, Amsterdam 1: 44-48.

van Rijn, P. C. J., van Houten, Y. M., and Sabelis, M. W., 2002, How plants benefit from providing food to predators even when it is also edible to herbivores. Ecology 83: 2664-2679.

Wagner, G. J., 1991, Secreting glandular trichomes: more than just hairs. Plant Physiology 96: 675-679.

Walde, S. J., 1995, How quality of host plant affects a predator-prey interaction in biological control. Ecology 76: 1206-1219.

Wheeler, A.G., 2001, Biology of the Plant Bugs. Cornell University Press. Ithaca, NY.

White, T.C.R., 1984, The abundance of invertebrate herbivores in relation to the availability of nitrogen in stressed food plants. Oecologia 63: 90-105.

Wurst, S., and Hefin Jones, T., 2003, Indirect effects of earthworms (*Aporrectodea caliginosa*) on an above-ground tritrophic interaction. Pedobiologia 47: 91-97.

Yasui, H., 2001, Sequestration of host plant-derived compounds by geometrid moth, *Milionia basalis*, toxic to a predatory stink bug, *Eocanthecona furcellata*. Journal of Chemical Ecology 27: 1345-1353

Zwahlen, C., Nentwig, W., Bigler, F., and Hilbeck, A., 2000, Tritrophic interactions of transgenic *Bacillus thuringiensis* corn, *Anaphothrips obscurus* (Thysanoptera: Thripidae), and the predator *Orius majusculus* (Heteroptera: Anthocoridae). Environmental Entomology 29: 846-850.

TROPHIC AND GUILD INTERACTIONS AND THE INFLUENCE OF MULTIPLE SPECIES ON DISEASE

Matt B. Thomas[1,2], Steve P. Arthurs[1] and Emma L. Watson[1]

[1]NERC Centre for Population Biology, Imperial College London, Silwood Park Campus, Ascot, Berks. SL5 7PY, UK; [2]Current address: CSIRO Entomology, GPO Box 1700, Canberra ACT 2601, Australia

Abstract. Most studies of host-parasite systems consider the interaction between one host and one parasite species. The aim of this chapter is to illustrate how the presence of other interacting species can influence the outcome of particular host-pathogen/parasite interactions. We begin with a brief review of the disease literature considering three broad categories of interactions: (i) cases with one pathogen/parasite and multiple shared hosts; (ii) cases with one host and multiple shared pathogens and/or parasites; (iii) cases with generally one host and pathogen but considering the influence of other interacting species such as natural enemies or competitors. We then consider some specific examples drawn from investigations into the use of fungal entomopathogens for biological control of locusts and grasshoppers. These case histories serve to highlight how an understanding of more complex trophic or guild interactions could help in improving the safety and effectiveness of biological control using pathogens.

1. INTRODUCTION

The vast majority of host-pathogen/parasite studies consider the direct interaction between one disease agent and one host species. As apparent throughout this book, however, interactions between species rarely occur in isolation. Our aim in this chapter is to highlight how the presence of other species within a system can influence the outcome of a focal host-pathogen interaction. In line with the main theme of the book we consider trophic and intraguild interactions involving pathogens, but also extend our study to more complex multi-species interactions. This broader perspective is justified by the fact that studies on strict intraguild predation involving pathogens, for example, are relatively rare (especially if we restrict ourselves to just biocontrol). Moreover, if we are to improve our understanding of disease ecology, then we need to consider the full ecological context in which host-pathogen interactions are played out and not just a subset of potential influencing factors. We begin with a brief review of the disease literature breaking down more complex, biodiverse systems into three broad categories: (i) cases with one pathogen/parasite and multiple shared hosts; (ii) cases with one host and multiple shared pathogens and/or parasites (this can be considered as a special class of intraguild interactions with two or more pathogens interacting within a single host); (iii) cases with generally one host and pathogen but considering the influence of other 'third party' species such as competitors or natural enemies

J. Brodeur and G. Boivin (eds.), Trophic and Guild Interactions in Biological Control, 101–122.

(which includes intraguild predators). Our aim is to be illustrative rather than exhaustive and so we pick examples from a diverse range of systems and consider both theoretical and empirical studies. Building on this broad overview, we then focus in on biological control and consider some specific examples drawn from our own investigations into the use of fungal entomopathogens for biological control of locusts and grasshoppers. These empirical case histories are again illustrative but serve to highlight how an understanding of more complex trophic or guild interactions could help in improving the safety and effectiveness of biological control using pathogens

2. OVERVIEW OF MULTI-SPECIES EFFECTS ON HOST-PATHOGEN/PARASITE INTERACTIONS

2.1. Multiple hosts and a single pathogen/parasite

Recent years have seen increased interest in the role of biodiversity in ecosystem functioning and the provision of ecosystem services (Loreau *et al.* 2001). Pest and disease control has been identified as a valuable ecosystem service delivered by biodiversity (Pimentel 1961; Altieri 1991; Naylor and Ehrlich 1997), and there is considerable evidence that as agricultural production systems are intensified they tend to lose biodiversity and become destabilised, with increased frequency and extent of pest and disease outbreaks (Pimentel 1961; Kruess and Tscharntke 1994; Swift *et al.* 1996; Knops *et al.* 1999). For example, the adoption of monocultures characteristic of modern intensive agriculture has acted to increase the density of a particular plant species or cultivar, while reducing the numbers of crop species, varieties within species, and genetic differences within varieties (Wolfe 2000). The risk this creates is that if a pest or disease is able to exploit the one dominant variety, then it has almost unlimited potential to spread throughout the field and landscape. The potential to redress this balance and reduce disease incidence by increasing plant diversity has been recognized for many years. For example, Elton (1958) argued that disease should be reduced in more complex plant systems (though this increased complexity could apply to the addition of host or non-host species). More recently in a study in Yunnan Province in China, rice farmers were able to control a key fungal disease (rice blast) through the use of variety mixtures, interplanting one row of a susceptible glutinous rice variety to every four or six rows of a more resistant commercial variety (Zhu *et al.* 2000). This simple increase in intra-specific diversity led to a substantial reduction in prevalence of rice blast and an increase in yield of the susceptible variety. However, there are also cases where an increase in host plant diversity can lead to an increase in disease prevalence. In a study of barley yellow dwarf virus, for example, Power and Mitchell (2004) showed that more diverse plant communities could lead to higher rates of infection through the addition of highly competent reservoir species for the virus.

Host diversity has also been shown to play a role in the dynamics of animal diseases. For example, in the US, the white-footed mouse (*Peromyscus leucopus*) is the main reservoir for the Lyme disease spirochete. Following feeding on these mice, ticks can transmit the bacterium to other hosts. However, while some host species (such as the white-footed mouse) are effective reservoirs, many are incompetent reservoirs of Lyme disease (Mather 1993). This creates a situation where high levels of host diversity (effectively reducing the representation of white footed mice relative to other vertebrate hosts) could reduce infection prevalence via a 'dilution effect' (Norman *et al.* 1994; Ostfeld and Keesing 2000; LoGuidice *et al.* 2003). That is, increasing the frequency of encounters with hosts who are incompetent reservoirs relative to highly competent white-footed mice should reduce infection prevalence of ticks leading to a lower risk of exposure to Lyme disease in humans. Empirical support for this dilution effect was provided by Ostfeld and Keesing (2000) who demonstrated significant negative relationships between species richness of small mammals and lizards and the per capita number of Lyme disease cases for each state along the eastern seaboard of the United States from Maine to Florida. Interestingly, a contrasting positive association was identified with species richness of ground-dwelling birds. This suggests that birds may be acting as competent reservoirs contributing to a 'rescue effect' whereby increased diversity helps maintain the disease agent at a relatively constant prevalence by buffering against fluctuations of individual host populations (Ostfeld and Keesing 2000).

Regarding biocontrol, most studies that consider multiple host effects approach this from the perspective of evaluating host range and the possible impact of a biocontrol agent on the non-target community. There appears limited research considering the influence of multiple hosts on the impact of disease on a target pest. This mirrors a much broader situation of a relatively good understanding of the population biology of single-host pathogens, yet much poorer understanding of more complex multi-host systems, in spite of the fact that most pathogens appear to be generalists (Woolhouse *et al.* 2001). Recent theoretical studies by Holt *et al.* (2003) have demonstrated the potential for additional host species to potentially inhibit establishment of infectious diseases. This negative association with diversity is limited to situations where certain host species are more resistant and applies especially to free-living pathogens or those transmitted by vectors (*c.f.* the 'dilution' effect discussed previously). In relation to biocontrol, this raises the possibility for infection of non-target hosts to restrict the impact of an agent against the target. In contrast, Holt *et al.* (2003) also identified situations where there is a positive association with diversity with parasite persistence enhanced by the presence of multiple host species (e.g. if the density of one host falls, a pathogen or parasite may still be sustained through interspecific transmission with another host). This is relevant since the aim of biocontrol is to drive a pest species down to low densities. If the pathogen or parasite is highly host specific, the low pest densities may cause the biocontrol agent to drop out of the system, whereas is if there are alternate hosts, the agent might persist.

2.2. Multiple pathogens and parasites and a single host

From above it can be seen that presence of additional host genotypes or species can strongly influence the outcome of specific host-parasite interactions, with the effects dependent on the specific nature of the interacting elements. The possible effects of altering pathogen/parasite diversity are no less complicated.

Evidence from a range of systems indicates that 'concomitant' or 'mixed' infections involving two or more parasite species or genotypes within a host are common (Petney and Andrews 1998; Cox 2001). These multi species/genotype infections have been examined in numerous hosts including mammals (e.g. Nilssen *et al.* 1998; Petney and Andrews 1998; Behnke *et al.* 2001; Cox 2001), birds (e.g. Forbes *et al.* 1999), reptiles (Schall and Bromwich 1994; Lainson 2002) fish (Sousa *et al.* 1996; Barker *et al.* 2002) and invertebrates (e.g. Tang *et al.* 2003; Thomas *et al.* 2003).

In mixed infections, complex interactions between parasites and an individual host may arise such that the burden of one or both of the infectious agents may be increased, one or both may be suppressed, or one may be increased and the other suppressed (Cox 2001). For example, Tang *et al.* (2003) demonstrated that preinfection with infectious hematopeoitic and hypodermal necrosis virus (IHHNV) reduced viral load and mortality of shrimps subsequently exposed to white spot syndrome virus. Similarly, pre-exposure of rainbow trout to a non-pathogenic virus has been shown to reduce impact of a pathogenic virus (Hedrick *et al.* 1994), possibly through a host-derived immune response.

With particular reference to biological control, a number of studies have examined (intraguild) interactions between pathogens within insect hosts (e.g. Ritter and Tanada, 1978 and references therein; Fuxa, 1979; Barbercheck and Kaya, 1991; Malakar *et al.*, 1999) to investigate the potential for improving the efficacy of biocontrol or to explore the effect of applying a biocontrol agent where indigenous entomopathogens are present. These studies have shown that concomitant infections can affect both the qualitative (which pathogen wins) and quantitative (number of propagules and speed of kill) outcome of an infection (Barbercheck and Kaya, 1990, 1991; Thurston *et al.*, 1994; Inglis *et al.*, 1997, 1999; Ishii *et al.*, 2002).

2.3. Effects of predators and other non-host species

It might be expected that the addition of more hosts or pathogens into a particular host-pathogen system can alter the outcome of the interaction. Perhaps less obvious is the role that other species can play through more indirect mechanisms affecting host density, disease transmission, virulence etc.. For example, the majority of viruses that are transmitted by insects tend to be found at lower incidences in polycultures (i.e. mixed species plantings) than monocultures (Power and Flecker 1996). This is because greater plant species diversity tends to reduce the abundance of individual insect vectors. The diversity of crop species in an agroecosystem has a much less predictable effect on microbial pathogens that do not rely on insect vectors, such as most fungi (Matson *et al.*, 1997). Whereas fungal diseases can be less severe in polycultures than monocultures (Boudreau, 1993), the opposite effect is also seen (Boudreau and Mundt 1997). Generalizations are difficult because the effects of intercropping on disease dynamics depend on a variety of factors, including microclimate effects and the spatial scale of pathogen dispersal relative to the pattern of planting (Boudreau and Mundt 1997).

Predators are often amongst the first species to disappear in the process of biodiversity loss. When predators are lost to ecosystems, their prey may increase in abundance, leading to increased disease transmission efficiency (Packer *et al.* 2003). Many zoonotic reservoir species are rodents (Mills and Childs 1998) whose dynamics can be strongly affected by predators (Ostfeld and Holt 2004 and refs therein). While these dynamic effects have rarely been linked to increases in zoonotic disease, in general terms, if rodents are maintained at low density away from humans, then disease transmission will be reduced (Ostfeld and Holt 2004). It is proposed that one generic effect of reducing predators could be to increase the equilibrial abundance of infected reservoirs and the fraction of reservoirs infected (Ostfeld and Holt 2004). The loss of vertebrate predators via habitat destruction and environmental degradation could, therefore, lead to a trophic cascade increasing transmission of rodent-borne disease to humans (Ostfeld and Holt 2004).

Other indirect interactions include, for example, the role of disease in mediating the interaction between competitors. In the UK, grey squirrels from America have largely displaced native red squirrels with the aid of a parapox virus, brought by resistant grey squirrels, that decimates red squirrels (Tompkins *et al.* 2003). Similarly, movement of the crayfish plague (*Aphanomyces astaci*)-resistant North American crayfish into the UK and European mainland resulted in extirpation of native crayfish species throughout much of Europe (Alderman 1993; Nylund and Westman 1995; Liley *et al.* 1997).

Focusing more on biological control of insects, intraguild interactions (both antagonistic and mutualistic) between diseases, such as entomopathogenic fungi, and other natural enemies have been shown to play a role in the outcome of insect host-pathogen interactions, at least under laboratory conditions. Most reports concern host-fungus-parasitoid interactions (e.g. Powell *et al.* 1986; Brobyn *et al.* 1988; Akalach 1992; Furlong and Pell 1996). In these studies, the pathogen was normally considered as the intraguild predator, as dual infection and parasitism commonly resulted in disrupted parasitoid development. However, foraging coccinellid predators of aphids have been shown to be potential intraguild predators of the entomophthoralean fungus, *Pandora* (*Erynia*) *neoaphidis* (Pell *et al.* 1997; Roy *et al.* 1998; Roy *et al.*, 2003) although evidence for the 'vectoring' of fungal conidia to uninfected hosts by the predator suggests the extent of the antagonistic interaction may be reduced (Pell *et al.* 1997; Roy *et al.* 1998).

2.4. Discussion

The common elements of a host-pathogen/parasite system are a host, a pathogen or parasite and some mechanism of disease transmission between pathogen/parasite and susceptible hosts. Based on the examples given above we can see that through their influence on either host or parasite abundance (which broadly includes effects such as reduced pathogen reproductive fitness or altered virulence through co-infection), or transmission efficiency, the addition of further species can affect disease in complex ways. Increased parasite or host diversity can increase or decrease the impact of individual disease agents. The presence of other interacting species or larger scale changes in habitat and ecosystem diversity can also affect disease risk and spread. The complex nature of these multispecies interactions makes generalisations difficult, but what is clear is that it is essential to consider not just the single host and pathogen/parasite of interest, but also the environmental context in which the interactions are played out. The examples presented above notwithstanding, relatively few studies of disease consider these broader trophic or guild interactions. This applies particularly to the use of pathogens in biological pest control where the focus is almost always on the performance one particular agent against a specific pest. Below we present some of our own research to illustrate how a broader perspective considering trophic or guild interactions can provide a better understanding of both the impact of a microbial biocontrol agent on non-target species, and also the influence of other species on the performance of the biocontrol agent.

3. MULTI-SPECIES INTERACTIONS AND THE SAFETY AND EFFECTIVENESS OF BIOLOGICAL CONTROL: SOME CASE HISTORIES FROM THE BIOLOGICAL CONTROL OF LOCUSTS AND GRASSHOPPERS USING A FUNGAL ENTOMOPATHOGEN

The biopesticide, Green Muscle^{TM}, has been developed for biocontrol of locusts and grasshoppers in Africa (Lomer *et al.* 2001). This biopesticide is based on the mitosporic fungus *Metarhizium anisopliae* var. *acridum,* strain IMI 330189 (= *flavoviride* (Driver *et al.*, 2000)), which is a strain with high virulence to acridids (Orthoptera). Extensive environmental safety testing indicates negligible direct effects of the pathogen on non-target arthropods following field applications (Prior, 1997; Peveling *et al.*, 1999a,b; Stoltz, 1999). In the context of our broader review above, this type of research fits within the framework of a single pathogen/parasite with multiple shared hosts. However, these and most other conventional studies on host range tend to be limited to measuring direct non-target mortality and do not consider longer-term dynamic implications. Moreover, relatively few studies have considered broader interactions such as those with the indigenous microbial community or with other natural enemies, which fit within our broader framework of multiple pathogens or shared natural enemies. In this next section, therefore, we present the results of some empirical research aimed at evaluating the risk potential of *M. anisopliae* var. *acridum* sprayed as a biopesticide considering some of these more complex trophic/guild interactions.

3.1. Effects of biopesticide on indigenous microbial competitors – multiple pathogens and multiple hosts

As indicated above, initial development and testing of Green Muscle^{TM} was directed at locust and grasshopper control in Africa. However, interest in use of this product for sustainable locust and grasshopper control has recently extended to Europe with trials conducted against species such as Moroccan locust and Italian grasshopper in Spain (Thomas and Kooyman 2004). The *acridum* variant has a wide geographical distribution but has not, to date, been isolated in Spain. It is of interest, therefore, whether spray applications of exotic *M. anisopliae* var. *acridum* could impact on non-targets in this new range. In particular, the local locust and grasshopper communities have been shown to support a relatively high prevalence of a second fungal species, *Beauveria bassiana* (Hernandez-Crespo and Santiago-Alvarez, 1997). The potential exists, therefore, for competition between the newly introduced and indigenous fungal pathogens.

The extent of inter-specific competition and risk of a negative effect such as displacement of a native pathogen depends in part on the relative host ranges of the fungal pathogens. Given the specificity of *M. anisopliae* var. *acridum*, direct competition between *B. bassiana* and *M. anisopliae* var. *acridum* will tend to be restricted to acridids and potential for displacement will be mitigated if the indigenous pathogens have a broader host range than that of the introduced pathogen. To help determine this risk we conducted a study examining the diversity of the indigenous fungal entomopathogen community and the likely specificity of particular isolates to acridid hosts. In brief, indigenous fungal entomopathogens were isolated from acridids, other insects and soil samples, from field sites in Spain where Green Muscle^{TM} was being trialled. The fungi were identified to species level using morphological characteristics and found to consist primarily of *Beauveria bassiana*. A random subset of these *B. bassiana* isolates (n=30) were then characterised using PCR amplification of ssreps (simple sequence repeats). The standard *M. anisopliae* var. *acridum* isolate (IMI 330189) was included in the analysis as an out-group and a *B. bassiana* isolate from the USA included as a positive control.

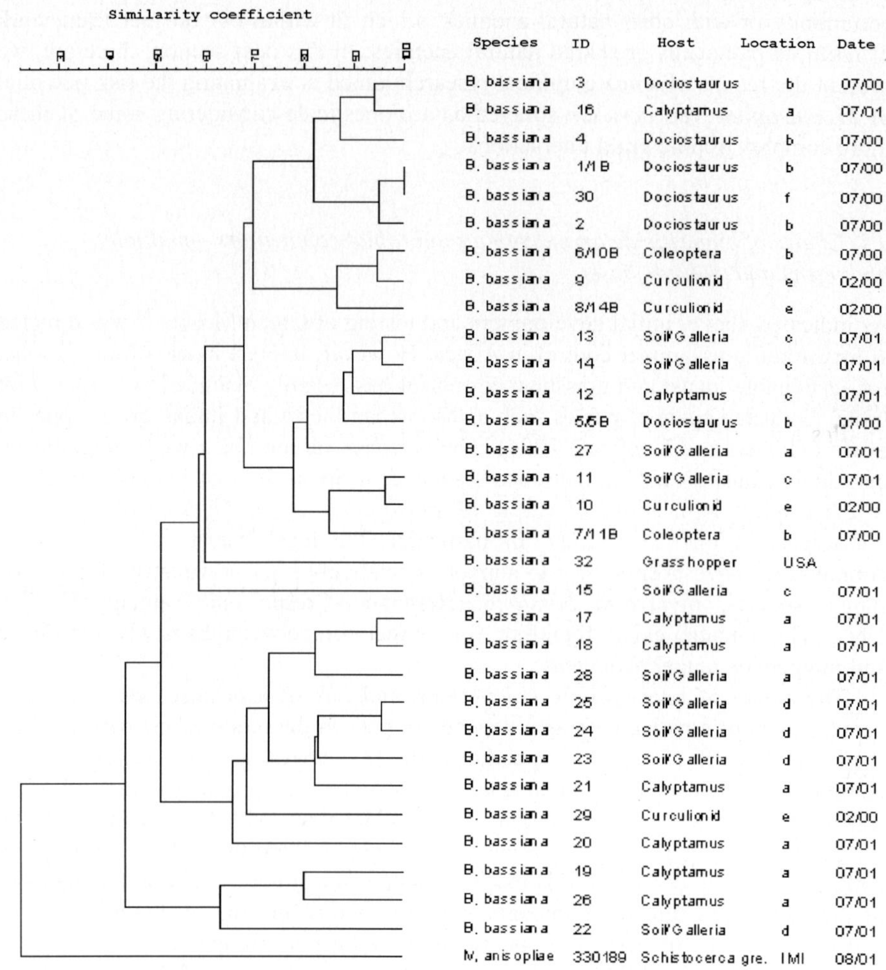

Figure 1. Dendrogram of B. bassiana isolates collected from La Serena and Ciudad Real, Spain, control Beauveria isolates from the UK and USA, and one M. anisopliae isolate. Data from PCR of simple sequence repeat regions using primers AAC (5'-(AAC)5-3'), MR (5'-GAGGGTGGCGGTTCT-3') and RY (5'-(CAG)5-3'), were analysed using UPGMA and DICE similarity coefficients. Isolate identification numbers indicate individual isolates. Location letters a-f indicate specific field sites within Spain.

Based on the molecular data, a dendrogram of similarity relationships was produced by unweighted pair-group means analysis (UPGMA) of band classes using DICE similarity coefficients. This dendrogram (Fig. 1) indicates considerable diversity amongst the *B. bassiana* isolates. There is a degree of clustering apparent depending on location of origin (e.g. isolates 17, 18 and 28 from a common field location 'a') although each location also includes genetically diverse isolates (e.g. isolates 16, 27, 21, 20, 19 and 26 from location 'a'). For most hosts, there is little clustering apparent; with isolates originating from Curculionids, for example, showing only ~40% similarity, and those from the Italian grasshopper, *Calliptamus italicus* (Orthoptera: Acrididae) being evenly distributed throughout the dendrogram. In contrast, five of the six Moroccan locust-originating isolates cluster together with ~80% similarity, and the sixth shows ~65% similarity with this cluster. The out-group, *M. anisopliae* var. *acridum*, shows less than 30% similarity with the *B. bassiana* isolates.

In addition to the molecular characterisation, the pathogenicity of five indigenous *B. bassiana* isolates was measured against the Moroccan locust, *Dociostaurus maroccanus* (Acrididae). The five isolates had different genotypic profiles and included two that had been isolated from the locust host species, and three from different insect hosts. Despite their diverse host origins and genetic diversity, all five of the *B. bassiana* isolates examined were pathogenic to Moroccan locust (Fig. 2) and exhibited high levels of virulence with LT_{50} values between 5 and 6 days.

Survival analysis (age-specific hazard with Weibull distribution and censoring) showed that the two *Beauveria* isolates originating from locusts (1 and 5) and the two from Coleoptera (10 and 11) had a similar level of virulence under the test conditions. The isolate originating from a Curculionid host (14) and the *M. anisopliae* var *acridum* isolate showed a significantly lower virulence compared to 1 ($P<0.005$, z value = 2.907 and $P<0.001$, z value = 4.484, respectively). The dendrogram (Fig. 1) indicates that these five isolates (designated B1, B5, B10, B11 and B14) are genetically diverse, and yet all show a high level of virulence to *D. maroccanus*.

The results of the soil and insect surveys indicate that *B. bassiana* is widespread and abundant in the locust-affected area. These results are in agreement with previous surveys in the region, which show relatively high prevalence of *B. bassiana* but few other entomopathogenic fungi (Hernandez-Crespo and Santiago-Alvarez, 1997). Molecular analysis revealed a considerable level of genetic variation amongst isolates from soil and different host insects, suggesting an assemblage of genotypes that might represent a species complex. Whilst these genotypes could exhibit differences in host specificity, the five genotypes tested, which came from a range of hosts, were all highly virulent to the Moroccan locust. Although not definitive, these results do suggest that the indigenous *B. bassiana* species complex consists of isolates with low host specificity. The detection of a large number of isolates in the soil also suggests the species has saprophytic ability. These characteristics are in sharp contrast to *M. anisopliae* var. *acridum* which has very little variability across its geographic range (Bridge *et al.*, 1997), and an ecological host range that is restricted to the Orthoptera (Peveling *et al.*, 1999a). This broader diversity and host range within the *B. bassiana* species-complex is likely to mitigate any direct competition for Orthopteran hosts from the exotic *Metarhizium* isolate. That is, even if *B. bassiana* is out-competed by *M. anisopliae* var. *acridum* for Orthopteran hosts, its ability to infect other taxa should provide refuges from competition. It is of course possible that a *B. bassiana* genotype exists which is highly specific to Orthoptera and therefore at a greater risk of competitive exclusion. This intra-host competition between coinfecting entomopathogens is explored below.

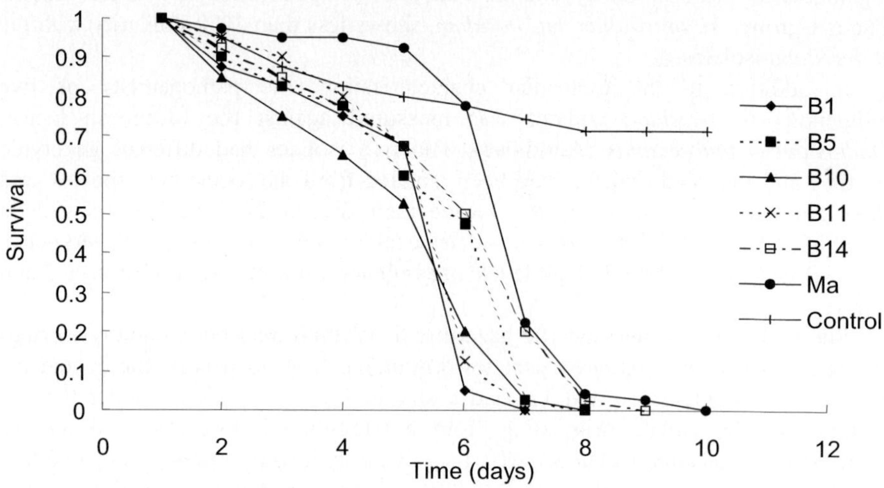

Figure 2. Survival curves of D. maroccanus following inoculation with 104 conidia per insect of one M. anisopliae var. acridum (Ma) and five B. bassiana strains (B1, B5, B10, B11, B14), compared with uninfected controls.

3.2. Effects of biopesticide on indigenous microbial competitors – mixed infections within a single host

Natural populations of insects, including locusts, are frequently infected with pathogens and mixed infections are expected to be common (Hernandez-Crespo and Santiago-Alvarez, 1997; Cox, 2001). This is especially likely following inundative application of a biocontrol agent where a high proportion of targets and non-targets will be artificially challenged by disease, above and beyond any existing background infection.

As indicated in section 2, intra-host interaction between two pathogens can result in: either or both pathogens increasing in virulence (synergism); neither pathogen being affected (independence); or either or both being inhibited and reducing in virulence (antagonism) (Tanada and Kaya, 1993; Cox, 2001). Which of these possible outcomes transpires will depend on numerous factors including the growth rates and virulence levels of the two pathogens, the order of, and time-lag between infections, the specific niches and host resources utilised by each, as well as many factors relating to the host (e.g. age and physiological status) and environment (reviewed for parasite co-infections by Cox, 2001). Inglis *et al.* (2001) review studies where two different entomopathogens have been applied to a host and report results of both increased efficacy and no effect, whilst other studies have reported a decrease in virulence of one or other pathogen (Ritter and Tanada, 1978; Fuxa, 1979; Gothama *et al.*, 1995).

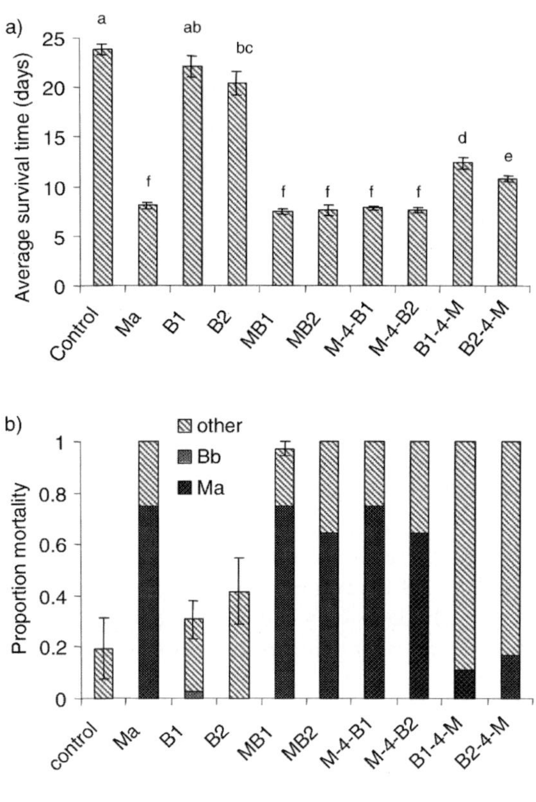

Figure 3. The effect of isolates M. anisopliae var. acridum (Ma) and B. bassiana (B1 and B2) inoculated alone, in combination (MB1 and MB2), and consecutively with a four day time lag between (M-4-B1, M-4-B2, B1-4-M, B2-4-M) on Schistocerca gregaria. *Single treatments were at 103 spores insect-1 and combination treatments 103 spores of each isolate insect-1 (total of 106 spores insect-1). A) Mean (± SE) survival time (days) of treated* S. gregaria *incubated at a constant 30°C. Different letters designate significant differences between treatments (P<0.05). B) Mean (± SE) proportion mortality and cause of death of the same insects.*

Whilst it is difficult to make generalisations about the outcome of a concomitant infection, the range of possible effects have clear implications for biocontrol. For example, if prior infection of the host by an indigenous pathogen precludes development and sporulation of the biocontrol pathogen, secondary transmission would be reduced, reducing persistence and overall levels of control (Thomas *et al.*, 1995). On the other hand, if a biocontrol agent is highly virulent, competitive exclusion of the indigenous pathogen might occur (although as argued above, this will also depend on whether the indigenous isolates have alternate hosts, for example).

To investigate this further, we conducted an experiment to examine the effects of mixed infection considering the interactions between *M. anisopliae* var. *acridum* (denoted below as Ma) and two of the *B. bassiana* isolates (denoted as B1 and B2) isolated from the Moroccan locust in Spain as reported above. The experiment used the desert locust as a test organism and involved six treatments covering the various combinations of single and mixed infections, plus controls: i) Ma alone; ii) B1 alone; iii) B2 alone; iv) Ma + B1 combined; v) Ma + B2 combined; vi) control. In addition, we also explored the effects of sequential rather than simultaneous infections by inoculating locusts with the following combination treatments with a gap of four days between the first and second inoculation: i) Ma + B1 (M-4-B1), ii) Ma + B2 (M-4-B2), iii) B1 + Ma (B1-4-M), iv) B2 + Ma (B2-4-M). Full details of the methodologies are presented in Thomas *et al.* (2003)

The main results of the study are presented in Fig. 3. Survival analysis and ANOVA of mortality (at 25 days) of treated locusts showed a significant treatment effect (for mortality: $F_{9,29} = 25.92$, $P<0.001$). Almost all of the treatment effects on mortality were due to Ma ($F_{1,29} = 225.1$, $P<0.001$), producing ~100% mortality, whilst neither *Beauveria* isolate had a significant effect (on mortality) over the control. *Metarhizium* either alone (Ma), concomitantly applied with *Beauveria* (MB1, MB2), or applied prior to *Beauveria* (M4B1, M4B2), reduced average survival time to *ca.*8 days. Of the two *Beauveria* isolates, only B2 had any effect on host survival (Fig. 3a); although only reducing average survival time from *ca.*24 days to *ca.*20 ($P<0.05$), with no significant effect on overall mortality compared to the control (Fig. 3b). Only one case of *Beauveria* cadaver colonisation was observed, and this was due to B1.

When locusts were inoculated with Ma four days after B1 (B1-4-M), average survival time was lengthened by a corresponding four days, suggesting that all mortality was due to Ma. However, infection with B2 four days before Ma (B2-4-M) resulted in a significantly shorter average survival time (by one day), suggesting that prior infection with this isolate had an enhancing effect on *Metarhizium* virulence (Fig. 3a). In addition, both B1-4-M and B2-4-M treatments reduced the overall sporulation of *Metarhizium* on cadavers ($F_{1,20}= 71.21$, $P<0.001$) compared to the other Ma treatments (Fig. 3b).

As expected from other studies run under similar conditions (e.g. Prior *et al.*, 1995; Blanford and Thomas, 1999; Elliot *et al.*, 2002), locusts inoculated with *M.*

anisopliae var. *acridum* rapidly succumbed to infection with average survival time around 7-8 days. In contrast, the *B. bassiana* isolates showed little or no virulence to desert locust, with only the B2 isolate having any significant effect on locust survival time (and this just 1 day) and only one cadaver exhibiting *Beauveria* sporulation.

Following simultaneous infections of the two species, or with *Metarhizium* inoculated first, virulence was as high as for the *Metarhizium*-alone infections. This is in line with work by Malakar *et al.* (1999) and Ishii *et al.* (2002) who found that simultaneous infections of entomopathogens resulted in the same level of virulence as the most virulent pathogen. However, there was evidence of an interaction when *Beauveria* was inoculated into the host first and allowed to establish for 4 days prior to inoculation with *Metarhizium*. Whilst the B1-4-M treatment resulted in an average survival time of four days more than the *Metarhizium* treatments, as would be expected if virulence and mortality were due to *Metarhizium*, the B2-4-M treatment showed a higher virulence, suggesting a synergistic effect of the two isolates. Moreover, there was also a significant reduction in colonisation of cadavers by *Metarhizium* for both B1-4-M and B2-4-M treatments (Fig. 3) from ~70% to <20%. Although *Beauveria* alone is not virulent enough to produce mortality over this time scale, in four days it appears to establish within the host sufficiently to have a significant effect on a subsequent *Metarhizium* infection. Malakar *et al.* (1999) found similar results when coinfecting gypsy moths (*Lymantria dispar*) (Lymantriidae) with a nucleopolyhedrosis virus and *Entomophaga maimaiga*. When inoculated simultaneously, virulence was due to *E. maimaiga* due to its shorter incubation time. When the NPV was inoculated ten days before *E. maimaiga*, however, at low doses of NPV there was a significantly increased virulence and higher mortality, in this case, due to the NPV, but lower progeny production.

As mentioned earlier, a reduction in propagules will have consequences for host-to-host transmission and secondary cycling, thereby potentially affecting the outcome of control. Of particular note in our study is that avirulent pathogens, which hitherto have been given little attention and may go largely undetected in the field, can still impact on fitness of another pathogen and so could play a significant role in mediating the outcome of host-pathogen interactions and contribute unexpected variability in host-pathogen dynamics over time and/or space. A similar effect has recently been shown for certain facultative bacterial endosymbionts, which appear to alter susceptibility of pea aphids to the fungal pathogen *Pandora neoaphidis* and also alter the reproductive fitness of the fungal pathogen should it infect an aphid host (Scarborough *et al.* 2005). In terms of biocontrol, such effects could translate to variable efficacy of an agent depending on the nature of the resident microbial community. In Spain, for example, the background prevalence of *B. bassiana* in locust populations has been observed to range from < 2 to > 50% in different sites and years (Hernandez-Crespo and Santiago-Alvarez 1997; Authors' unpublished data). Accordingly we would anticipate considerable variation in prevalence of mixed infections following biopesticide spray treatments of *M. anisopliae* and, given the effects of pre-infection with *B. bassiana* on reproductive fitness of *M. anisopliae* (plus a small effect on virulence), different patterns of mortality and pathogen establishment and spread.

3.3. Effects of biopesticide applications on intra-guild interactions

Green Muscle™ is based on a naturally occurring fungal pathogen with potential for horizontal transmission between infected and non-infected hosts after a spray application (Baker *et al.*, 1994; Lomer *et al.*, 1997). The importance of this disease cycling for overall control has been explored using theoretical population models based on empirically-derived transmission parameters (Thomas *et al.*, 1995; Thomas 1999). These models predict that horizontal transmission can provide substantial additional control by the end of the season following a single spray application. Cycling of the pathogen can, therefore, provide a biological substitute for chemical persistence and reduce the need for repeated spraying. However, although the pathogen can cycle and natural epizootics do occasionally occur, the frequency and extent of horizontal transmission following spray applications in field trials has been far less than predicted by these models. A partial explanation for this is that assessments in many trials rarely continue for long enough and are often not precise enough to quantify the effects of horizontal transmission (the difficulty being that horizontal transmission acts over many weeks on an ever decreasing population, and is also often confounded by movement of insects into and out of treated areas over the season). In addition, however, there is now experimental evidence to suggest some of the underlying assumptions in the models themselves, may also be leading to inaccurate predictions. That is, one of the assumptions in the basic models is that infected insects go on to die from the disease and become infective cadavers, providing new sources of inoculum. However, field observations from large-scale field trials in the Sahel reveal an apparent shortfall in the number of cadavers in the field of > 95% relative to the reduction in live grasshoppers (Arthurs *et al.* 2003). One possible explanation for this shortfall is that cadavers are rapidly removed by scavengers and so are difficult to find in the field. To investigate this we conducted a number of studies placing cadavers in the field and assessing removal rates (e.g. see Arthurs *et al.* 2003). The results of these investigations revealed a significant difference in the patterns of persistence of grasshopper cadavers depending on whether they were infected with *M. anisopliae* var *acridum* or were uninfected (Fig. 4). Non-infected cadavers were removed rapidly by the numerous ground dwelling scavengers, mainly tenebrionid beetles and ants. On the other hand, infected cadavers (whether simply internally mycosed, or sporulating externally) were found to be avoided by scavengers and to persist for several days, with the pattern of decay largely mediated by heavy rainfall.

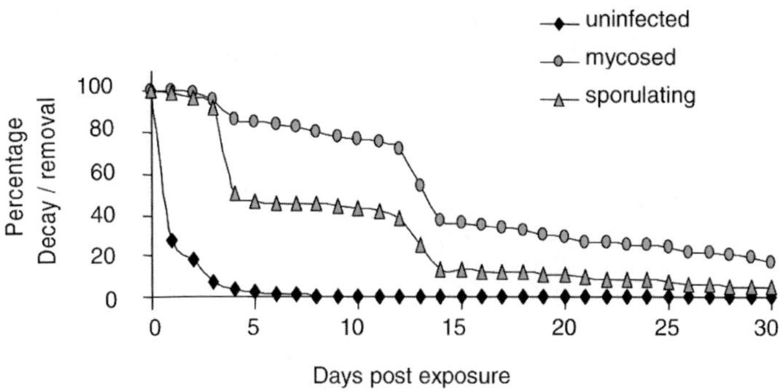

Figure 4. Comparison of persistence of infected and uninfected cadavers in the field during the wet season in the Sahel. Data are expressed as a percentage score representing level of damage or removal. Different symbols represent initial infective state of the cadavers. Uninfected cadavers were rapidly scavenged while the persistence of infected cadavers was determined largely be rainfall events, hence the step-wise decay pattern (modified from Arthurs et al. 2003).

Given that infected cadavers appear to be avoided by scavengers and to persist in the field for several days this suggests that many grasshoppers are being removed before they are killed by the biopesticide proper, with predation being the likely mechanism. This suggestion is supported by a range of empirical studies in lab and field, which indicate grasshoppers and locusts suffer increased susceptibility to predation during the disease incubation period (Thomas *et al.* 1998; Arthurs and Thomas 1999, 2001). For example, as part of our investigations into the effects of Green Muscle on Brown locust in South Africa, we examined comparative predation rates of infected and non-infected locusts by tethering individual locusts in the field using fine nylon. Following tethering, locusts were monitored at 24 h intervals for 3 days and classified according to the following categories: i) unaffected, ii) missing, or iii) predated (insect dead/missing but with signs of having been eaten by predators, indicated by the presence of limbs, head capsule or wings). The distribution of these different categories between infected and healthy locusts across the study period is shown in Fig. 5. Analysis for each observation period revealed a significant difference in class distributions between treated and control insects on day 2 and 3, treated insects being less likely to be observed (Pearson chi-square day 1 = 5.58 (P =0.06); day 2 = 7.85; (P< 0.05); day 3 = 7.0; (P< 0.05)).

These results indicate increased susceptibility to predation due to infection. Placing this in a more general ecological framework, this demonstrates intraguild predation that is predominantly unidirectional (see Polis and Holt 1992; Rosenheim *et al.* 1995) with the pathogen serving as a primary but not secondary consumer, and indigenous natural enemies serving as the intraguild predators.

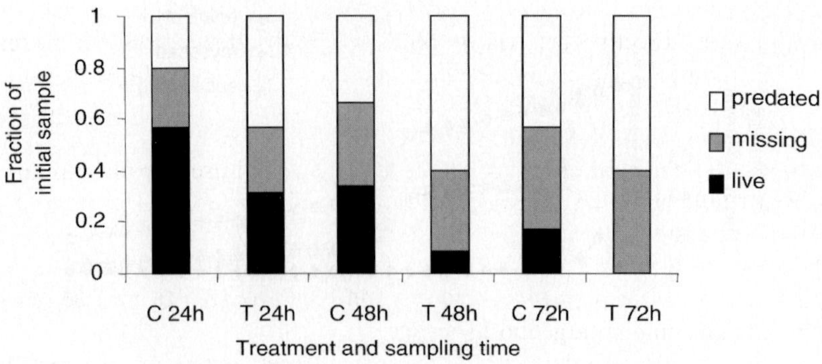

Figure 5. Data describing daily condition of adult brown locusts tethered in the field. Locusts were either treated with M. anisopliae var acridum (T) or untreated controls (C) and their condition assessed daily over 72 hours.

For our locust/grasshopper system, an important consequence of this type of intraguild predation concerns its repercussions for longer-term host-pathogen dynamics. Deuteromycete fungi infect new hosts via asexual spores (conidia) that are produced on the host following its death (Hajek and St. Leger 1994). Predation of infected but non-infective hosts, will reduce the ultimate density of cadavers and limit production of further inoculum. This may have a significant impact on the overall mortality caused by spray applications, with important implications for long-term control strategies (Thomas *et al.* 1995; Thomas *et al.* 1999). It may also help explain why horizontal transmission following spray applications is generally not as apparent as predicted and, in part, why natural epizootics are rare. Few theoretical or empirical studies have considered the extent to which predators (or, as indicated in the previous section, co-infecting pathogens) could act as biotic constraints to infection or how their impact could vary across time and/or space.

4. CONCLUSIONS

In this chapter we have tried to highlight some of the ways that multi-species interactions (either within or across trophic levels) can influence the outcome of individual host-pathogen/parasite interactions and hence, disease dynamics. We feel that further understanding of these multi-species interactions and the diversity of mechanisms involved would enhance our knowledge of disease in natural systems and increase our ability to manage disease in agricultural systems. Moreover, moving away from the conventional host-pathogen paradigm to consider the broader range of interactions and mechanisms that prevail in nature could create new opportunities for utilising pathogens in biological control. For example, most microbial control programmes (whether classical or inundative) consider the use of virulent pathogens for direct control of insect pests. However, there could be considerable scope for using less virulent, sub-lethal pathogens that impact not by killing the pest directly, but via trophic or guild effects such as increasing susceptibility to predation or competition. Indeed, given the emphasis on virulence and direct mortality, many potentially useful pathogens are probably overlooked because they do not kill the target, yet the most common effects of disease are sub-lethal. Similarly, potential effects of mixed infections and use of pathogen mixtures has received little attention. The possibility exists, for instance, to combine two or more largely avirulent pathogens to produce a much stronger lethal effect. Alternatively, if there are antagonistic effects of co-infection then it might be possible to use a second pathogen as a means of restricting the impact of a generalist virulent pathogen on non-target host species.

Extending the idea of mixed infections further, recent studies have demonstrated considerable potential for using pathogens for biocontrol of vector borne diseases. Blanford *et al.* (2005), for example, demonstrated that the number of mosquitoes potentially able to transmit malaria could be substantially reduced with fungal entomopathogens used as biological pesticides. This reduction resulted from two complementary effects of fungal infection. The primary effect was high-level mosquito mortality before mosquitoes were able to transmit malaria with more rapid mortality in mosquitoes infected with fungus and malaria. The secondary effect was that significantly fewer surviving mosquitoes had malaria sporozoites in their mouthparts (i.e. an impact on the malaria parasite as a consequence of co-infection). Other sub-lethal effects, such as reduced propensity to feed (Blanford *et al* 2005; Scholte *et al.* in press), are likely to further reduce malaria transmission. Such indirect effects of fungal infection raise the possibility of using pathogens to reduce malaria transmission without imposing selection for fungal resistance in the vector population. Indeed, influencing vector capacity need not even require successful fungal infection; simply provoking an additional non-specific immune response, for example, could potentially affect successful development and multiplication of the vector-borne disease. Whether this holds and to what extent these more subtle effects of co-infection could be exploited for control other vector-borne diseases warrants further investigation.

Regarding biocontrol safety, then our case histories illustrate the need to consider an array of trophic and guild interactions if we are to fully evaluate risks to non-target biodiversity. First, non-target effects extend beyond possible alternative hosts and include the microbial community; determining impacts on potential microbial competitors requires an understanding of microbial diversity and extent of overlap in host use. Second, short-term impact on non-target hosts may itself be mediated by the presence of other pathogens (i.e. effects of co-infection increasing or decreasing virulence). Third, risks of longer-term effects, as likely influenced by the establishment potential of the biocontrol agent, can be further affected by other trophic or guild interactions (e.g. our example of co-infection reducing reproductive fitness of *M. anisopliae* var *acridum*, or invertebrate natural enemies preventing production of infective cadavers). Furthermore, the possibility also exists for reciprocal effects of non-target species on biocontrol. The theoretical studies of Holt *et al.* (2003) referred to in section 2, for example, identify situations where parasite persistence can be enhanced by the presence of multiple host species. This creates an interesting tension between the growing demands for highly specific agents so as to minimize possible risks to the environment, and possible beneficial effects of a broader host range that could maximize long-term effectiveness of a microbial control agent. Managing potential operational conflicts such as this requires that we better understand the influence of multiple species in the ecology of pathogens and parasites.

5. ACKNOWLEDGEMENTS

This article derives from a presentation at a joint meeting of the IOBC-NRS/Canada Biocontrol Network held in Quebec on 8-11 May 2005. We are most grateful to Guy Boivin and Jacques Brodeur for providing the opportunity to participate in this meeting. MBT is also grateful to Carolyn Friedman and Kevin Lafferty for providing many useful insights into the effects of biodiversity on disease ecology.

6. REFERENCES

Akalach, M., Fernandez-Garcia, E., and Moore, D.,1992, Interaction between *Rastrococcus* invadens (Hom.: Pseudococcidae) and two natural enemies, Entomophaga 37: 99-106.

Alderman, D.J., 1993, Crayfish plague in Britain, the first twelve years. Freshwater Crayfish IX., University of Southwestern Louisiana, Lafayette, LA (USA), pp. 266-272.

Altieri, M.A., 1991, Increasing biodiversity to improve insect pest management in agro-ecosystems, in: The Biodiversity of Microorganisms and Invertebrates: its Role in Sustainable Agriculture, D. Hawksworth, ed., CAB International, Wallingford, UK.

Arthurs, S.P. and Thomas, M.B., 1999, Factors affecting horizontal transmission of entomopathogenic fungi in locusts and grasshoppers, Aspects Appl. Biol. 53, Challenges in Applied Population Biology, 89-98.

Arthurs, S.P. and Thomas, M.B., 2001, Investigation into behavioural changes in *Schistocerca gregaria* following infection with a mycoinsecticide: implications for susceptibility to predation, Ecol. Entomol. 26: 227-234.

Arthurs, S.P., Thomas, M.B. and Langewald, J., 2003, Field observations of the effects of fenitrothion and *Metarhizium anisopliae* var. *acridum* on non-target ground dwelling arthropods in the Sahel, Biol. Control 26: 333-340.

Baker, G. L., Milner, R.J., Lutton, G.G. and Watson, D., 1994, Preliminary field trial on control of *Phaulacridium vittatum* (Sjostedt) (Orthoptaera: Acrididae) populations with *Metarhizium flavoviride* Gams and Rozsypal (Deuteromycotina: Hyphomycetes), J. Australian Entomol. Soc. 33: 190-192.

Barbercheck, M.E. and Kaya, H.K., 1990, Interactions between *Beauveria bassiana* and the entomogenous nematodes, *Steinernema feltiae* and *Heterorhabditis heliothidis*, J. Invertebr. Pathol. 55: 225-234.

Barbercheck, M.E. and Kaya, H.K., 1991, Competitive interactions between entomopathogenic nematodes and *Beauveria bassiana* (Deuteromycotina, hyphomycetes) in soilborne larvae of *Spodoptera exigua* (Lepidoptera, noctuidae), Environ. Entomol. 20: 707-712.

Barker, D.E., Cone, D.K. and Burt, M.D.B., 2002, *Trichodina murmanica* (Ciliophora) and *Gyrodactylus pleuronecti* (Monogenea) parasitising hatchery-reared winter flounder, *Pseudopleuronectes americanus* (Walbaum): effects on host growth and assessment of parasite interaction, J. Fish Dis. 25: 81-89.

Behnke, J.M., Barnard, C.J., Bajer, A., Bray, D., Dinmore, J., Frake, K., Osmond, J., Race, T. and Sinski, E., 2001, Variation in the helminth community structure in bank voles (*Clethrionomys glareolus*) from three comparable localities in the Mazury Lake District region of Poland, Parasitology 123: 401-414.

Blanford, S. and Thomas, M.B., 1999, Host thermal biology: the key to understanding insect-pathogen interactions and microbial pest control? Agric. Forest Entomol. 1: 195-202.

Blanford, S., Chan, B.H.K., Jenkins, N.E., Sim, D., Turner, R.J., Read, A.F. and Thomas, M.B., 2005, Fungal pathogen reduces potential for malaria transmission, Science 308: 1638-1641.

Boudreau, M.A., 1993, Effects of intercropping beans with maize on angular leaf spot of bean in Kenya, Plant Pathol. 42: 16-25.

Boudreau, M.A. and Mundt, C.C., 1997, Ecological approaches to disease control. In: Environmentally Safe Approaches to Crop Disease Control, Rechcigl, J. and Rechcigl, N., eds., CRC/Lewis Press, Boca Raton, FL.

Bridge, P.D., Prior, C., Sagbohan, J., Lomer, C.J., Carey, M., and Buddie, A., 1997, Molecular characterization of isolates of *Metarhizium* from locusts and grasshoppers, Biodivers. Conserv. 6: 177-189.

Brobyn, P.J., Clark, S.J. and Wilding, N., 1988, Effect of fungus infection of *Metopolophium dirhodum* (Hom, Aphididae) on the oviposition behavior of the aphid parasitoid *Aphidius rhopalosiphi* (Hym, Aphidiidae), Entomophaga 33: 333-338.

Cox, F.E.G., 2001, Concomitant infections, parasites and immune responses, Parasitology 122: S23-S38.

Driver, F., Milner, R.J., and Trueman, W.H., 2000, A taxonomic revision of *Metarhizium* based on phylogenetic analysis of rDNA sequence data, Mycol. Res. 104: 134-150.

Elliot, S.L., Blanford, S. and Thomas, M.B., 2002, Host-pathogen interactions in a varying environment: temperature, behavioural fever and fitness Proc. R. Soc. Lond. Ser. B-Biol. Sci. 269: 1599-1607.

Elton, C.C., 1958, The Ecology of Invasions by Animals and Plants, Methuan & Co. London.

Forbes, M.R., Alisauskas, R.T., McLaughlin, J.D. and Cuddington, K.M., 1999, Explaining co-occurrence among helminth species of lesser snow geese (*Chen caerulescens*) during their winter and spring migration, Oecologia 120: 613-620.

Furlong, M.J. and Pell, J.K., 1996, Interactions between the fungal entomopathogen *Zoophthora radicans* Brefeld (Entomophthorales) and two hymenopteran parasitoids attacking the diamondback moth, *Plutella xylostella* L, J. Invertebr. Pathol. 68: 15-21.

Fuxa, J.R., 1979, Interactions of the microsporidium *Vairimorpha necatrix* with a bacterium, virus and fungus in *Heliothis zea*, J. Invertebr. Pathol. 33: 316-323.

Gothama, A.A.A., Sikorowski, P.P. and Lawrence, G.W., 1995, Interactive effects of *Steinernema carpocapsae* and *Spodoptera exigua* nuclear polyhedrosis virus on *Spodoptera exigua* larvae, J. Invertebr. Pathol. 66: 270-276.

Hajek, A. E., and St. Leger, R. J., 1994, Interactions between fungal pathogens and insect hosts, Ann. Rev. Entomol. 39: 293-321.

Hedrick, R.P., La Patra,, S.E., Yun, S., Lauda, K.A., Jones, G.R., Congleton, J.L., and de Kinkelin, P., 1994, Induction of protection from infectious hematopoietic necrosis virus in rainbow trout *Oncorhynchus mykiss* by pre-exposure to the avirulent cutthroat trout virus (CTV), Dis. Aquat. Organ. 20: 111-118.

Hernandez-Crespo, P. and Santiago-Alvarez, C., 1997, Entomopathogenic fungi associated with natural populations of the Moroccan locust *Dociostaurus maroccanus* (Thunber) (Orthoptera: Gomphocerinae) and other Acridoidea in Spain, Biocontrol Sci. Techn. 7: 357-363.

Holt, R.D., A.P. Dobson, M. Begon, R.G.Bowers, and E.M. Schauber, 2003, Parasite establishment in host communities, Ecol. Letts. 6: 837-842.

Inglis, G.D., Johnson, D.L., Cheng, K.J., and Goettel, M.S., 1997, Use of pathogen combinations to overcome the constraints of temperature on entomopathogenic hyphomycetes against grasshoppers, Biol. Control 8: 143-152.

Inglis, G.D., Duke, G.M., Kawchuk, L.M., and Goettel, M.S., 1999, Influence of oscillating temperatures on the competitive infection and colonization of the migratory grasshopper by *Beauveria bassiana* and *Metarhizium flavoviride*, Biol. Control 14: 111-120.

Inglis, G.D., Goettel, M.S., Butt, T.M., and Strasser, H., 2001, Use of hyphomycetous fungi for managing insect pests, in: Fungi as Biocontrol Agents, Butt, T.M., Jackson, C.W. and Magan, N., eds., CAB International, Wallingford, UK, pp. 23-69.

Ishii, T., Takatsuka, J., Nakai, M., and Kunimi, Y., 2002, Growth characteristics and competitive abilities of a nucleopolyhedrovirus and entomopoxvirus in larvae of the smaller tea tortrix, *Adoxophyes honmai* (Lepidoptera: Tortricidae), Biol. Control 23: 96-105.

Knops, J.M.H., Tilman, D., Haddad, N.M., Naeem, S., Mitchell, C.E., Haarstad, J., Ritchie, M.E., Howe, K.M., Reich, P.B., Siemann, E., and Groth, J., 1999, Effects of plant species richness on invasion dynamics, disease outbreaks, insect abundances and diversity, Ecol. Letts. 2: 286-293.

Kruess, A. and Tscharntke, T., 1994 Habitat fragmentation, species Loss, and biological-control, Science 264: 1581-1584.

Lainson, R., 2002, Intestinal coccidia (Apicomplexa: Eimeriidae) of Brazilian Lizards. *Emeria carmelinoi* n.sp, from *Kentropyx calcarata* and *Acroeimeria paraensis* n.sp from *Chemidophorus lemniscatus lemniscatus* (Lacertilia: Teiidae), Mem. I. Oswaldo Cruz. 97: 227-237.

Liley, J.H., Cerenius, L., and Soderhall, K., 1997, RAPD evidence for the origin of crayfish plague outbreaks in Britain, Aquaculture 157: 181-185.

LoGiudice, K., Ostfeld R.S., Schmidt, K.A., and Keesing, F., 2003, The ecology of infectious disease: effects of host diversity and community composition on Lyme disease risk, Proc. Natl. Acad. Sci. U.S.A. 100: 567-571.

Lomer, C.J., Prior, C. and Kooyman, C., 1997, Development of *Metarhizium* spp. for the control of locusts and grasshoppers, Memoirs Entomol. Soc. Canada 171: 265-286.

Lomer, C.J., Bateman, R.P., Johnson, D.L., Langewald, J., and Thomas, M.B, 2001, Biological control of locusts and grasshoppers, Ann. Rev. Entomol. 46: 667-702.

Loreau, M., Naeem, S., Inchausti, P., Bengtsson, J., Grime, J.P., Hector, A., Hooper, D.U., Huston, M.A., Rafaelli, D., Schmid, B., Tilman, D. and Wardle, D.A., 2001, Biodiversity and ecosystem functioning: current knowledge and future challenges, Science 294: 804-808.

Malakar, R., Elkinton, J.S., Hajek, A.E. and Burand, J.P., 1999, Within-host interaction of *Lymantria dispar* (Lepidoptera: Lymantriidae) nucleopolyhedrosis virus and *Entomophaga maimaiga* (Zygomycetes: Entomophthorales), J. Invertebr. Pathol. 73: 91-100.

Mather, T.N., 1993, The dynamics of spirochete transmission between ticks and vertebrates, in: Ecology and Environmental Management of Lyme Disease, H.S. Ginsberg, ed., Rutgers University Press, New Brunsqick, New Jersey, pp. 43-60.

Matson, P.A., Parton, W.J., Power, A.G. and Swift, M.J., 1997, Agricultural intensification and ecosystem properties, *Science* 277: 504-509.

Mills, J.N. and Childs, J.E. 1998. Ecological studies of rodent reservoirs: their relevance for human health, Emerg. Infect. Dis. 4: 529-537.

Naylor, R.L. and Ehrlich, P.R.,1997, Natural pest control services and agriculture, in: Nature's Services, G.C. Daily, ed., Island press, Washington, pp. 151-174..

Nilssen, A.C., R.E. Haugerud, and I. Folstad, 1998, No interspecific covariation in intensities of macroparasites of reindeer, *Rangifer tarandus* (L.), Parasitology 117: 273-281.

Norman, R., Begon, M. and Bowers, R. G., 1994, The population dynamics of Macroparasites and vertebrate hosts: the importance of recovery and immunity, Theor. Pop. Biol. 46: 96-119.

Nylund, V. and Westman, K., 1995, Frequency of visible symptoms of the crayfish plague fungus (*Aphanomyces astaci*) on the signal crayfish (*Pacifastacus leniusculus*) in natural populations in Finland in 1979-1988. Eight International Symposium on Astracology, Louisiana State university printing Office, Baton Rouge, LA (USA), pp. 577-588.

Ostfeld, R.S. and Keesing, F., 2000, Biodiversity and disease risk: the case of Lyme disease, Conserv. Biol. 14: 722-728.

Ostfeld R.S., and Holt R.D., 2004, Are predators good for your health? Evaluating evidence for top-down regulation of zoonotic disease reservoirs, Frontiers Ecol. Environ. 2: 13-20.

Packer, C., R. D. Holt, P. J. Hudson, K. D. Lafferty, and A. P. Dobson, 2003, Keeping the herds healthy and alert: implications of predator control for infectious disease, Ecol. Letts. 6: 797-802.

Pell, J.K., Pluke, R., Clark, S.J., Kenward, M.G. and Alderson, P.G., 1997, Interactions between two aphid natural enemies, the entomopathogenic fungus *Erynia neoaphidis* Remaudiere & Hennebert (Zygomycetes: Entomophthorales) and the predatory beetle *Coccinella septempunctata* L (Coleoptera: Coccinellidae). J. Invertebr. Pathol. 69: 261-268.

Petney, T.N. and R.H. Andrews, 1998, Multiparasite communities in humans and animals: frequency, structure and pathogenic significance, Int. J. Parasitol. 28: 377-393.

Peveling, R., Attignon, S., Langewald, J. and Ouambama, Z., 1999a, An assessment of the impact of biological and chemical grasshopper control agents on ground-dwelling arthropods in Niger, based on presence/absence sampling, Crop Prot. 18: 323-339.

Peveling, R., Rafanomezantsoa, J.-J., Razafinirina, R., Tovonkery, R. and Zafimaniry, G., 1999b, Environmental impact of the locust control agents fenitrothion, fenitrothion-esfenvalerate and triflumuron on terrestrial arthropods in Madagascar, Crop Prot. 18: 659-676.

Pimentel, D., 1961, Species diversity and insect populations outbreaks, Ann. Entomol. Soc. Amer. 54: 76-86.

Polis, G.A. and Holt, R.D., 1992, Intraguild predation – the dynamics of complex trophic interactions, Trends Ecol. Evol. 7: 151-154.

Powell, W., Wilding, N., Brobyn, P.J. and Clark, S.J., 1986, Interference between parasitoids (Hym, Aphidiidae) and fungi (Entomophthorales) attacking cereal aphids, Entomophaga 31: 293-302.

Power, A.G. and Flecker, A.S., 1996, The role of biodiversity in tropical managed ecosystems, in: Biodiversity and Ecosystem Processes in Tropical Forests, Orians, G.H., Dirzo, R. and Cushman, J.H., eds., Springer-Verlag, New York, pp173-194.

Power, A.G. and Mitchell, C.E., 2004, Pathogen spillover in disease epidemics, Amer. Nat. 164: S79-S89.

Prior, C., Carey, M., Abraham, Y.J., Moore, D. and Bateman, R.P., 1995, Development of a bioassay method for the selection of entomopathogenic fungi virulent to the desert locust, *Schisotcerca gregaria* (Forskål), J. Appl. Entomol. 119: 567-573.

Prior, C., 1997, Susceptibility of target acridoids and non-target organisms to *Metarhizium anisopliae* and *M. flavoviride*, in: New Strategies in Locust Control, Krall, S., Peveling, R. and Ba Diallo, D., eds., Birkhauser Verlag, Basel/Switzerland, pp. 369-375.

Ritter, K.S. and Tanada, Y., 1978, Interference between two nuclear polyhedrosis viruses of the armyworm, *Pseudaletia unipuncta* [Lep. : Noctuidae], Entomophaga 23: 349-359.

Rosenheim, J.A., Kaya, H.K., Ehler, L.E., Marois, J.J. and Jaffee, B.A., 1995, Intraguild predation among biological-control agents: theory and evidence, Biol. Control 5: 303-335.

Roy, H.E., Pell, J.K., Clark, S.J. and Alderson, P.G., 1988, Implications of predator foraging on aphid pathogen dynamics, J. Invertebr. Pathol. 71: 236-247.

Roy, H.E., Alderson, P.G. and Pell, J.K., 2003, Effect of spatial heterogeneity on the role of *Coccinella septempunctata* as an intra-guild predator of the aphid pathogen *Pandora neoaphidis*, J. Inverterbr. Pathol. 82: 85-95.

Scarborough, C.L., Ferrari, J. and Godfray, H.C.J., 2005, Aphid protected from pathogen by endosymbiont, Science 310: 1781.

Schall, J.J. and Bromwich, C.R., 1994, Interspecific interactions tested – 2 species of malarial parasite in a West African lizard, Oecologia 97: 362-332.

Scholte, E-J, Knols, B.G.J. and Takken, W., in press, Infection of the malaria mosquito *Anopheles gambiae* with the entomopathogenic fungus *Metarhizium anisopliae* reduces blood feeding and fecundity. J. Invertebr. Pathol..

Sousa, J.A., Romalde, J.L., Ledo, A., Eiras, J.C., Barja, J.L. and Toranzo, A.E., 1996, Health status of two salmonid aquaculture facilities in North Portugal: characterisation of the bacterial and viral pathogens causing notifiable diseases, J. Fish Dis. 19: 83-89.

Stoltz, I., 1999, The effect of *Metarhizium anisopliae* (Metsch.) Sorokin (= *flavoviride*) Gams and Rozsypal var. *acridum* (Deutermycotina: Hyphomycetes) on non-target Hymenoptera. PhD, Univ. Basel, Switzerland.

Swift, M.J., Vandermeer, J., Ramakrishnan, P.S., Anderson, J.M., Ong, C.K., and Hawkins, B.A., 1996, Biodiversity and agroecosystem function, in: Functional Roles of Biodiversity: a Global Perspective, H.A. Mooney, J.H. Cushman, E. Medina, O.E. Sala and E.D. Schulze, eds., Wiley, New York, pp. 261-297.

Tanada, Y. and Kaya, H.K., 1993, Insect Pathology, Academic Press, London.

Tang, K.F.J., Durand, S.V., White, B.L., Redman, R.M., Mohney, L.L., and Lightner, D.V., 2003, Induced resistance to white spot syndrome virus infection in *Penaeus stylirostris* through pre-infection with infectious hypodermal and hematopoietic necrosis virus-a preliminary study, Aquaculture 216: 19-29.

Thomas, M.B., 1999, Ecological approaches and the development of 'truly integrated' pest management, Proc. Natl. Acad. Sci. U.S.A. 96: 5944-5951.

Thomas, M.B., Blanford, S., Gbongboui, C. and Lomer, C.J., 1988, Experimental studies to evaluate spray applications of a mycoinsecticide against the rice grasshopper, *Hieroglyphus daganensis*, in northern Benin, Entomol. Exp. Appl. 87: 93-102.

Thomas, M.B., Wood, S.N. and Lomer, C.J., 1995, Biological control of locusts and grasshoppers using a fungal pathogen: the importance of secondary cycling, Proc. R. Soc. Lond. Ser. B-Biol. Sci. 259: 265-270.

Thomas, M.B., Watson. E.L. and Valverde-Garcia, P., 2003, Mixed infections and insect-pathogen interactions, Ecol. Letts. 6: 183-188.

Thomas, M.B. and Kooyman, C., 2004, Locust biopesticides: a tale of two continents, Biocontrol News Info. 25: 47N-51N.

Thurston, G.S., Kaya, H.K., and Gaugler, R., 1994, Characterising the enhanced susceptibility of milky disease-infected scarabaeid grubs to entomopathogenic nematodes, Biol. Control 4: 67-73.

Tompkins, D. M., A. R. White, and M. Boots, 2003, Ecological replacement of native red squirrels by invasive greys driven by disease, Ecol. Letts. 6: 189-196.

Wolfe, M.S., 2000, Crop strength through diversity, Nature 406: 681-682.

Woolhouse, M.E.J., Taylor, L.H. and Haydon, D.T., 2001, Population biology of multi-host pathogens, Science 292: 1109-1112.

Zhu, Y., H. Chen, J. Fan, Y.Wang, Y. Li, J. Chen, J.-X. Fan, S. Yang,L. Hu, H. Leung, T.W. Mew, P.S. Teng, Z. Wang and C.C. Mundt, 2000, Genetic diversity and disease control in rice, Nature 406: 718-722.

INTRA- AND INTERSPECIFIC INTERACTIONS AMONG PARASITOIDS: MECHANISMS, OUTCOMES AND BIOLOGICAL CONTROL

Guy Boivin[1] and Jacques Brodeur[2]

[1]CRDH, Agriculture and Agrifood Canada, Saint-Jean-sur-Richelieu, Québec, Canada; [2]Institut de Recherche en Biologie Végétale, Université de Montréal, Montréal, Canada

Abstract. Insect parasitoids are a major mortality factor for their herbivore hosts. Parasitoids typically interact, within their guild and between trophic levels, with several organisms from their community. Both intra- and interspecific competitions occur and the intensity of this competition influences the population dynamics of parasitoids. A female parasitoid may face several types of competition from females of her own species or from different species or from either conspecific or heterospecific females. The strategies evolved by parasitoids to cope with competition have implications both for the population dynamics of these species and for their use as biological control agents. The response of parasitoids to the type and intensity of competition varies at the level of the species, population and individual. In addition, a female parasitoid may change her response to competition depending on the conditions under which she developed and the experience she gained.

1. INTRODUCTION

Insect parasitoids are key components of terrestrial ecosystems both in terms of diversity and abundance. They may exert strong mortality on herbivore populations (Hawkins *et al.* 1997) and have stimulated extensive research in behavioral ecology, population dynamics, and biological control (Godfray and Shimada 1999; Hassell 2000). Comprehensive laboratory studies have been carried out on their anatomy, development, reproduction, host specificity, foraging behaviour, and molecular, biochemical and physiological interactions with the host. In the field, researchers have explored their distribution, movement, seasonal ecology and impact on host populations. An increasing number of studies on the natural enemies and competitors of parasitoids have led to a reconsideration of the most significant processes shaping the ecology and evolution of parasitoids including trophic and guild interactions.

In this chapter, we consider the trophic and guild relationships of parasitoids to natural enemies from their community. We briefly describe the nature and outcome of potential interactions between parasitoids and predators, entomopathogens, antagonistic parasitoids and hyperparasitoids. We then focus our attention to competitive interactions between parasitoids, including facultative hyperparasitoids, and attempt to review aspects of intra- and interspecific competition, as well as the

J. Brodeur and G. Boivin (eds.), Trophic and Guild Interactions in Biological Control, 123–144.

variability in the response of parasitoids to competitors. Throughout the chapter we aim to identify the impact that higher trophic and guild interactions may have on the use of natural enemies in biological control.

2. NATURAL ENEMIES AND COMPETITORS OF PARASITOIDS

Parasitoids are mostly viewed as primary carnivores (developing on herbivores), but they can also be secondary carnivores, by attacking predators or parasitoids (hyperparasitism), or even herbivores when, as adults, they feed and acquire energy from plant resources such as nectar. The community surrounding a parasitoid can be complex and includes taxonomically dissimilar species of antagonistic parasitoids, predators, entomopathogens, and hyperparasitoids. The interactions involve predation and competition processes.

We will limit the scope of this chapter to the major direct interactions within simple parasitoid-natural enemy/competitor associations because more complex indirect interactions involving three or more organisms are poorly understood. Of course, arthropod communities are structured by more intricate biotic connections than pairwise interactions between species (Holt and Lawton 1994, Müller *et al.* 1999), and the overall impact of multi-species interactions on populations must be assessed within a food-web perspective. For example, competitive interactions between parasitoids can be mediated via a third member of the community. Among the few documented examples are the work of van Nouhuys and Hanski (2000) who showed how short-term apparent competition between two parasitoid species might reduce each other's population growth rates via a shared generalist hyperparasitoid.

Figure 1 describes the potential trophic and guild connections between a parasitoid and its natural enemies and competitors. For each of the interactions illustrated we will briefly provide information about the nature of the interaction and the consequences to parasitoid populations. Competitive interactions between the parasitoid and antagonist parasitoids or facultative hyperparasitoids will be described in further details below (section 3).

2.1. Parasitoid-predator interactions

Arthropod communities are rich in species of predators. Whether they share the guild of the parasitoid or belong to a higher trophic level and are considered as higher-order predators (i.e. generalist predators of herbivores that also attack other predators), insect predators are often labile in their prey choices and may feed on parasitoids. All developmental stages of the parasitoid are vulnerable to predation. Immatures can be killed by a predator feeding on parasitized hosts, whereas adults can be captured when foraging (reviewed by Rosenheim *et al.* 1995, Brodeur and Rosenheim 2000).

Within a given guild, parasitoid-predator interactions are mostly asymmetric, in favor of the predators (Fig. 1). Implicitly, a parasitoid is not adapted to attack non-host insects, and is thus disadvantaged during confrontations with generalist predators.

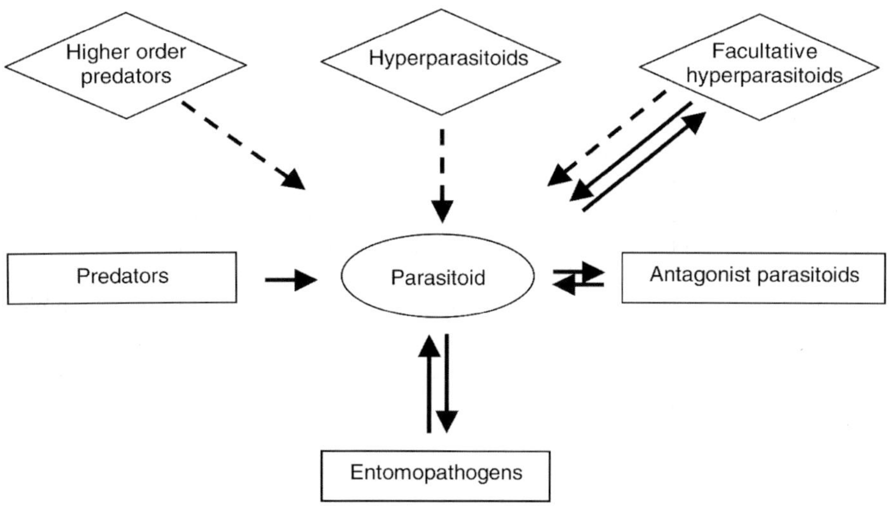

Figure 1. Schematic illustration of paired, direct trophic and guild interactions between a parasitoid and its natural enemies or competitors. Solid lines are guild interactions, dashed lined are trophic ones. Guild and higher trophic level protagonists of the parasitoid are enclosed in rectangles and diamonds, respectively. The arrowhead indicates a negative effect.

2.2. Parasitoid-entomopathogen interactions

Although interactions between parasitoids and entomopathogens (virus, bacteria, fungi) have rarely been described and quantified in nature, they seem likely to occur in most ecosystems, to be complex and to play an important role in community ecology (Hochberg and Lawton 1990). Flexner *et al.* (1986) and Brooks (1993) summarized a number of studies describing the susceptibility of immature and adult parasitoids to infection and examined the consequences on parasitoid fitness. As with most host-pathogen interactions, infection may have lethal or sublethal effects on parasitized hosts or adult parasitoids, the costs of an infection being typically dose-related. Furthermore, the competitive outcome in parasitized hosts depends on the timing of infection. A common observation is that entomopathogens usually outcompete parasitoids, except when parasitoid development is initiated sufficiently in advance of the infection.

As a general rule, parasitoids and entomopathogens compete asymmetrically, in the sense that fungi, bacteria and viruses typically benefit at the expense of the parasitoid (Fig. 1). However, in some cases, the relationship may operate in the other direction, and be considered to be mutual as, for example, when a parasitoid consumes fungal bodies when developing in infected hosts (Askary and Brodeur 1999).

2.3. Parasitoid-parasitoid interactions

In addition to predation and infection, the lateral effects of competition may hinder the development of parasitoid populations. Competition involves any mutually negative interactions between two or more species within the same guild (Morin 1999). In the broad sense, parasitoid-parasitoid interactions can frequently be viewed as a case of mutual predation when one individual kills and most often eats the other one (Brodeur and Rosenheim 2000). Interactions between parasitoids involve protagonists from the same species or from different species that have evolved similar strategies to exploit a common resource, in the same habitat. Their intraguild links are therefore likely to be much stronger than with a predator or an entomopathogen. Essentially three types of competition may occur between parasitoids. First, when two foraging parasitoids encounter each other, the interactions cause them to stop exploiting a patch and leave. Second, one species defends a host patch and excludes competitors through behavioural or chemical means. Third, one species restrains another by exploiting a shared host.

Guild interactions between parasitoids are often mutual (Fig. 1) and several factors may determine the direction of the interaction (Brodeur and Rosenheim 2000). Competitive interactions between parasitoids, the mechanisms behind them, and the consequences to biological control are further discussed below.

2.4. Parasitoid-hyperparasitoid interactions

In trophic studies, hyperparasitoids have a special meaning as they are the only natural enemies specific to parasitoids. Hyperparasitoids are obligate (species that can only develop at the expense of primary parasitoids) or facultative (species that can develop on either the primary parasitoids or its hosts). The evolutionary transitions from primary parasitism to facultative hyperparasitism, and from facultative hyperparasitism to obligate hyperparasitism, have generated a variety of often complex trophic and competitive interactions within host-parasitoid-hyperparasitoid associations (Brodeur 2000).

Obligate hyperparasitoids always develop at the expense of parasitoids and may directly affect the distribution, establishment and abundance of their hosts. Facultative hyperparasitoids may either act as higher-order natural enemies or guild competitors (Fig. 1), interactions can therefore be asymmetrical or mutual.

Ecological factors that determine the dynamics of host use by facultative hyperparasitoids will be discussed below.

3. INTRAGUILD COMPETITION AMONG PARASITOIDS

When resources are limited, as it is almost always the case, individuals have to compete for food, shelter, nesting sites, and mates to insure their survival, growth or reproduction. Competition can occur between individuals of the same species (intraspecific competition) or between individuals of different species (interspecific competition). Because a single host can be used by different female parasitoids, both types of competition can be experienced by parasitoids. Under conditions of strong competition, the contribution of an individual to the next generation will be reduced. Both intra- and interspecific competition thus affects the population dynamics of parasitoid species and may be important selective forces shaping behaviors linked to resource finding and exploitation. However most of the data available on the impact of competition on behavior are from laboratory experiments and the importance of the behavioral mechanisms expressed in interspecific competition in shaping parasitoid coexistence in the field has been questioned (Hawkins 2000).

3.1. Direct competition

Direct competition occurs either when two or more parasitoid females of the same or different species simultaneously exploit a host patch (i.e. extrinsic competition) or when larvae sharing a host compete for that resource (i.e. intrinsic competition (Demoraes *et al.* 1999)). Indirect competition refers to cases when a female enters a patch that has been previously parasitized and reacts to that previous competition. When detecting indirect competition females typically change their progeny and sex allocation and their patch residence time. Indirect competition is covered in the next section on superparasitism. In species where it occurs, direct competition can influence both the level of parasitism that a female can achieve in a habitat and the potential fitness gain obtained per host.

When more than one female parasitoid exploits a host patch, they can interfere directly, through aggressive behaviors. Fights between females have been observed in several parasitoid species and can be contests for access to the patch, or for defense of hosts (Pérez-Lachaud *et al.*, 2002; Batchelor *et al.*, 2005; Goubault *et al.*, 2005). These aggressive interactions often take the form of host guarding where guarding females attempt to repel conspecific females or even females of a different species. Aggressive intra- and interspecific behaviors consist mostly of striking the opponent with the antennae, the wings or the legs, biting and stinging. In some cases, protagonists may be injured or even killed (Lawrence, 1981). Guarding the parasitized host may last until the first immature to be deposited can defend itself against a competitor or last for almost the complete immature development, as in some Bethylidae (Griffiths and Godfray, 1988; Hardy and Blackburn, 1991).

Several factors influence the occurrence of aggressive behaviors between female parasitoids exploiting a host patch. The size of the female is an obvious parameter, the larger female generally initiating and winning the contest (Lawrence, 1981; Petersen and Hardy, 1996). In this case it is not the absolute size of the female that is important but being larger than the opponent. The prior ownership of the host, or the host patch, also influences the result of the fight. Females that first found a host or that have started ovipositing in it generally win the contest, even against intruding larger females (Petersen and Hardy, 1996; Field and Calbert, 1999; Batchelor *et al.*, 2005). The physiological status of a female also influences its capacity to defend a host. A high egg load increases the motivation of females to fight, either to defend their host or to drive the owner away, as a high egg load increases the value of the host (Hughes *et al.*, 1994; Stokkebo and Hardy, 2000). Other factors such as the age of the female and the availability of hosts in the habitat may influence the motivation of the female to fight.

Defending a host increases the probability that a female will gain some fitness from that host, and we would then expect that direct competition between parasitoid females should be common. However, guarding and fighting have been described in less than 30 species in the Ichneumoidea, Chrysidoidea, Proctotrupoidea and Chalcidoidea (Goubault, 2003). Its relative rarity could be explained by two factors. First, defending a host increases the probability that a female will gain fitness from that host, but it also decreases opportunities of finding and parasitizing other hosts. Secondly, there is a long delay between the investment of a parasitoid female in a host and the gain of fitness that occurs when the immatures emerge successfully as adults. In many cases this period, corresponding to the developmental time of the immature, is longer than the adult lifetime of the female. Defending a host may thus be mostly advantageous in long-lived species whose hosts are rare and where the parent can control to some extent the amount of risk to offspring.

Because females defending hosts must invest time, direct competition normally results in an increase in patch residence time. However, these females are busy defending parasitized hosts and not searching for hosts, and this increase in patch residence time does not necessarily result in an increase in the level of parasitism within a patch. From a biological control point of view, the result is negative as fewer patches are likely to be discovered in a habitat and the level of parasitism within a patch does not increase. Clearly, for the same reason, when females defending hosts detect a level of within-patch intraspecific competition, they may stay longer in the patch and invest less time searching for new patches. The result is also a decrease in the proportion of host patches found in the habitat. However, when competition occurs between females of different sizes, the smaller females can be driven away by the large females and therefore can exploit other host patches (Lawrence, 1981). The advantage for a female of defending a patch decreases with the probability of finding a new host patch and it is not surprising that strong host guarding behaviors are found in species where the likelihood of finding hosts is low.

Direct competition also occurs between females of different parasitoid species. Such encounters might be expected when a host species is attacked by several parasitoid species. In general, the same factors present in intraspecific competition apply, with the addition of the level of aggressiveness of the species. The more aggressive species almost always displaces the less aggressive ones (Mills, 1991). Direct competition between parasitoid species should be taken into account when selecting natural enemies, especially in classical biological control programmes. When the guild of natural enemies attacking the bark beetle *Leperisinus varius* (Coleoptera: Scolytidae) was evaluated, it was found that females of the most abundant species with the highest attack and net reproductive rates, the Braconidae *Coeloides filiformis*, were attacked and displaced while ovipositing by other species of Pteromalidae (*Cheiropachus quadrum*) and Eurytomidae (*Eurytoma morio*) (Mills, 1991). These aggressive species seemed to have lower host searching ability and searched actively for ovipositing females of *C. filiformis*, suggesting that they act as cleptoparasitoids. This study concludes that the more efficient Braconidae should be selected for introduction but not the Pteromalidae and Eurytomidae.

Intra- and interspecific competition also occurs between parasitoid immatures that compete to exploit the same host. More than one individual of a given species exploiting a host is called superparasitism whereas two or more species exploiting the same host is called multiparasitism (Godfray 1994). The presence of more than one larva in a host may change the value of that host for a female and the decrease in expected fitness return for the female is due to the fact that the immatures compete in several ways. In order to respond to this change of potential host value females have evolved strategies that are described in the following sections.

In solitary parasitoids, superparasitism or multiparasitism implies that only one individual remains alive to exploit the entire resource. It is thus a contest competition (Brodeur and Boivin 2004) that can occur through direct fighting or physiological suppression (Mackauer 1990). When competition occurs principally through larval fights, the first instar is generally the fighting instar and it often presents morphological adaptations, is mobile (with a caudal appendage or setae) and is mandibulated (van Baaren *et al.* 1997). Morphologically, the following instars are often unable to fight and can be killed by the first instar when a second oviposition occurs soon after the first one. Physiological suppression regroups several mechanisms resulting in the death of competitors that were either present before or arrived later than the individual expressing these mechanisms. The most common type of physiological suppression is asphyxiation as oxygen availability is limited for parasitoid immatures (Quicke 1997). Older larvae can reduce the oxygen level in the hemolymph therefore causing the eggs or younger larvae to asphyxiate.

Competition among larvae of gregarious species is more of a scramble-type competition. All larvae must share the resources available in the host and when these resources are insufficient for the survival of all larvae, all or part of the brood cannot complete its development. Although the majority of larvae of gregarious species do not have functional mandibles, they have nonetheless evolved competitive adaptations. In most *Trichogramma* species (Hymenoptera: Trichogrammatidae), the emerging larva consumes the host egg content at a rapid pace, completely consuming it within 8h after hatching (Jarjees and Merritt 2002). Such rapid ingestion of food prevents other larvae to develop within the same host, unless of course they develop as facultative hyperparasitoid as in *Anaphes victus* (Hymenoptera: Mymaridae) (van Baaren *et al.* 1995a). When several larvae of *Trichogramma* hatch simultaneously in a host, they ingest food rapidly. Video recording of developing *Trichogramma* larvae also showed that when a larva faces a competitor during this rapid feeding period, it can kill a competitor and absorb its content (Heslin and Merritt 2005). There are cases where first-instar larvae of gregarious species have functional mandibles, especially when they are likely to face multiparasitism from a solitary species whose larvae are mandibulate. Acquisition or retention of such morphological characteristics enable them to compete effectively with a solitary species (Boivin and van Baaren 2000).

3.2. Indirect competition (superparasitism)

Unlike predation, once a parasitoid female has exploited a host, that host remains in the habitat. It can subsequently still be parasitized either by the same female (self-superparasitism) or by a different female (conspecific superparasitism). The females thus do not interfere directly with each other but rather indirectly through cues used by the second female to detect previously parasitized hosts (host discrimination). The advantages and costs of superparasitism vary for solitary or gregarious parasitoids.

Self-superparasitism is normally disadvantageous for a solitary parasitoid as the supernumerary individuals compete, either through physical fights or physiological suppression, with only one progeny surviving (i.e. intrinsic competition (Demoraes *et al.*, 1999)). This behavior is nonetheless frequently found in nature and was first attributed to the failure of the female to recognize that a host was previously parasitized (van Alphen and Visser, 1990). Since then, it has been shown that self-superparasitism by solitary parasitoids can be optimal when additional eggs are: 1) deposited as an insurance against infertility or immature mortality (Parker and Courtney, 1984, Godfray and Shimada, 1999): 2) to increase the probability that one of the female's progeny will win in a competition with offspring of other con or hetero-specific females (Godfray, 1994); 3) to decrease the probability that a second female will be able to estimate the number of eggs already present and thus adjust its progeny allocation (van Alphen and Visser, 1990); or 4) to overcome the encapsulation capacity of the host (Blumberg and Luck, 1990). There are also cases where the female parasitoid does not possess host discrimination capacity and thus may self-superparasitize unknowingly as in *Aphidius rhopalosiphi* (Hymenoptera: Braconidae), an aphid parasitoid whose females rather adjust their patch residence time in order to decrease the risk of self-superparasitism (Outreman *et al.*, 2001).

Self-superparasitism by gregarious parasitoids does not imply that surnumerary individuals will be killed by fighting but rather that the immatures will share the resources within the host. Of course, the more immatures within a host, the smaller the emerging adults will be. The female must then allocate her progeny within a host to optimize her total fitness gain per host. Several factors are involved in the decision of the female to superparasitize. As the number of immatures per host increases, we can expect an increased mortality of the immatures, a gradual decrease in the fitness gain per immature and, in the case of species where the female attempts to reduce competition for her progeny by killing previous eggs or larvae, a longer host handling time that reduces the instantaneous rate of fitness gain of the female.

Conspecific superparasitism presents other constraints as the immatures sharing the host are from different females. In solitary parasitoids, females should superparasitize when the probability of their immature winning the contest exceeds the cost of searching for a better quality host. The fitness gained by superparasitism is lower than depositing an egg in an unparasitized host, both because that second progeny may lose the competition, and bring a fitness of 0, and because even if it wins the competition and develops alone in the host, it may bring a fitness gain lower than if it developed in an unparasitized egg (van Baaren *et al.*, 1995a). In general, females tend to superparasitize when unparasitized hosts are rare, when the density of host patch in the habitat is low, or when their life expectancy is approaching 0. For gregarious parasitoids, adding an extra progeny to a host does not generally implies that mortality occurs, unless the carrying capacity of the host is exceeded. Except in cases where a minimum number of immatures is necessary for successful development, for each additional egg deposited, the fitness gain per progeny decreases and thus the value of the host for a female also decreases. As expected, females will optimize the number of progeny deposited per host based on the number of progeny deposited by previous females. In conspecific superparasitism, as in multiparasitism, the total fitness produced by the host is not considered by the second female that attacks the host, as the previous immatures are from other females. Females that superparasitize in this situation thus generally decrease their egg clutch size and increase the proportion of males deposited.

The impact of all types of superparasitism is important to biological control. When mass rearing parasitoids, superparasitism results in a loss of eggs for solitary parasitoids and in the production of smaller adults and more males in gregarious parasitoids. By adjusting the ratio of hosts to females and the duration of contact, the level of superparasitizing can be maintained at a reasonable level. However, producers of natural enemies also want to reach high levels of parasitism to avoid wasting hosts. Obtaining a high level of parasitism with a low level of superparasitism can often be contradictory, as females will switch to superparasitism when the proportion of unparasitized hosts decreases in the patch. Once a parasitoid is released in a biological control program, any factors increasing the probability of superparasitism decrease the efficiency of the parasitoid.

3.3. Competition for mates

In all organisms that reproduce sexually, mate searching and acquisition are major factors of their life history. Insect parasitoids are no exception although competition for mates has received less attention than other aspects of parasitoid life history. The parasitoid Hymenoptera reproduce either through arrhenotoky, where the males develop from unfertilized eggs and the females from fertilized eggs, or through thelytoky, where only females are produce parthenogenetically. In arrhenotokous species, females should mate, as unmated females are constrained to produce only males, which is costly in terms of fitness.

The importance of mate competition varies with the spatial and temporal distribution of mating opportunities. In species that disperse from the native patch before mating and where mating occurs later in life, often after a maturation period, intense mate competition is unlikely. In these species, males and females normally find each other with the aid of sexual pheromones and it is generally the first acceptable male to reach the female that mates with her (Cormier *et al.* 1998, Fauvergue *et al.* 1998). The situation is very different with species that mate on the native patch, soon after emergence. In these species, several males and females are present simultaneously and both sexes can express mate selection. Competition for mates is probably as important for males as host acquisition for females, as it is only through mate acquisition that males can gain fitness. Few data are available on the influence of mate competition on the males' fitness. In *Trichogramma turkestanica*, males leaving the native patch differ greatly in their mating history. While some have almost exhausted their sperm supply others did not mate at all before leaving the patch. This points to differences in the males' capacity to acquire mate; some "sexy" males acquire a disproportionally large proportion of the available females while others fail to mate (Martel and Boivin, unpublished data). Details on the actual competition are still lacking but both sexes of that species have been shown to be able to choose their mates based on size or mating history (Boivin and Lagacé, 1999, Martel, Damiens and Boivin unpublished data).

3.4. Multiparasitism

Multiparasitism occurs when parasitoid females from different species attack the same host. This type of relationship is often much more asymmetric than superparasitism as the differences between parasitoid species are likely to be more important than intraspecific differences. The capacity to recognize that a host has been parasitized by a different species (interspecific discrimination) is not common in parasitoids (Vinson and Ables, 1980; van Strein-van Liempt and van Alphen; 1981, Hagvar and Hofsvang; 1988, van Alphen and Visser, 1990; Godfray and Shimada, 1999). As parasitoids are expected to encounter hosts parasitized by conspecifics more frequently than by a different species, the selection pressure to evolve interspecific discrimination is weak. Because the asymmetry between species can be important, some species may always win when in competition with an intrinsically weaker species and therefore are under no selection to recognize already parasitized hosts. Finally it may be more difficult for a species to evolve the capacity to recognize an alien marking than a conspecific marking (Godfray, 1994). In certain ecosystems, the level of multiparasitism can be significant (Hawkins, 2000), even reaching 44% for parasitized gypsy moth larvae (Eichhorn, 1996).

Multiparasitism generally results in the death of one of the immatures (Strand *et al.*, 1990). Not surprisingly, it is often the species whose immature is at a disadvantage that develop interspecific host discrimination (Chow and Mackauer, 1986; McBrien and Mackauer, 1991; Weber *et al.*, 1996; Quicke, 1997). In other cases, a species can detect the marking of another species where the two have recently speciated, as between the two mymarid species *Anaphes victus* (= *A.* n.sp.) and *A. listronoti* (= *A. sordidatus*) where the first species avoid hosts parasitized by the second species (van Baaren *et al.*, 1994). These two species are closely related and probably only relatively recently speciated (Landry *et al.*, 1993).

Following egg deposition, the relative advantage of each of the two species involved in multiparasitism can change over time, and the female has an advantage if she can estimate the time since parasitization by the first female. Both the Trichogrammatidae *Trichogramma pretiosum* and the Scelionidae *Telenomus heliothidis* attack the eggs of *Heliothis virescens*. The larva of *T. pretiosum* consumes the entire host within 10h after eclosion (Wu *et al.*, 2000). When a larva of *T. pretiosum* hatches 6h before the larva of *T. heliothidis*, it wins the competition because it has monopolized most of the host resources. However, *T. heliothidis* larvae win the competition if they hatch before, or less than 6 h after, *T. pretiosum* larvae (Strand, 1986).

There are a few cases where multiparasitism results in the emergence of two and even three different parasitoid species from the same host (Hawkins, 2000). When eggs of the Diopsidae *Diopsis macrophthalma* are parasitized by different species of Trichogrammatidae (*Trichogramma kalkae*, *T. pinneyi*, *Trichogrammatoidea simmondsi* and *Paracentrobia* sp.) a small percentage (2.8 %) of double emergence occurs (Feijen and Schulten, 1981). Similar cases have been reported when the Braconidae *Apanteles yakutatensis* and the Tachinidae *Madremya saundersii* attack the Noctuidae *Autographa californica* (Miller, 1982) or when the Tachinidae *Voria ruralis* and the Encyrtidae *Copidosoma truncatellum* attack the Noctuidae *Trichoplusia ni* (Browning and Oatman, 1984). However these cases of successful multiparasitism are the exception rather than the rule.

From a biological control perspective, multiparasitism has always a negative impact, except when the species of interest is intrinsically superior in competition. Even then, the immature winning the fight against the inferior species may inherit a host of reduced quality as some resources may have already been used by the other species. The presence of competition from other parasitoid species for the same targeted host should be evaluated prior to release just as the presence of hyperparasitoids should be. Although the impact would likely be less than for hyperparasitoids, the presence of multiparasitism may influence the establishment and the field efficacy of parasitoids used in biological control programs (see Mills this volume).

3.5. *Facultative hyperparasitism*

Facultative hyperparasitism is widespread among families of parasitic wasps and probably the most common form of hyperparasitism (Brodeur 2000). Facultative hyperparasitoids have evolved specific attributes enabling them to exploit hosts from both the primary and secondary consumer trophic level. Acceptance of alternate hosts from the same trophic level may reduce competition by excluding primary parasitoids from the habitat. For example, recent observational and experimental studies by Pérez-Lachaud and her colleagues suggest that within a guild of bethylid wasps attacking the coffee berry borer, *Cephalonomia hyalinipennis* outcompetes via facultative hyperparasitism the other parasitoid species, as expressed by its higher intrinsic rate of increase (Pérez-Lachaud *et al.* 2002, 2004, Batchelor *et al.* 2005). *Cephalonomia hyalinipennis* has the capacity to develop hyperparasitically on two other bethylid species and also to perform conspecific and allospecific ovicide and larvicide. This study indicates that mechanisms evolved by immature and adult facultative hyperparasitoids to avoid or face competition are similar to those described above for primary parasitoids with whom they share common evolutionary origins and life-history strategies.

There are very few hypotheses and models about the dynamics of host use by facultative hyperparasitoids. Clearly, several ecological processes, operating at different spatial and time scales are involved. Within parasitoid communities, the strength of competitive interactions is density-dependent and the expression of facultative hyperparasitism is likely to be also determined by variations in primary host density (Brodeur 2000). We therefore expect facultative hyperparasitoids to perform as hyperparasitoids and be better competitor than primary parasitoids when inter-guild hosts are rare (Brodeur 2000).

We are not aware of any laboratory or field studies testing the ability of facultative hyperparasitoids to adapt their trophic host selection in response to host abundance. Host use may be determined by the nutritional quality of the host as fitness gains can be different for a facultative hyperparasitoid developing either in a parasitoid host or a primary (herbivorous) host (Brodeur 2000). The few studies that have compared the suitability of hosts from different trophic levels for facultative hyperparasitoids have yielded opposite conclusions. Grandgirard *et al.* (2002) showed that the pteromalid *Pachycrepoideus dubius*, prefers to parasitize the cabbage maggot, *Delia radicum* (Anthomyiidae) over the primary parasitoid, *Trybliographa rapae* (Hymenoptera: Figitidae) and that, concurrently, fitness costs (higher mortality between oviposition and emergence, smaller adult size) were associated with hyperparasitism. This observation concurs with a pattern described by Kfir and Rosen (1981) about a reduction in the size of tertiary and quaternary parasitoids following depletion of host resources along with an increase in trophic level. On the other hand, the size of adult females of *C. hyalinipennis* reared on two out of three parasitoid species was similar to those of conspecifics developing as primary parasitoids of the coffee berry borer (Pérez-Lachaud *et al.* 2004).

Practitioners of biological control have always been hesitant to release hyperparasitoids because they pose risk to primary parasitoids. Ecologists are still unsure about the ecological consequences of obligate and facultative hyperparasitism for the dynamics of arthropod communities and biological control. This issue was discussed comprehensively by Rosenheim *et al.* (1995) and Rosenheim (1998) who reviewed empirical and analytical models of host-parasitoid-hyperparasitoid interactions. Because obligate hyperparasitoids always develop at the expense of parasitoids, they are likely to limit top-down control of herbivores. In contrast, some models (e.g. Luck *et al.* 1981, Briggs 1993) predict that obligate hyperparasitoids may play a role to regulate herbivore populations by stabilizing host-parasitoid interactions. In any event, obligate hyperparasitoids have never been considered and should by no means be introduced in a biological control program.

The situation is somewhat different for facultative hyperparasitoids. They are fierce competitors (Godfray 1994) and a number of species, including quite a few heteronomous parasitoids, have been released for classical and inundative biological control. Assessing the role of a species that can either act as a primary parasitoid, a competitor and a hyperparasitoid presents a huge challenge that, to our knowledge, remains to be unraveled theoretically and empirically. Using competition models to infer the role of facultative hyperparasitoids, Rosenheim *et al.* (1995) concluded that 'intrinsically superior parasitoids can either enhance or disrupt biological control; the relative likelihood of these effects depends on a large number of assumptions regarding the biology of competing parasitoids, including their relative searching abilities and degree of niche differentiation'. Pedersen and Mills (2004) reached similar conclusions and urged biological control practitioners to evaluate searching efficiency, patterns of aggregation and niche separation before dismissing *a priori* facultative hyperparasitoids. Furthermore, the impact of a facultative hyperparasitoid as a biological control agent may be determined by its tendency to exploit herbivorous pests or primary parasitoids. If a facultative hyperparasitoid has a preference for herbivores rather than for primary parasitoids, then biological control can be enhanced. Assessing the nutritional quality of hosts from different trophic levels (see above), as well as the relationship between herbivore (pest) density and facultative hyperparasitism, might help determining the propensity of facultative hyperparasitoids to develop as primary parasitoid or hyperparasitoid, and thereby their suitability as biological control agents.

4. IMPORTANCE OF VARIABILITY IN RESPONSE

Insect parasitoids, like all organisms, show an important variability in their response to internal and external factors, including their perception of competition, either direct or indirect. Anyone who has done behavioral experiments with parasitoids can testify to the large differences in the response of parasitoids to environmental or ecological situations. While biological control scientists have documented interspecific differences in the biology, host preference and capacity to parasitize a given host, fewer data exist on the variability found within a species. Variability is also expressed in the response of parasitoids to competition. Genetic variability has been shown in the strategies used by parasitoids to respond to competition such as sex allocation (Wajnberg, 1993), superparasitism (Wajnberg *et al.*, 1989) and patch time allocation (Wajnberg *et al.*, 1999).

Three levels of variability can be recognized. First, different populations of a given species can respond differently to a number of factors. These populations can be found in separate geographical situations or can be sympatric but use different host species. The fact that distinct populations respond differently to environmental conditions is well documented but differences in biological-based factors (such as competition) are less known. Second, within a population, inter-individual differences are expected, as genetic variability must exist for selection to operate. Finally, a given individual may also show change in its response to factors because of the conditions under which it developed or because of previous experience in his lifetime.

4.1. Between populations

The existence in most parasitoid species of "strains", "biotypes" or "ecotypes" has long been known. The differences found between populations are part of the genotypic diversity (Lewis *et al.*, 1990). In addition to the usual differences found in the morphology or physiology of the populations studied, there is some evidence that genetic variation between populations of parasitoids correlates with variation in host-use patterns (Baker *et al.*, 2003). In aphid parasitoids, species may consists of genetically distinct populations that show differences in their host range, some being more specialist while other are more generalist. These differences could be mediated through changes in the response to semiochemicals involved in host recognition (Powell and Wright, 1992). There are few data available on differences in the response to competition between populations of parasitoids. However, two populations of *Pachycrepoideus vindemmiae* (Hymenoptera: Pteromalidae), a solitary pupal parasitoid of cyclorraphous dipteran species, showed different strategies in their host discrimination (Goubeault *et al.* 2004). While one population discriminated against parasitized hosts following internal examination, the other population relied on external examination, a more rapid but less precise behavior. These results suggest a trade-off between the speed and the accuracy of host discrimination by *P. vindemmiae*.

The presence of populations within a species, each exhibiting a different set of morphological, physiological and behavioral characteristics, suggests that these populations have been selected according to the conditions where they evolved. In a sense, this level of variability is a significant advantage from a biological control point of view. It is possible, through careful selection to select a population well adapted to the specific needs of a biological control program. There are numerous studies on the selection of the best population for biological control purposes (Hassan, 1994, Liu and Smith, 2000) although the selection is often left to chance (Pak, 1992). Using an inappropriate population for release accounts for about 12% of all documented failure in classical biological control programs (Stiling, 1993).

4.2. Between individuals

Within a population, each individual differs slightly based on its genotype. Differences due to the environment under which an individual has developed are considered below (Section 4.3). Obviously inter-individual variability is an important component of the global variability in diploid and haplodiploid species but in Hymenoptera parasitoids that reproduce through thelytokous parthenogenesis, this level of variability is much reduced, all individuals within a population being more or less identical clones.

Variability in the response of different individuals to cues originating from the host or its host plant has been well documented (Lewis *et al.*, 1990; Vinson, 1998). For example, responses to volatile allelochemicals by *Microplitis croceipes* (Braconidae) (Prévost and Lewis, 1990) and response to host distribution in *Trichogramma* spp. (Chassain *et al.*, 1988) are inherited. Although less documented, similar differences exist between individuals in the response to stimuli related to competition. When isofemale lines are produced from a single population, differences found in the response to different factors reflect the inter-individual variability in a population. In *Trichogramma*, when such isofemale lines are tested, significant differences in the level of superparasitism (Wajnberg *et al.*, 1989) and sex allocation (Wajnberg, 1993) are found. In the Scelionidae *Telenomus busseolae,* when females exploit a host patch, each host rejection increases their tendency to leave the patch (Wajnberg *et al.*, 1999). Rejected hosts are often already parasitized hosts, an indication of the level of competition within a patch. Isofemale lines differ with respect to the impact that rejecting a host has on the patch leaving tendency (Wajnberg *et al.*, 1999).

4.3. Within an individual

The response of an individual parasitoid to cues from its surroundings, including cues indicating competition, may also vary over time, reflecting phenotypic plasticity. Phenotypic plasticity is the ability of a genotype to produce distinct phenotypes when exposed to different environments throughout its ontogeny (Pigliucci, 2005). This type of variability has to be within the limits of the genotype, but still may be significant. Three factors may influence the phenotypic expression of an individual. A behavior may be expressed or changed through previous experience, learning being a major component of the behavioral ecology of parasitoids (Jaenike and Papaj, 1992). The environmental or biological conditions experienced generally during immature development may influence several aspects of the morphology and behavior of parasitoids (Gandolfi *et al.*, 2003). Several ecological and physiological factors, such as host habitat, egg load and host quality are known to affect behaviors linked to host location and acceptance (Duan and Messing, 1999) including progeny and sex allocation (Ueno, 1999). Age, sex, nutritional state, species, size are all host characteristics that may change the fitness and behaviour of the parasitoid, although the latter has been less examined. The physiological state of an individual may also change its behavior, thereby its response to competition. Nutritional status, mating status, health, and changes brought by the perception of stimuli will change the perception of a female of the level of competition occurring in a patch.

Levels of intra-individual variability in host discrimination have been thoroughly examined in the Mymaridae *Anaphes victus*, a parasitoid of Curculionidae eggs. Females of *A. victus* discriminate (van Baaren *et al.*, 1994) against hosts parasitized either by themselves, other females of the same species or different, closely related, species. These females can assess the degree of genetic proximity and adjust their level of acceptance accordingly (van Baaren and Boivin, 1998a). Female *A. victus* can reject parasitized hosts at two occasions during the oviposition process. Upon contact with the egg, the female touches the host egg for a few seconds and, if external marking pheromone is detected, host rejection can occur. If the host is considered acceptable at this stage, the female drills the chorion and inserts her ovipositor. The female can then also reject the host, probably after detecting the presence of another parasitoid immature within the host egg (van Baaren *et al.*, 1995b). Rejections following an antennal contact are much faster (ca. 2 sec) than rejections following ovipositor insertion (ca. 48 sec). When females of *A. victus* contact several parasitized hosts, they first reject them after ovipositor insertion. However, as the female progresses in a sequence of rejection, she starts to reject these parasitized hosts more and more after only an antennal contact (van Baaren and Boivin, 1998b). Eventually, the female relies almost only on antennal rejection. This change in behavior occurs through associative learning where the female learns to associate the presence of an external marking pheromone with the presence into a host of a parasitoid immature and, doing so, she decreases the amount of time necessary to reject a low quality host.

This illustrates how a female changes her behavior as she gains experience when exploiting host patches. Obviously, by learning how to discriminate against parasitized hosts faster, these females gain an advantage when several females simultaneously exploit a patch. In addition, females that can learn will probably decrease their patch residence time, for a similar level of exploitation, compared to females not expressing learning. We could thus expect in biological control programs that such females will reduce competition within a patch and exploit a larger number of host patches in the habitat.

Comparison of the patch exploitation strategy of *A. victus* between habitats of different quality showed that the response of females to a given patch quality strongly depended on their past experiences. Females allocated more time and more eggs in a mixed quality patch after experiencing a poor quality patch than after experiencing a good quality patch. In a poor quality patch, females superparasitize more frequently after experiencing a poor quality patch than after experiencing a good quality patch. In a good quality patch, *A. victus* females laid more eggs after having visited poor quality patches than after visiting good quality patches. Recent foraging experiences are used to estimate both the availability and spatial distribution of hosts in the environment and adjust foraging decisions accordingly (Boivin *et al.* 2004). The observed variability in the patch-leaving rules within the same species stresses the importance of previous experience when describing behaviors of female parasitoids (van Baaren *et al.* 2005a).

The conditions experienced by the parasitoid immature during its development can also change its adult behavior, therefore adding to the individual variability. Adults of *A. victus* can learn olfactory and gustatory cues, particularly during host location and patch exploitation. When second instar larvae are exposed to low temperature during increasing periods of time, the emerging adult females oviposited fewer eggs, were less able to learn external marks, oviposited more in already parasitized hosts and used different patch-leaving rules. The change in behavior was positively correlated with the duration of cold exposure. Low-temperature exposure thus affected the reproductive success, expression of learning, host discrimination ability and patch-leaving decision rules of the parasitic wasps (van Baaren *et al.* 2005b).

Such changes in parasitoid behavior should be taken into account in biological control. Cold storage of parasitoids is currently done for several species in order to spread the cost of production over time. The impact of low temperature storage is normally measured on morphological (size) and physiological (survival, longevity, fecundity) characteristics of the parasitoids but rarely on the behavior of the emerging adults. The results obtained on *A. victus* indicate clearly that individuals stored at low temperature as immatures express behaviors that are not optimal and that result in underexploitation of good host patches and overexploitation of low quality patches.

5. CONCLUSIONS

Insect parasitoids live in a complex natural world. They struggle with conspecific and allospecific parasitoids, and are exposed to predation, hyperparasitism and infection. Researchers have barely begun to assess the ecological significance of these species interactions. For most parasitoids, trophic and guild interactions with competitors and natural enemies are significant. At a higher level of species organization, these interactions influence the structure, diversity and stability of parasitoid populations and communities (Askew and Shaw 1986, Hawkins 2000).

From an applied perspective, the potential of competitors and natural enemies of parasitoids to limit topdown control of herbivore pests by these parasitoids, and even prevent the establishment of introduced parasitoid species, has become a hot topic in ecology and biological control (Rosenheim *et al.* 1995).

Figure 1 clearly illustrates a marked asymmetry in the relationships between parasitoids and their protagonists. Except for competitive interactions, all interactions are detrimental to parasitoids. The outcomes are mainly determined by the contest ability of the protagonists and the prior ownership when competition occurs with antagonist parasitoids, facultative hyperparasitoids or entomopathogens. Because of their mode of development and relatively high degree of host specialization, parasitoids are disadvantaged during confrontations with generalist natural enemies. As a consequence, Brodeur and Rosenheim (2000) hypothesized that parasitoid population dynamics are in large part governed by higher trophic and guild interactions.

Nevertheless, parasitoids are successful organisms. They represent one of the most diverse and abundant group of arthropods in terrestrial ecosystems. They have evolved a range of morphological, physiological and behavioral attributes to avoid or reduce competition. The frequency and intensity of trophic and guild interactions also led to the evolution of various mechanisms to cope with natural enemies. Establishment and survivorship of parasitoids in natural and managed ecosystems may result from life-history characteristics (high fecundity, high developmental rate), great searching efficiency and realized rates of oviposition in the field, as well as different strategies of resource partitioning (spatial and temporal segregation, exploitation of alternative host species).

While important progresses have been made in the understanding of individual responses of parasitoids to intra- and interspecific competition, much remain to be done in transferring this information to the population level. Even when the response of an individual facing competition can be predicted, the inherent variability of animal behavior complicates the transition from the individual to the population. These considerations are particularly important when these organisms are to be used in biological control. For instance, parasitoids released in inundative programs face intense competition when mass reared, with consequences both for the productivity of rearing system and the efficacy of the parasitoids once released in the field. Creating models linking individual responses to competition to the capacity of a given species to impact its host population remains an important challenge in the future.

6. REFERENCES

van Alphen, J. J. M., and Visser, M. E., 1990, Superparasitism as an adaptive strategy for insect parasitoids. Annu. Rev. Entomol. 35: 59-79.

Askary, H., and Brodeur, J., 1999, Susceptibility of larval stages of the aphid parasitoid *Aphidius nigripes* to the entomopathogenic fungus, *Verticillium lecanii*. J. Invert. Pathol. 73: 129-132.

Askew, R. R., and Shaw, M. R., 1986, Parasitoid communities: their size, structure and development. In: Insect Parasitoids, J. Waage, and D. Greathead, eds, , Academic Press, San Diego, pp. 225-264.

van Baaren, J., and Boivin, G., 1998a, Genotypic and kin discrimination in a solitary Hymenopterous parasitoid: implications for speciation. Evol. Ecol. 12: 523-534.

van Baaren, J., and Boivin, G., 1998b, Learning affects host discrimination behavior in a parasitoid wasp. Behav. Ecol. Sociobiol. 42: 9-16.

van Baaren, J., Boivin, G., Le Lannic, J., and Nénon, J. P., 1997, The male and female first instar larvae of *Anaphes victus* and *A. listronoti* (Hymenoptera: Mymaridae). Zoomorphology 117: 189-197.

van Baaren, J., Boivin, G., and Nénon, J. P., 1994, Intra- and interspecific host discrimination in two closely related egg parasitoids. Oecologia 100: 325-330.

van Baaren, J., Boivin, G., and Nénon, J. P., 1995a, Intraspecific hyperparasitism in a primary hymenopteran parasitoid. Behav. Ecol. Sociobiol. 36: 237-242.

van Baaren, J., Boivin, G., and Outreman, Y., 2005a, Patch exploitation strategy by an egg parasitoid in constant or variable environment. Ecol. Entomol. 30: 502-509.

van Baaren, J., Nénon, J. P., and Boivin, G., 1995b, Comparison of oviposition behavior of a solitary and a gregarious parasitoid (Hymenoptera: Mymaridae). J. Ins. Behav. 8: 671-686.

van Baaren, J., Outreman, Y., and Boivin, G., 2005b, Effect of low temperature exposure on oviposition behaviour and patch exploitation strategy in parasitic wasps. Anim. Behav. 70: 153-163.

Baker, D. A., Loxdale, H. D., and Edwards, O. R., 2003, Genetic variation and founder effects in the parasitoid wasp, *Diaeretiella rapae* (M'intosh) (Hymenoptera: Braconidae: Aphidiidae), affecting its potential as a biological control agent. Mol. Ecol. 12: 3303-3311.

Batchelor, T. P., Hardy, I. C. W., Barrera, J. F., and Pérez-Lachaud, G., 2005, Insect gladiators II: competitive interactions within and between bethylid parasitoid species of the coffe berry borer, *Hypothenemus hampei* (Coleoptera: Scolytidae). Biol. Contr. 33: 194-202.

Blumberg, D., and Luck, R. F., 1990, Differences in the rate of superparasitism between two strains of *Comperiella bifasciata* (Howard) (Hymenoptera: Encyrtidae) parasitizing California red scale (Homoptera: Diaspididae): an adaptation to circumvent encapsulation? Ann. Entomol. Soc. Am. 83: 591-597.

Boivin, G. and van Baaren, J., 2000, The role of larval aggression and mobility in the transition between solitary and gregarious development in parasitoid wasps. Ecol. Lett. 3: 469-474.

Boivin, G., Fauvergue, X., and Wajnberg, E., 2004. Optimal patch residence time in egg parasitoids: innate versus learned estimate of patch quality. Oecologia 138: 640-647.

Boivin, G., and Lagacé, M., 1999, Effet de la taille sur la fitness de *Trichogramma evanescens* (Hymenoptera: Trichogrammatidae). Ann. Soc. Entomol. Fr. 35 (suppl.): 371-378.

Briggs, C. J., 1993, Competition among parasitoid species on a stage-structured host and its effect on host suppression. Am. Nat. 141: 372-397..

Brodeur, J., 2000, Host specificity and trophic relationship of hyperparasitoids. In: Parasitoid Population Biology, M. E. Hochberg and A.R. Ives, eds, Princeton University Press, pp. 163-183.

Brodeur, J., and Boivin, G., 2004, Functional ecology of immature parasitoids. Annu. Rev. Entomol. 49: 27-49.

Brodeur, J., and Rosenheim, J. A., 2000, Intraguild predation in aphid parasitoids: a review. Entomol. Exp. Appl. 97: 93-108.

Brooks, W.M., 1993, Host-parasitoid-pathogen interactions. In: Parasites and pathogens of insects, N.E. Beckage, S.N. Thompson and B.A. Federici, eds, Vol. 2, Academic Press, San Diego, pp. 231-272.

Browning, H. W., and Oatman, E. R., 1984, Intra- and interspecific relationships among some parasites of *Trichoplusia ni* (Lepidoptera: Noctuidae). Environ. Entomol. 13: 551-556.

Chassain, C., Boulétreau, M., and Fouillet, P., 1988, Host exploitation by parasitoids: local variations in foraging behaviour of females among populations of *Trichogramma* species. Entomol. Exp. Appl. 48: 195-202.

Chow, F. J., and Mackauer, M., 1986, Host discrimination and larval competition in the aphid parasite, *Ephedrus californicus*. Entomol. Exp. Appl. 41: 243-254.

Cormier, D., Royer, L., Vigneault, C., Panneton, B., and Boivin, G., 1998, Effect of female age on daily cycle of sexual pheromone emission in gregarious egg parasitoid *Anaphes listronoti*. J. Chem. Ecol. 24: 1595-1610.

Demoraes, C. M., Cortesero, A. M., Stapel, J. O., and Lewis, W. J., 1999, Intrinsic and extrinsic competitive interactions between two larval parasitoids of *Heliothis virescens*. Ecol. Entomol. 24: 402-410.

Duan, J. J., and Messing, R. H., 1999, Effect of origin and experience on patterns of host acceptance by the opiine parasitoid *Diachasmimorpha tryoni*. Ecol. Entomol. 24: 284-291.

Eichhorn, O., 1996, Experimental studies upon the parasitoid complex of the gypsy moth (*Lymantria dispar* L.) (Lep., Lymantriidae) in lower host populations in eastern Austria. J. Appl. Entomol. 120: 205-212.

Fauvergue, X., Fouillet, P., Mesquita, A. L. M., and Boulétreau, M., 1998, Male orientation to trail sex pheromones in parasitoid wasps: does the spatial distribution of virgin females matter? J. Ins. Physiol. 44: 667-675.

Feijen, H. R., and Schulten, G. G. M., 1981, Egg parasitoids (Hymenoptera: Trichogrammatidae) of *Diopsis macrophthalma* (Diptera: Diopsidae) in Malawi. Neth. J. Zool. 31: 381-417.

Field, S. A., and Calbert, G., 1999, Don't count your eggs before they're parasitized: contest resolution and the trade-off during patch defense in a parasitoid wasp. Behav. Ecol. 10: 122-127.

Flexner, J. L., Lighthart, B., and Croft, B.A., 1986, The effects of microbial pesticides on non-target, beneficial arthropods. Agric. Ecosys. Environ. 16: 203-254.

Gandolfi, M., Mattiacci, L., and Dorn, S., 2003, Mechanisms of behavioral alterations of parasitoids reared in artificial systems. J. Chem. Ecol. 29: 1871-1887.

Godfray, H. C. J., 1994, Parasitoids: Behavioral and Evolutionary Ecology. Princeton University Press, Princeton.

Godfray, H. C. J., and Shimada, M., 1999, Parasitoids: a model system to answer questions in behavioral, evolutionary and population ecology. Res. Popul. Ecol. 41: 3-10.

Goubault, M., 2003, Exploitation des hôtes chez un Hyménoptère parasitoïde solitaire: conséquences adaptatives de la compétition intraspécifique sur les stratégies comportementales des femelles. PhD Thesis, Université de Rennes 1, Rennes.

Goubault, M., Krespi, L., Boivin, G., Poinsot, D., Nénon, J.-P., and Cortesero, A. M., 2004, Intraspecific variations in host discrimination behavior in the pupal parasitoid *Pachycrepoideus vindemmiae* Rondani (Hymenoptera: Pteromalidae). Environ. Entomol. 33: 362-369.

Goubault, M., Outreman, Y., Poinsot, D., and Cortesero, A. M., 2005, Patch exploitation strategies of parasitic wasps under intraspecific competition. Behav. Ecol. 16 693-701.

Grandgirard, J., Poinsot D., Krespi L., Nénon J.-P., and Cortesero, A. M., 2002, Costs of secondary parasitism in the facultative hyperparasitoid *Pachycrepoideus dubius*: does host size matter? Entomol. Exp. Appl. 103: 239-248.

Griffiths, N. T., and Godfray, H. C. J., 1988, Local mate competition, sex ratio and clutch size in bethylid wasps. Behav. Ecol. Sociobiol. 22: 211-217.

Hagvar, E. B., and Hofsvang, T., 1988, Interspecific competition between the aphid parasitoids *Aphidius colemani* Viereck and *Ephedrus cerasicola* Stary (Hymenoptera: Aphidiidae). Entomol. Exp. Appl. 59: 145-150.

Hardy, I. C. W., and Blackburn, T. M., 1991, Brood guarding in a bethylid wasp. Ecol. Entomol. 16: 55-62.

Hassan, S. A., 1994, Strategies to select *Trichogramma* species for use in biological control. In: Biological Control with Egg Parasitoids, E. Wajnberg and S. A. Hassan, eds, CAB International, Wallingford, pp. 55-71.

Hassell, M.P., 2000, The Spatial and Temporal Dynamics of Host-Parasitoid Interactions. Oxford University Press, Oxford..

Hawkins, B. A., 2000, Species coexistence in parasitoid communities: does competition matter? In: Parasitoid Population Biology, M. A. Hochberg and A. R. Ives, eds, Princeton University Press, Princeton, pp. 198-213.

Hawkins, B.A., Cornell, H.V., and Hochberg, M.E., 1997, Predators, parasitoids, and pathogens as mortality agents in phytophagous insect populations. Ecology 78: 2145-2152.

Heslin, L. M., and Merritt, D.J., 2005, Cannibalistic feeding of larval *Trichogramma carverae* parasitoids in moth eggs. Naturwissenschaften 92: 435-439.

Hochberg, M.E., and Lawton, J. H., 1990, Competition between kingdoms. TREE 5: 367-371.

Holt, R.D., and Lawton, J. H., 1994, The ecological consequences of shared natural enemies. Annu. Rev. Ecol. Syst. 25: 495-520.

Hughes, J. P., Harvey, I. F., and Hubbard, S. F., 1994, Host-searching behavior of *Venturia canescens* (Grav.) (Hymenoptera: Ichneumonidae): Interference- The effect of mature egg load and prior behavior. J. Ins. Behav. 7: 433-454.

Jaenike, J., and Papaj, D. R., 1992, Behavioral plasticity and patterns of host use by insects, In: Insect Chemical Ecology: an Evolutionary Approach, B. D. Roitberg and M. B. Isman, eds, Chapman & Hall, New York, pp. 245-264.

Jarjees, E. A., and Merritt, D. J., 2002, Development of *Trichogramma australicum* Girault (Hymenoptera: Trichogrammatidae) in *Helicoverpa* (Lepidoptera: Noctuidae) host eggs. Austr. J. Entomol. 41: 310-315.

Kfir, R., and Rosen D., 1981, Biology of the hyperparasite *Pachyneuron concolor* (Förster) (Hymenoptera: Pteromalidae) reared on *Microterys flavus* (Howard) in brown soft scale. J. Entomol. Soc. South. Afric. 44: 151-163.

Landry, B. S., Dextrase, L., and Boivin, G., 1993, Random amplified polymorphic DNA markers for DNA fingerprinting and genetic variability assessment of minute parasitic wasp species (Hymenoptera: Mymaridae and Trichogrammatidae) used in biological control programs of phytophagous insects. Genome 36: 580-587.

Lawrence, P. O., 1981, Interference competition and optimal host selection in the parasitic wasp, *Biosteres longicaudatus*. Ann. Entomol. Soc. Am. 74: 540-544.

Lewis, W. J., Vet, L. E. M., Tumlinson, J. H., van Lenteren, J. C., and Papaj, D. R., 1990, Variations in parasitoid foraging behavior: essential element of a sound biological control theory. Environ. Entomol. 19: 1183-1193.

Liu, F.-H., and Smith, S. M., 2000, Measurement and selection of parasitoid quality for mass-reared *Trichogramma minutum* Riley used in inundative release. Biocontr. Sci. Tech. 10: 3-13.

Luck, R. F., Messenger, P. S., and Barbieri, J. F., 1981, The influence of hyperparasitism on the performance of biological control agents. In D. Rosen, ed., The Role of Hyperparasitism in Biological Control: A Symposium, Berkeley, U.S.A.: Division of Agricultural Science, University of California, pp. 34-42..

Mackauer, M., 1990, Host discrimination and larval competition in solitary endoparasitoids. In: Critical Issues in Biological Control, M. Mackauer, L. E. Ehler, and J. Roland, eds, Intercept/VCH, Andover, pp. 41-62.

McBrien, H., and Mackauer, M., 1991, Decision to superparasitise based on larval survival: Competition between larval parasitoids *Aphidius ervi* and *Aphidius smithi*. Entomol. Exp. Appl. 59: 145-150.

Miller, J. C., 1982, Life history of insect parasitoids involved in successful multiparasitism. Oecologia 54: 8-9.

Mills, N. J., 1991, Searching strategies and attack rates of parasitoids of the ash bark beetle (*Leperisinus varius*) and its relevance to biological control. Entomol. Entomol. 16: 461-470.

Morin, P.J. 1999, Community ecology. Blackwell Science, Malden..

Müller, C. B., Adriaanse, I. C. T., Belshaw, R., and Godfray, H. C. J., 1999, The structure of an aphid-parasitoid community. J. Anim. Ecol. 68: 346-370.

van Nouhuys, S., and Hanski, I., 2000, Apparent competition between parasitoids mediated by a shared hyperparasitoid. Ecol. Lett. 3: 82-84.

Outreman, Y., Le Ralec, A., Wajnberg, E., and Pierre, J. S., 2001, Can imperfect host discrimination explain partial patch exploitation in parasitoids? Ecol. Entomol. 26: 271-280.

Pak, G. A., 1992, Inundative release of *Trichogramma* for the control of cruciferous Lepidoptera: preintroductory selection of an effective parasitoid. In: Diamondback moth and other crucifer pests,

N. S. Talekar, eds, Asian Vegetable Research and Development Center, AVRDC Publication; v. No. 92-368, Tainan, Taiwan, pp. 297-308.

Parker, G. A., and Courtney, S. P., 1984, Models of clutch size in insect oviposition. Theor. Popul. Biol. 26: 27-48.

Pedersen, B.S., and Mills, N.J., 2004, Single vs. multiple introduction in biological control : the roles of parasitoid efficiency, antagonism and niche overlap. J. Appl. Ecol. 41: 973-984.

Pérez-Lachaud, G., Bachelor, T. P., and Hardy, I. C. W., 2004, Wasp eat wasp: facultative hyperparasitism and intra-guild predation by bethylid wasps. Biol. Contr. 30: 149-155.

Pérez-Lachaud, G., Hardy, I. C. W., and Lachaud, J. P., 2002, Insect gladiators: competitive interactions between three species of bethylid wasps attacking the coffe berry borer, *Hypothenemus hampei* (Coleoptera: Scolytidae). Biol. Contr. 25: 231-238.

Petersen, G., and Hardy, I. C. W., 1996, The importance of being larger: parasitoid intruder-owner contests and their implications for clutch size. Anim. Behav. 51: 1363-1373.

Pigliucci, M., 2005, Evolution of phenotypic plasticity: where are we going now? TREE 20: 481-486.

Powell, W., and Wright, A. F., 1992, The influence of host food plants on host recognition by four aphidiine parasitoids (Hymenoptera: Braconidae). Bull. Entomol. Res. 81: 449-453.

Prévost, G., and Lewis, W. J., 1990, Heritable differences in the response of the Braconid wasp, *Microplitis croceipes* to volatile allelochemicals. J. Ins. Behav. 3: 277-287.

Quicke, D. L. J., 1997, Parasitic Wasps. Chapman & Hall, London..

Rosenheim, J.A., 1998, Higher-order predators and the regulation of insect herbivore populations. Annu. Rev. Entomol. 43: 421-447.

Rosenheim, J.A., Kaya, H. K., Ehler, L. E., Marois, J. J., and Jaffee, B. A., 1995, Intraguild predation among biological control agents: theory and evidence. Biol. Contr. 5: 303-335.

Stiling, P., 1993, Why do natural enemies fail in classical biological control programs? Am. Entomol. 39(1): 31-37.

Stokkebo, S., and Hardy, I. C. W., 2000, The importance of being gravid: egg load and contest outcome in a parasitoid wasp. Anim. Behav. 59: 1111-1118.

Strand, M. R., 1986, The physiological interactions of parasitoids with their hosts and their influence on reproductive strategies, in: Insect Parasitoids, J. K. Waage and D. Greathead, eds., Academic Press, London, pp. 97-136.

Strand, M. R., Johnson, J. A., and Culin, J. D., 1990, Intrinsic interspecific competition between the polyembryonic parasitoid *Copidosoma floridanum* and the solitary endoparasitoid *Microplitis demolitor* in *Pseudoplusia includens*. Entomol. Exp. Appl. 55: 275-284.

van Strien-van Liempt, W. T. F. H., and van Alphen, J. J. M., 1981, The absence of interspecific host discrimination in *Asobara tabida* Nees and *Leptopilina heterotoma* (Thompson) coexisting larval parasitoids of *Drosophila* species. Neth. J. Zool. 31: 701-712.

Ueno, T., 1999, Host-size-dependent sex ratio in a parasitoid wasp. Res. Popul. Ecol. 41: 47-57.

Vinson, S. B., 1998, The general host selection behavior of parasitoid Hymenoptera and a comparison of initial strategies utilized by larvaphagous and oophagous species. Biol. Contr. 11: 79-96.

Vinson, S. B., and Ables, J. R., 1980, Interspecific competition among endoparasitoids of tobacco budworm larvae (Lep.: Noctuidae). Entomophaga 25: 357-362.

Wajnberg, E., 1993, Genetic variation in sex allocation in a parasitic wasp: variation in sex pattern within sequences of oviposition. Entomol. Exp. Appl. 69: 221-229.

Wajnberg, E., Pizzol, J., and Babault, M., 1989, Genetic variation in progeny allocation in *Trichogramma maidis*. Entomol. Exp. Appl. 53:177-187.

Wajnberg, E., Rosi M. C., and Colazza, S., 1999, Genetic variation in patch time allocation in a parasitic wasp. J. Anim. Ecol. 68: 121-133.

Weber, C.A., Smilanik, I. M., Ehler, L. E., and Zalom, F. G., 1996, Ovipositional behavior and host discrimination in three scelionid egg parasitoids of stink bugs. Biol. Contr. 6: 245-252.

Wu, Z. X., Cohen, A. C., and Nordlund, D. A., 2000, The feeding behavior of *Trichogramma brassicae*: new evidence for selective ingestion of solid food. Entomol. Exp. Appl. 96: 1-8.

INDIRECT EFFECTS, APPARENT COMPETITION AND BIOLOGICAL CONTROL

F. J. Frank van Veen[1], Jane Memmott[2] and H. Charles J. Godfray[1]

[1]NERC Centre for Population Biology and Division of Biology, Imperial College London, Silwood Park Campus, Ascot, Berkshire, SL5 7PY, United Kingdom; [2]School of Biological Sciences, University of Bristol, Woodland Road, Bristol, BS8 1UG, United Kingdom

Abstract. In biological control in its simplest form only direct interactions between the control agent and the pest and potential non-targets are considered. Ecologists are however amassing an ever increasing body of evidence for the importance of indirect effects in ecological communities. Indirect effects are the effects of one species on another mediated by at least one intermediate species. An example is so-called apparent competition which is the negative indirect effect that prey species have on each other when they share natural enemies. This effect is thought to play a particularly significant role in phytophagous insect communities where the scope for resource competition is limited. We show that there is experimental evidence for apparent competition amongst phytophagous insects. We describe a community of aphids and their parasitoids, predators and pathogens that we have been studying for over 10 years. We discuss how this species-rich community in a relatively natural environment may be structured by indirect effects. Returning to biological control we discuss how these ideas from community ecology can be applied to enhance pest control and to assess the ecological risks of the introduction of control agents. Introducing or encouraging species that share natural enemies with a target pest may lead to increased pest control through an apparent competition effect by boosting the natural enemy population. We conclude that although occasional attempts are made, such techniques are currently still much underutilised. Equally, we show how indirect effects may cause or increase the impact of introduced control agents on native flora and fauna but that these possible effects are rarely taken into account.

1. INTRODUCTION

1.1. Background

When attempting to understand and predict the dynamics of a pest and its control agent, efforts are generally focussed on the direct interactions between the two. However, host-parasitoid or predator-prey pairs rarely operate in isolation but are often embedded within complex communities of interacting species. Indirect effects propagating through such a network of interactions may significantly influence a pest and its control agent. Phytophagous insect species interact with each other in numerous ways, but the two most important are through competition for resources and through sharing natural enemies. In both cases, the commonest type of interaction is indirect, mediated in the first instance by changes in the level of the shared resource, and in the second case by changes in the distribution and abundance of a predator, parasitoid or pathogen. In the past, much of community ecology

J. Brodeur and G. Boivin (eds.), Trophic and Guild Interactions in Biological Control, 145–169.

concentrated on how resource competition can affect the population dynamics of species and how this determines community structure. This theory has been applied to understanding the structure of phytophagous insect communities on particular host plants, but the overall importance of resource competition in structuring assemblages of herbivorous insects is limited by the high degree of host specificity shown by many species (which may, of course, itself be a consequence of the ghost of competition past) (Lawton and Strong 1981, Strong *et al.* 1984).

But as was first realised by Holt (1977), many of the classical community ecology patterns that can be generated by resource competition can also be generated by interactions mediated by natural enemies. Indeed, the similarities are such that Holt called this type of interaction apparent competition. At least in theory, apparent competition can cause species exclusion (very relevant to biological control!), lead to segregation along the equivalent of a niche axis, give rise to character displacement, and result in mosaic distributions of species that cannot co-exist at one site (Jeffries and Lawton 1984, Holt and Lawton 1994, Wootton 1994, Abrams and Chen 2002).

In this contribution we shall explore apparent competition and related phenomena, and discuss their implications for biological control. In the rest of this introduction we shall introduce some of the terminology used in the field In the second section we briefly review some of the few experimental studies of apparent competition among phytophagous insects. In the third section we review a series of experiments we have done exploring indirect effects in a community of aphids and their natural enemies. Although this particular community resides in a relatively natural environment, we feel it illustrates very well the sort of processes that may also play an important part in agricultural environments. The fourth section is more speculative and in it we explore indirect effects and biological control. We ask both how a consideration of these effects may help the design of biological control programmes, and also whether there may be negative consequences of biological control, mediated by indirect effects, for future research.

1.2. Types of indirect effects

Apparent competition is a type of indirect interaction between two species that is mediated by a third species. The last couple of decades has seen a great increase in interest in indirect interactions among ecologists (Kerfoot and Sih 1987, Schmitt 1987, Wootton 1993, 1994, Menge 1995, Müller and Godfray 1999a, Werner and Peacor 2003), and it has become customary to classify them as either density (or trophically) mediated or trait (or behaviourally) mediated (Wootton 1993, Abrams 1995, Werner and Peacor 2003). In the former it is a change of density in a third species that transmits the indirect effect, while in the latter one species interacts with another by altering the behaviour or another trait of a third species that interacts directly with the second. Density mediated indirect effects include trophic cascades where an increase in density of a top predator benefits a prey species by decreasing the abundance of an intermediate predator, and many forms of exploitation competition where one species reduces the density or abundance of a shared resource. Trait-mediated indirect effects include instances where prey benefit from the presence of a top predator, not because the density of the intermediate predator changes, but because its behaviour changes, perhaps because it alters its habitat preference to inhabit safer micro-habitats.

Apparent competition is defined as a negative effect between two species mediated by a shared natural enemy. The simplest models of apparent competition suggest that other things being equal two prey/hosts cannot co-exist at the same site if they share a natural enemy (Holt and Lawton 1993, Holt *et al.* 1994). This result is the exact equivalent of the competitive exclusion principle that says no two competitors can exist on a common resource (Lotka 1925, Volterra 1926, Tilman 1977). In the competition case, the species that survives is the one that can persist on the lowest equilibrium density of resources. In the apparent competition case the winner is the species that, at equilibrium, supports the highest density of natural enemies. The second species goes extinct because the natural enemy pressure is sufficient to prevent it replacing itself. Thus one consequence of species exclusion through natural enemy effects is that a parasitoid or predator may appear to be monophagous not because of any physiological or morphological constraint, but because of the emergent dynamics of the community. Holt and Lawton (1993) thus named the phenomenon dynamic monophagy.

It is useful to distinguish the time scale over which the reciprocally negative effects of apparent competition operate. In a long-term interaction the presence of shared natural enemies will affect the equilibrium densities of all species concerned, or more precisely the distribution of densities observed over time. These effects are termed long-term apparent competition, and are distinguished from short-term apparent competition which describes the effects of shared natural enemies on the immediate trajectory of the population densities of the species concerned (Holt and Lawton 1994). By its nature, short-term apparent competition is easier to study and demonstrate experimentally. Because of the transient nature of crops, it is probably also short term apparent competition that is of most relevance in biological control although in some cases, in particular negative impacts on native species, long-term apparent competition may be important.

Indirect effects mediated by natural enemies may not only cause reciprocally negative consequences to the species involved. Suppose that natural enemy densities are effectively constant, their numbers being regulated by something other than resource supply. The introduction of an alternative prey for the predator or parasitoid may then actively benefit another prey species by diluting the natural enemy attack. Indeed, if the alternative prey is abundant or favoured by the natural enemy this effect can be very strong. Such reciprocally positive interactions are called apparent mutualism, and it is easy to think of interactions that may be asymmetric positive/negative, positive/neutral or negative/neutral (Chaneton and Bonsall 2000).

In addition to indirect effects mediated by shared natural enemies, we have already mentioned that exploitation competition can be viewed as an indirect effect mediated by a shared resource (Wootton 1994). There can also be indirect interactions mediated by mutualists (not to be confused with the apparent mutualism of the last paragraph). Perhaps the best examples involving herbivorous insects come from those honeydew-producing species, typically homopterans, which are tended by ants that protect them from predators. Several studies have shown that aphids benefit from the presence of ants, and that the benefits are reduced in the presence of other aphids, especially those more favoured by the ants (Addicott 1978, Cushman and Addicott 1989, Sakata 1995, Müller and Godfray 1999b, Sakata 1999, Fischer *et al.* 2001). The presence of homopterans may also boost ant populations in crops and so increase ant predation on other insect pests (see Eubanks and Styrsky in this volume). This effect can be considered to be a form of asymmetrical apparent competition.

We have concentrated in this brief introduction on indirect interactions between pairs of species, but of course in reality these species pairs will be imbedded in much more complex communities. These will be both harder to study, but also offer even greater scope for indirect interactions. At the level of interactions between the second and third trophic levels one may find diffuse apparent competition (Müller and Godfray 1999a) where a species is influenced by the nature of the natural enemy community supported by a broad range of herbivorous insects, and where identification of specific dyadic interactions may be difficult or impossible. But as Polis and Strong (1996) among others have stressed, the trophic structure of most communities involving phytophagous insects is more complex than a simple trophic layer cake, with many predators feed at a variety of trophic levels, and with even some herbivores not being adverse to a bit of meat (see Gillespie and Roitberg in this volume). Studies involving communities based on phytophagous insects have led in the development of the field of intra-guild predation (see Janssen *et al.*, Rosenheim and Harmon, Thomas *et al.* in this volume) and an exciting prospect is to explore how indirect effects involving shared intra-guild predators may affect community structure and dynamics (van Veen *et al.* 2006).

2. APPARENT COMPETITION IN INSECT HERBIVORES COMMUNITIES

2.1. Patterns in communities

In principle, apparent competition, like traditional competition, can be studied by looking at patterns in the distribution of organisms, or experimentally through manipulations in the field or laboratory (van Veen *et al.* 2006). Observations are normally the only practical means of studying the long-term effects of apparent competition, whether it be the imprint of past interactions on the current set of species found in a community (the equivalent of the ghost of competition past) or the evolutionary consequences of indirect interactions. While very important, great care must be taken in interpreting observational data as they lack the formal controls found in well-planned experiments, as well as often being bedevilled by the difficulties of reconstructing past evolutionary scenarios.

Non-random distributions of species across niche space have been used to infer the structuring role of resource competition, and equivalently non-random distributions of herbivore-natural enemy associations might reveal the action of apparent competition. Unfortunately, far fewer data have been collected at this higher trophic level, and formal tests of such hypotheses have yet to be performed. There are some intriguing hints of the possibility of community structure driven by natural enemies. For example, the leaf-mining moth genus *Phyllonorycter* (Gracillariidae) includes many species attacking the majority of genera of broad-leaved trees in Europe and Asia (Rott and Godfray 2000). It is attacked by a genus of parasitoid wasp, *Achrysocharoides* (Eulophidae), which appears to be relatively uniformly distributed across possible host plants, with at least in the UK one species tending to attack the miners found on a single host plant genus (Askew and Shaw 1974). However, there are exceptions (two species attacking miners on *Quercus*, none on *Populus*) and with the number of species available and the difficulties of defining a suitable null hypothesis, a formal test is not feasible. Note also in this case that the parasitoid host range appears to be more determined by host plant rather than host. This might suggest that resource competition among parasitoids rather than apparent competition among host is the more powerful force at work here. It also raises the question of why several *Phyllonorycter* species can co-exist on the same host plant heavily parasitised by the same species of wasp; a simple application of the idea of dynamic monophagy would lead one to expect that all but one would be driven to extinction by the shared parasitoid.

2.2. Enemy free space

Evolutionary ecology often postulates that species will evolve to colonise unoccupied niches. The idea that natural selection will also favour a move into areas of niche space where predation or parasitism is reduced also has a long history, though was crystallised by Jeffries and Lawton (1984) in their concept of enemy-free space. Though almost certainly important, testing this theory is difficult because once a species has occupied a patch of enemy-free space it is unlikely to remain enemy-free as resource competition selects for niche shifts in natural enemies. To give a concrete example, fig wasp parasitoids live in galls constructed by their fig wasp pollinator hosts, and are often themselves attacked by specialised hyperparasitoids that have very long ovipositors that are used to locate suitable larvae deep within the fig. In the Austro-Papuan *Ficus hispidioides* one species of fig wasp parasitoid (*Apocryptophagus* sp.) (Hymenoptera: Agaonidae) has evolved to produce its own larger galls deeper within the fig, protruding into the internal cavity of the synconium (Godfray 1988). In so doing the species has also shifted from being a parasitoid to feeding on gall tissue itself. Because the new large gall is further from the fig surface the normal hyperparasitoids can no longer attack it, and thus one might hypothesise that the niche shift has been driven, or partly driven, by selection to colonise enemy-free space. This may be true, but at least today the species experiences heavy parasitism by a monophagous specialist, *Apocrypta mega* (Hymenoptera: Pteromalidae) (its generic name is unfortunate as it feeds on rather than being fed on by *Apocryptophagus*). We cannot simply conclude whether *Apocryptophagus* colonised enemy free space to avoid its hyperparasitoids and that subsequently *Apocrypta* caught it up, or whether the niche shift was driven by a different process and the two species evolved into their new niches simultaneously. Conceivably molecular phylogenetics might resolve this issue if relative divergence times can be estimated.

2.3. *Experimental evidence for apparent competition*

Turning to experimental studies, the contemporary action of apparent competition can be demonstrated using natural or planned field manipulation experiments. In principle all that needs to be done is to manipulate the density of one species and observe what happens to other species that interact with the focal species only through their shared natural enemies. We review here the major studies of this type that have involved herbivorous insects, with the exception of work on aphids which is discussed in the following section.

Grape vine in California is attacked by the native leaf-hopper *Erythroneura elegantula* (Hemiptera: Cicadellidae). The eggs of this herbivore are attacked by *Anagrus epos*, minute parasitoid wasps in the family Mymaridae which contain the smallest known insects. In the 1980s the San Joaquin valley in California was invaded by a second species of *Erythroneura*, *E. variabilis* which also acted as a host for the egg parasitoid. Settle and Wilson (1990) observed that post-invasion the densities of *E. elegantula* were significantly reduced. They argued that the invasive *E. variabilis* was able to maintain itself at relatively high densities, and also to support a sizeable population of *A. epos*. The swollen parasitoid population was then responsible for the reduction in density of the native leaf hopper. Of course, it is possible that indirect effect is mediated not through the natural enemy but through the host-plant: *E. variabilis* is a more significant pest than *E. elegantula*, but Settle and Wilson also performed competition experiments which indicated that this was the less likely explanation. It would be very interesting to confirm this explanation with cage experiments. Most experiments involving herbivore manipulation provide information on only short-term apparent competition, and using deliberate or accidental introductions as in this study may be particularly useful in exploring long-term apparent competition.

Morris *et al.* (2004) also tried to look for medium to long-term effects of apparent competition. They worked with a community of leaf-miners in a tropical hurricane forest in Belize. A quantitative food web describing the interactions between the different hosts and parasitoids in the community was already available (Lewis *et al.* 2002) and they used this to predict the consequence of removing a common species of host plant that harboured two species of miner, one a hispine beetle and the other an agromyzid fly. They hypothesised that the removal of the fly would benefit a suite of other agromyzids in the same genera but feeding on different host plants which shared many common parasitoids, while the removal of the beetle would benefit just a single further hispine on a different host plant with which it shared parasitoids. The host plant was removed from six separate plots, and an equivalent amount of unmined foliage was removed from the same number of control plots. The plants were prevented from regrowing and the sites surveyed again after ten months, approximately six leaf-miner generations. As predicted, the abundance of the connected fly species increased significantly on experimental compared with control sites, while the same trend was seen in the beetle though the difference was not significant.

In exactly the same way that herbivores on different host plants can interact through shared natural enemies, so can parasitoids and predators through their shared natural enemies at higher trophic levels. Two studies have looked at apparent competition between parasitoids mediated by hyperparasitoids. van Nouhuys and Hanski (2000) studied the hyperparasitoids of the butterfly *Melitaea cinxia* (Nymphalidae) in Finland. This species is attacked by a locally monophagous primary parasitoid *Cotesia melitaearum* (Hymenoptera: Braconidae), which in turn is parasitized by the wingless hyperparasitoid *Gelis agilis* (Ichneumonidae), a species with a relatively broad host range including the well-known parasitoid, *Cotesia glomeratus* of cabbage butterflies *Pieris*. van Nouhuys and Hanski reared *C. glomeratus* in the laboratory and placed pupae in the field near colonies of the butterfly parasitised by *C. melitaearum*. Though their sample size was small, they found a strong tendency for rates of hyperparasitism to be higher at sites where *C. glomeratus* was present. Another study of apparent competition mediated by hyperparasitoids involves the parasitoids of aphids and is discussed in section 3.

2.4. Dynamic monophagy

We know of no demonstration of dynamic monophagy in the field, but mention here a laboratory study which although it does not concern herbivorous species does use insects with a rather similar life history. A number of moth groups have evolved to feed on dried organic manner and have become post-harvest pests of stored products. Some of the most significant pests are phycitine pyralids in the genera *Ephestia* and *Plodia*. These are attacked by a number of parasitoid groups, including ichneumonids such as *Venturia canescens* that in the wild are reared from phycitines feeding in decaying figs. In laboratory population cages, with wheat millings as a food source, *V. canescens* can persist in a stable interaction with either *Ephestia kuhniella* or *Plodia interpunctella*. In a clever experimental design, Bonsall and Hassell (1997) built a cage with two sections separated by a mesh. They placed *E.*

kuhniella and *P. interpunctella* separately in the two sections, the adults not being able to cross through the mesh. However, the parasitoid was able to transfer between the two compartments giving rise to a system in which the two moth species were coupled through a shared natural enemy but not a shared resource. The three species system was not stable with *E. kuhniella* being driven to extinction by the mortality imposed by the parasitoid population. Quantitative analysis of the interaction confirmed that of the two species *P. interpunctella* was the one that could survive in the presence of the higher density of the parasitoid.

2.5. Apparent competition in other systems

Our focus here has been on apparent competition as a potential force structuring insect herbivore communities, but we finish by briefly mentioning other systems in which apparent competition has been shown or postulated to play an important role. Invertebrate communities on rocky shores are one such example. Mobile gastropods and sessile bivalves are commonly attacked by a range of invertebrate predators. An experimental increase of either gastropods or bivalves has a negative effect on the other due to aggregation of predators (Schmitt 1987). Furthermore, in a comparison of interactions webs from 23 rocky shore intertidal habitats, Menge (1995) found that apparent competition was one of the most common types of indirect effect, while exploitation competition was rare. In vertebrates, pathogen mediated apparent competition has been put forward as a likely mechanism of declining populations of native species after introductions of aliens. Pheasants (*Phasianus colchicus*) released as game birds in Great Britain are the main source of infection of the wild grey partridge (*Perdix perdix*) with the nematode *Heterakis gallinarum*, thus contributing to the decline of the partridge (Tompkins *et al.* 1999, Tompkins *et al.* 2000). The replacement of native red squirrels with the North American grey squirrel throughout much of Great Britain is thought to be facilitated by a virus that is asymptomatic in greys but very virulent in reds, with greys thus likely being the major source of infection of reds (Tompkins *et al.* 2003).

3. APHIDS AND THEIR NATURAL ENEMIES

Aphids are common herbivores throughout the north temperate region with fewer species found in other parts of the world (Dixon 1985). They have received particular attention from ecologists interested in apparent competition, partly because of their economic importance as very significant pests of agriculture and to a slightly lesser extent forestry, but also because of the ease with which they can be manipulated in the field (Müller and Godfray 1997, Rott *et al.* 1998, Müller and Godfray 1999b, Morris *et al.* 2001, van Veen *et al.* 2001). In this section we describe a series of experiments, largely done by our group, designed to explore the role of apparent competition in structuring aphid communities in natural or semi-natural habitats. Discussion of the role of indirect effects specifically in the context of pest management is deferred to the next section.

3.1. Description of our study community

Biologists and farmers used to the depredations of aphids in highly fertilised crops are sometimes surprised at the hard life aphids have in old field and similar situations where their host plants are growing much more slowly. Here aphids sometime find it hard to persist caught between what McNeill and Lawton (1970) referred to as the devil of poor plant nutrition and the deep blue sea of natural enemy attack. For the last ten years or so we have been studying an assemblage of aphids and their natural enemies in such a habitat. We are particularly interested in the extent to which the species form an interacting community rather than a mere assemblage. As the majority of aphids feed on different host plants, direct and indirect resource competition is unlikely to be very important, and structuring, if it exists, is most likely to occur through the effects of shared natural enemies. The latter fall into three main categories: predators, many of which such as some species of ladybirds, hoverflies, midges, lacewings and bugs are aphid specialists; pathogens, six or so species of entomophthorales; and parasitoids, largely braconid wasps with relatively narrow host ranges. The parasitoids are attacked by specialised secondary parasitoids (of which more below).

To describe the aphid-based invertebrate community we have built quantitative food webs (Müller *et al.* 1999; unpublished data) to represent the disposition and strength of trophic links between aphids and each of the three main categories of natural enemies (for predators we considered only aphid specialists). For parasitoids and pathogens this is at least conceptually straightforward as infected aphids are mummified and are then relatively easy to sample. For a predator no such "body in the library" exists and the strength of trophic links must be calculated from the numbers of predator and prey present, and their estimated consumption rates. The three trophic webs show very different levels of connectance [the degree to which the community is joined together by trophic links (Martinez 1992, Bersier *et al.* 1999)] and this is reflected in the potential for apparent competition. By far the most strongly connected guild is the predators with most aphid species pairs sharing at least some common predators. The pathogen community shows considerably less connectance, though one species in particular has a broad host range and the potential for apparent competition. Finally, the primary parasitoids are least connected guild, with the large majority of species being highly specialised, and the few more oligophagous species being relatively rare (C. B. Müller and others, unpublished data).

We have also built quantitative webs to describe the community of secondary parasitoids. These fall into two distinct classes. The first is composed wholly of alloxystine figitids and are species that attack the primary parasitoid when it is a small larva within the still living host. The alloxystine larva suspends development until the primary larva has killed and mummified the host. We refer to these as hyperparasitoids. The second consists of a taxonomically heterogeneous group of species which attack the primary parasitoid (or possibly hyperparasitoid) after mummification. We call these mummy parasitoids. Hyperparasitoids tend to be quite specific, probably because of the intimacy of their association with the living aphid and primary parasitoid, while mummy parasitoids are more generalists. If apparent competition occurs at this trophic level then we would expect it to be mediated by mummy parasitoids not hyperparasitoids.

3.2. Experiments on this community

The quantitative food webs both allow us to make broad predictions about which guild of natural enemies is most likely to be involved in mediating indirect effects, as well as suggesting specific aphid pairs that it may be worth investigating experimentally. We have used both information from our food web as well as from other sources to design a series of experiments to look for chiefly short-term apparent competition and related phenomena. Most experiments have a conceptually similar design. The performance of an aphid species is compared in populations that are or are not in close proximity to another aphid species growing on a different host plant.

The first experiment we did was designed to test for indirect effects operating between the common nettle aphid (*Microlophium carnosum*) and aphids feeding on grass, the most important of which was *Rhopalosiphum padi*. *M. carnosum* populations, like those of many temperate aphids, peak in early summer and then decline in full summer, often with a smaller peak in the autumn. We asked whether the size and duration of the early summer peak would be reduced in the presence of high densities of the much more numerous grass aphids. The results were very clear: the grass aphids had a strong negative effect on nettle aphid numbers (Fig. 1). It was also clear what was responsible for the effect: coccinellids, both adults and larvae, were common predators on the grass aphids and migrated readily onto the nettle aphids, especially as grass aphid populations began themselves to decline. This was clear evidence of an indirect effect mediated by predators (Müller and Godfray 1997).

We were curious to see whether apparent competition might also occur mediated by parasitoids and conducted a second similar experiment, except this time using pea aphid (*Acyrthosiphon pisum*) as the target aphid and manipulating the presence or absence of nettle aphid. To maximise the chance of observing a parasitoid-mediated effect we removed predators whenever they appeared, patrolling the plots several times a day. We did get an effect though much weaker than in our first experiment: *A. pisum* populations declined more rapidly in the presence of nettle aphid. However, whatever was responsible it could not have been parasitoids: they were far too rare to influence aphid population dynamics. Instead we think the nearly-significant difference ($P = 0.06$) was due to predators. Even though we tried to remove all the predators that colonised the patch, we were aware that some were getting through and consuming aphids, especially very early in the morning. We suspect that were we not to have removed so many predators the effect would have been considerable stronger (Müller and Godfray 1997, Rott *et al.* 1998).

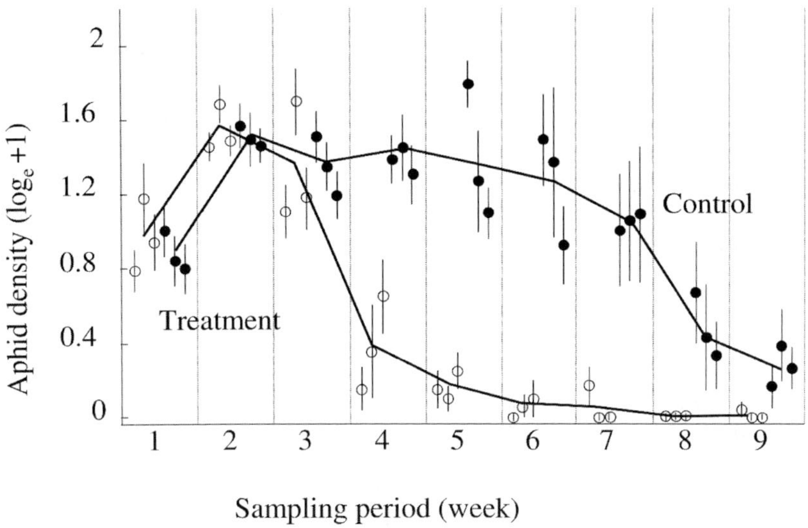

Figure 1. Apparent competition with grass aphids, mediated by ladybirds, strongly affects nettle aphid populations. Treatment aphids were on plants adjacent to fertilised grass plots which supported much higher numbers of grass aphids than unfertilised control plots. From Müller and Godfray (1997), redrawn with permission.

In addition to these experiments we have further evidence of apparent competition mediated by predators (this time between two aphid species on the same host, M. Ringel, unpublished data) and a second example of the lack of apparent competition mediated by parasitoids. Our working hypothesis is thus that aphid populations are dynamically linked by their predators, but that at least in our system parasitoids are seldom sufficiently common to have similar effects (we shall return to why this may be so below). We have also explored the possibility of apparent competition between aphids mediated by fungal pathogens, and while we failed to find a convincing effect, we are less certain about the role of these diseases (Pope *et al.* 2002). The problem here is that the severity of fungal diseases is strongly influenced by weather, which makes large-scale field experiments difficult. We think it most likely that interactions mediated by fungal pathogens will occur during periods of persistent warm and wet weather, and are likely to be much less important otherwise.

If predator-mediated apparent competition is important in this community then might it cause some species of aphids to be excluded? To test this we set up a suction trap and collected and identified all the winged aphids in the aerial plankton above our field site. This allowed us to determine several common species whose host plant occurred at the site but which themselves were never present or were extremely rare. Two of these, *Aphis jacobaeae* and *Brachycaudus cardui*, fed on ragwort, *Senecio jacobae*. We artificially set up replicate colonies of each of these species in our site under three different conditions: (i) controls with no protection from predators; (ii) colonies protected from ground predators by a ring of fruit-tree grease at the base of the plant; and (iii) colonies largely protected from flying predators as well as ground predators by being placed in a chicken wire cage with the wire coated in fruit-tree grease (this excludes most predators without influencing the micro-climate as a gauze barrier would; a few predators do get in but these can be removed by hand).

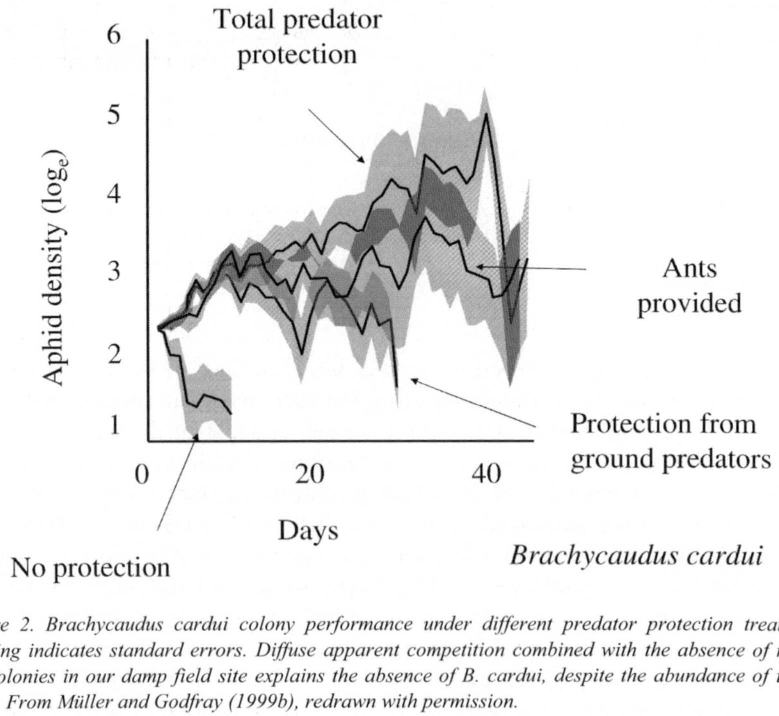

Figure 2. Brachycaudus cardui colony performance under different predator protection treatments. Shading indicates standard errors. Diffuse apparent competition combined with the absence of natural ant colonies in our damp field site explains the absence of B. cardui, despite the abundance of its host plant. From Müller and Godfray (1999b), redrawn with permission.

Artificial colonies with no protection from predators survived at most a few days; those protected from crawling predators survived a little longer, but few contained any aphids after a fortnight. However, colonies protected from aerial predators, which included the major aphid-specialist groups such as syrphids and coccinellids grew and persisted until the end of the season (Fig. 2). It appears that the community of predators at our field site is preventing the establishment of two species of aphids that otherwise would flourish. Possibly one particular predator is responsible, but we think it more likely that the sum total of predators is to blame, a form of diffuse apparent competition (Müller and Godfray 1999b).

This finding of course begs the question of why it is that the 15-odd regular resident species are able to persist. We do not know the answer to this, but suspect that each has some particular adaptation that provides it a refuge from predation. Some species are able to escape predation by throwing themselves off the plant and climbing back later on; others live on plants such as *Cirsium* that have leaves clustered at the growing tips where they can hide from attackers; some are poisonous to predators (Dixon 1958); still others live at low densities dispersed across their food plant where many may never be discovered. The experiment also raises the issue of why *A. jacobaeae* and *B. cardui* persist at other sites. For example, though *B. cardui* is absent from our study site it is quite common nearby in dry grassland at slightly higher elevation. Here the aphid is tended by ants and we tested whether it was the absence of mutualists at our damper field site that prevented this aphid from invading. We did this by in a fourth treatment providing our artificial *B. cardui* colonies with their own ants nest in a flower pot, connected to their food plant by string. Colonies tended by ants survived as well as those from which predators were excluded (Fig. 2), again demonstrating the importance of mutualists for aphid fitness, and suggesting how *B. cardui* may avoid apparent competition (Müller and Godfray 1999b).

In some situations, for example in many cereal crops, parasitoids can be very important natural enemies of aphids, whereas in our system they were relatively unimportant and seldom numerically abundant. The reason for this appears to be the levels of secondary parasitism they experience; typically 40-50% of mummies give rise to secondary rather than primary parasitoids. We believe in our system that secondary parasitoids tend to regulate primaries at population densities too low to cause major aphid mortality. But are the parasitoids also connected by indirect effects, but this time mediated by secondary parasitoids, and in particular the polyphagous mummy parasitoids? To test this Morris *et al.* (2001) designed an experiment in which mummies of pea aphid, *Acythosiphon pisum* parasitised by the braconid *Aphidius ervi*, were placed in the field beside nettle (*Urtica dioica*) plants that (i) had no aphids on them; (ii) were fed on by unparasitised nettle aphid *Microlophium carnosum*; or (iii) had colonies of nettle aphids parasitised by *Aphidius microlophii*. The experiment was run twice and in the first run the probability of mummy parasitism of the pea aphids was significantly higher in the presence of nettle aphid mummies compared with the other two treatments (Fig. 3).

The same trend was found in the second run, though here it was not significant. Primary parasitoid species may thus be dynamically linked by shared secondary parasitoids.

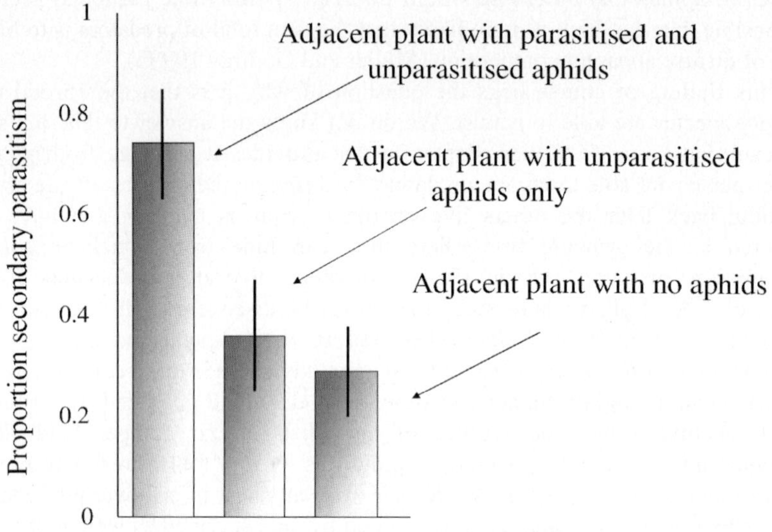

Figure 3. Apparent competition among primary parasitoids, mediated by secondary parasitoids. Bars show the proportion secondary parasitism of parasitised pea aphids on broad bean adjacent to nettles without aphids, with only unparasitised netle aphids or parasitised and unparasitised nettle aphids. From Morris et al (2001), redrawn with permission.

3.3. Summary of the structure of the aphid community

To conclude Fig. 4 illustrates our current ideas about how our aphid-based community is structured. At the bottom are the plants which we assume interact together to determine their relative abundances as well as secular changes in community composition. We think there is little or no feedback from aphids on plant abundance or composition, the aphids are just not common enough (aphids may act as vectors of plant pathogens but we have not observed this in this community). The aphids are chiefly monophagous and largely only one species is found per host plant. We think it unlikely that resource competition is important here, though the quality of the resource will affect aphid performance and their interactions with natural enemies. We believe that the aphid species do interact indirectly through their shared natural enemies: apparent competition mediated by predators seems rife in this system. We do not think parasitoids mediate many indirect interactions between aphids, but are less sure about the role of fungal pathogens. Unlike predators, parasitoids are largely monophagous, and the reason they are not involved in indirect interaction is, we believe, because their densities are tightly controlled by secondary parasitoids. These are of two types, true hyperparasitoids which tend to have narrow host range and tightly coupled dynamics with their primary parasitoid hosts, and more polyphagous mummy parasitoids. We think the specific coupled primary/hyper-parasitoid modules may be dynamically linked by shared mummy parasitoids, though our evidence here is not as strong as we would wish.

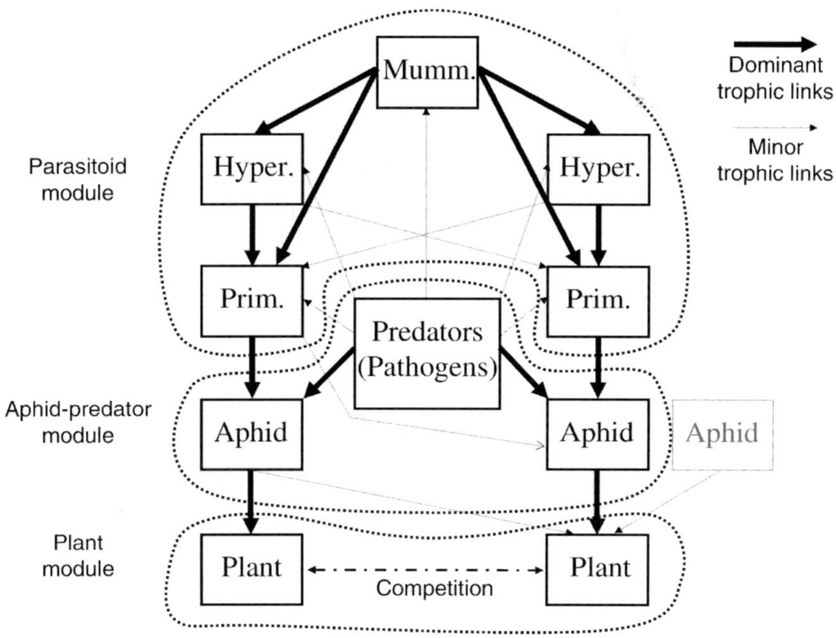

Figure 4. Diagram showing the interactions that we believe structure our aphid community. See text for explanation.

4. APPARENT COMPETITION AND BIOLOGICAL CONTROL

4.1. Improving biological control

Well before Holt coined the term apparent competition in 1977 many of its principles were evident to pest managers. Indeed, as Jeffries and Lawton (1984) have pointed out, ideas about apparent competition and enemy free space have permeated both pure and applied ecology since the birth of the subject. Many early ideas related to the importance of non-crop plants in providing alternative hosts for beneficial natural enemies. In the early 1970s it was noted for example that the presence of native *Rubus* and its leaf hoppers lead to higher rates of parasitism and lower densities of grape-pest leafhoppers in Californian vineyards (Doutt and Nakata 1973). Prompted by observations such as this, Starý (1993) devised a means of reducing pest aphids in greenhouses by establishing an aphid and its shared natural enemies on a non-crop plant before the economically important crop was grown. The pest in this case was the polyphagous *Myzus persicae* feeding on beans and the alternate host (of a pair of shared parasitoids) was the cereal aphid *Schyzaphis graminum*. Starý reported successful parasitism of the pest using this technique, though we are not aware of a formal replicated experimental demonstration of pest suppression. One of us wrote an unpublished report in 1980 advocating a very similar technique for the control of *Chrysanthemum* leafminers (*Chromatomyia syngenesiae*) (Diptera: Agromyzidae) in commercial greenhouses. This was prompted by the observation that low-light level grasses grew happily underneath greenhouse shelving and could be infested with leafminer species (*C. nigra* and *C. milii*) that shared five or six species of parasitoids with *C. syngenesiae* (including *Diglyphus isaea* (Hymenoptera: Eulophidae), used commercially as a biological control agent of greenhouse leafminers). Growers, however, were unwilling to take the risk in implementing the technique.

A similarly targeted approach was recently used in an orchard setting (Bribosia *et al.* 2005). Here, the objective was the control of the aphid *Dysaphis plantaginea*, which infects apple trees, by native populations of the parasitoid *Ephedrus persicae* (Hymenoptera: Aphidiidae). To achieve this Rowan trees (*Sorbus aucuparia*) adjacent to an orchard were artificially infested with the aphid *Dysaphis sorbi*. *E. persicae* naturally established on this alternative host and it was shown that these parasitoids were capable of parasitizing the pest aphid *D. plantaginea*. Thus there is a potential here for apparent competition between the two aphid species, although an effect on the population densities of the pest aphid has not yet been demonstrated.

A common problem with the natural biological control of pest insects is that the crops and the pests are only present for part of the year. Populations of parasitoids and predators can therefore not complete their annual life cycle *in situ* and have to reinvade the site every year, usually giving the pest a head start. The presence of alternative hosts on non-crop vegetation could promote the maintenance of resident natural enemy populations, and allow them to increase rapidly at the beginning of the season. In European cereal fields, parasitoids are probably the most important group of natural enemies controlling aphid densities (Schmidt *et al.* 2003). Langer and Hance (2004) found that wheat fields with strips of plants supporting alternative hosts had higher rates of parasitism of pest aphids than fields without strips. This was correlated with higher parasitoid activity early in the season suggesting that the strips and the insects they contain might be acting as a winter refuge for natural enemies. Unfortunately, the wheat strip experiment had only low replication and it is not possible formally to show that they were associated with increased pest control.

Complex landscapes, not exclusively used for arable crops, might be expected to support higher densities of natural enemies and thus benefit natural pest control through a form of diffuse apparent competition. This has been demonstrated most clearly for the rape pollen beetle (*Meligethes aeneus*) (Coleoptera: Nitidulidae) which suffers greater levels of parasitism in the presence of old field margins and fallow land, resulting in less crop damage (Thies and Tscharntke 1999). However, the approach of maintaining field margins and areas of fallow land or natural vegetation carries the danger of not only promoting apparent competition among herbivores of crops and natural vegetation but also among the vegetation and the crop mediated by the herbivores. For example, it has been found that although the rates of parasitism of wheat aphids in complex landscapes are higher, so too is the density of the wheat aphids themselves, probably because they maintain populations on grassland weeds. Landscape complexity in this case benefits both pest and natural enemies (Thies *et al.* 2005).

The examples just discussed all rely on the natural movement of parasitoids (and predators) from the alternative hosts to the pests. If this movement could be increased or encouraged, the negative effects on the pest might be enhanced. Chemical ecologists have made huge progress in the last several decades in discovering the cues that natural enemies use to locate their hosts, and this offers the prospect of manipulating insect behaviour. For example, among the cues aphid parasitoids use to locate oviposition sites is their host's sex pheromone. This can be synthesised and its application has been shown to increase the parasitism levels of cereal aphids near scent stations (Glinwood *et al.* 1998, Powell 1998, Glinwood *et al.* 1999a, Glinwood *et al.* 1999b). Plants damaged by herbivores can release volatiles that attract the natural enemies of the herbivore, signals that can be quite species-specific (Dicke *et al.* 1990, Turlings *et al.* 1990, Janssen *et al.* 2002). For example, parasitoids seem to be particularly attracted to plants damaged by host aphids and not to plants damaged by other aphids (Guerrieri *et al.* 1999). Infected plants can also induce nearby plants to produce synomones through root-root communication (Guerrieri *et al.* 2002). The signalling compound involved in this plant-to-plant communication might be employed to induce crops to attract parasitoids and predators and deter aphids (Powell and Pickett 2003).

Any attempt to attract natural enemies into a crop necessarily depends on the presence of areas of vegetation with suitable alternative hosts or prey to sustain sufficiently large populations of the beneficial insects. It thus involves the manipulation of an interaction involving apparent competition. Chemical ecology has promised much to pest management over the last quarter century and while it has had undoubted successes and take-up by the industry, these have perhaps been fewer than expected or hoped. We suspect that for this approach to prosper further it must become much more ecological and pay greater attention to the size, source and spatial distribution of the population of natural enemies that is to be brought into the crop.

4.2. Negative indirect effects

There has been much recent debate on the safety of biological control (Simberloff and Stiling 1996, Strong 1997, Thomas and Willis 1998, Ewel *et al.* 1999, Boettner *et al.* 2000, Follett and Duan 2000, Strong and Pemberton 2000) with particular concerns about the interactions between biological control agents and 'non-target' native species (Howarth 1991, Louda *et al.* 1997, Henneman and Memmott 2001). Such interactions can occur either directly, if an agent attacks a non-target host, or indirectly, when the agent affects non-target species via shared natural enemies. An insect herbivore introduced to control a weed could be attacked by generalist native parasitoids which also have native hosts (Lawton 1990, Cornell and Hawkins 1993, Hawkins and Marino 1997). If the weed biological control agent is common, then there is the potential for apparent competition between the agent and native herbivores, mediated via shared native parasitoids. Thus, even the introduction of an entirely host-plant specific biological control agent, assumed to be completely safe from non-target effects, still could have a community-wide impact. Only by understanding how invasive species interact within the context of the entire community can we hope to assess the risks to native species, whether they be direct effects on single species, or indirect effects on several species across trophic levels. While some data exists on direct impacts of biological control agents on non target hosts (Elliott *et al.* 1996, Hawkins and Marino 1997, Henneman and Memmott 2001), there is a paucity of field data on the impact of indirect effects, despite their potential to significantly influence community diversity and dynamics.

The ecological impact of intentionally introduced biological control agents of insect pests remains controversial. Some ecologists blame the practice for extinctions of native species (Howarth 1991), and others call for more regulation (Strong and Pemberton 2000). Biological control practitioners tend instead to believe that biological control is largely safe (Thomas and Willis 1998) although some are beginning to investigate non target effects themselves (Barron *et al.* 2003). Currently ecologists do not know whether the observed direct non-target effects are isolated cases, and they know even less about the frequency and impact of indirect effects. We discuss below a series of studies where a combination of the direct and indirect effects of the release of biological control agents have raised concerns about novel species introductions.

Introduced European thistles *Cirsium* spp. are a major pest of rangelands in the United States, and also invade natural habitats where they can be a threat to endemic plant diversity. A weevil, *Rhinocyllus conicus* (Coleoptera: Curculionidae), has been introduced as a biological control agent to control the exotic thistles. Unfortunately, the weevil is not host-specific and also attacks native *Cirsium* species, some of which are rare and threatened (Louda *et al.* 1997). For example, Rand and Louda (2004) reported that attack by *R. conicus* on the native thistle, *Cirsium undulatum*, increased 3-5 fold as exotic thistle density increased. This can be viewed as a direct effect of the introduced natural enemy, but also as apparent competition between native and exotic thistles. What is more, some of the endemic thistles support native seed predators which will suffer exploitation competition (an indirect effect) with *R. conicus*.

Over the years, a wide range of exotic parasitoids have been introduced into Hawaii to control a broad array of agricultural and forestry pests. This was especially true prior to 1945 before modern insecticides and then modern regulations changed practises. Henneman and Memmott (2001) asked to what extent these parasitoids had colonised more pristine habitats, such as the native forest on Kauai Island. To do this they constructed quantitative food webs to describe the interactions among plants, moths, and their parasitoids. Surprisingly, introduced (pre-1945) parasitoids were abundant in the forest, comprising 83% percent of the parasitoids reared from native moths. Many herbivore communities consist of species on different host plants that probably only interact through shared natural enemies (van Veen *et al.* 2006). These results suggest that the introduced parasitoids might be radically changing the way the community is structured.

In addition to introduced parasitoids, phytophagous insects have also been introduced to combat weeds, including the blackberry *Rubus argutus* which has infiltrated native habitats on Kauai. Several biological control agents have been introduced for the control of this species, including the moths *Croesia zimmermani* (Tortricidae) and *Schreckensteinia festaliella* (Heliodinidae) (Zimmerman 1978), both of which were present in Henneman and Memmott's (2001) food web. Indirect effects could ensue if native parasitoids attacked *C. zimmermani* or *S. festaliella* heavily, and were also found attacking native host caterpillars. However, no parasitoids, either alien or native, were reared from these agents indicating that at least in this case indirect effects are unlikely.

Boneseed or bitou bush, *Chrysanthemoides monilifera* ssp. *rotundata* (Asteraceae), is a South American plant that has invaded the southern states of Australia where it can form dense stands in forest understorey or bush. To control the weed, a seed-feeding tephritid fly, *Mesoclanis polana*, has been introduced, and this species is now heavily attacked by a guild of native parasitoids that normally feed on endemic tephritids. A quantitative food web has been built which shows that through its abundance *M. polana* now occupies a potential keystone position in the web of interactions connecting this guild of herbivores (Willis and Memmott In press). Thus a highly host-specific biological control agent such as *M. polana* has the potential to change community structure by increasing the abundance of native parasitoids and affecting native species through apparent competition. This is an excellent system to investigate these effects experimentally.

Redman and Scriber (2000) working in the USA studied indirect interactions between a native butterfly, the swallowtail *Papilio canadensis* (Papilionidae), and the alien gypsy moth, *Lymantria dispar* (Lymantriidae). They used laboratory and field experiments to look for evidence of apparent competition between the larvae of the two species. They found that the swallowtail was negatively affected by gypsy moths via both shared parasitoids and shared pathogens. Sterilizing the leaves of the host plant which destroys pathogen propagules significantly reduced the negative effects of the gypsy moth on the swallowtail caterpillars. In addition, swallowtail caterpillars had significantly increased rates of parasitism by the biological control agent *Compsilura concinnata* (Diptera: Tachninidae) among others when positioned near to gypsy moth infestations in the field.

5. CONCLUSIONS

In this chapter we have tried to review how insect communities may be structured by shared natural enemies in the same way that other communities are structured by competition for shared resources. Our emphasis on the higher trophic level should not be interpreted as implying that phytophagous insects are not influenced by bottom-up effects, by the quality of the host plant for example. There has been an unfortunate tendency to dichotomise bottom-up versus top-down effects, even though the dynamics of virtually every herbivorous insect will be strongly influenced by both. What is significant is not whether plant or natural enemy effects influence dynamics but how they influence them; for example is it bottom-up or top-down effects that have the density-dependent structure that can regulate populations, and through what processes, if any, do different species of insect interact to structure their communities?

Hairston, Smith and Slobodkin (1960) famously said the earth is green and hence herbivores do not compete for resources. This sparked the successful search for numerous counter-examples, and indeed many pairs of phytophagous insect species do compete for limiting resources. Yet we believe that for the vast majority of plant-feeding insects Hairston *et al.* are essentially correct: they are insufficiently abundant most of the time to deplete their resources. We suspect that these insects are regulated below resource carrying-capacity by natural enemies of one sort of another. (We digress briefly to note that often this regulation is only possible because of bottom-up effects: our aphid populations on wild grasses and herbs are much more easily controlled by predators than those in the nutritional cornucopia of a well-fertilised crop.)

So for us a major question is, are communities of phytophagous insects structured by something other than resource competition? They may not be, species numbers and their densities may fluctuate neutrally, influenced only by broad macro-evolutionary and macro-ecological processes. But if they are, then apparent competition and related phenomena seem to be the most likely candidate processes. The problem of course is how to demonstrate this.

There are two broad techniques for exploring processes in community ecology: analysis of large-scale patterns, and experimentation. The former has been used relatively little in investigating apparent competition because the data available are insufficiently resolved to address these questions. We are hopeful that new data sets, especially those employing quantitative food webs, may prove useful here. Given the antiquity of ideas concerning enemy-free space, and that Holt defined apparent competition nearly 30 years ago, it is perhaps surprising that there are still relatively few experimental studies in the field. The reason for this is that such experiments are often logistically challenging, especially if long-term rather than short-term competition is studied. We have found that perhaps 60% of our field experiments fail, not in the sense they give negative results but because something goes wrong with plants, insects or the weather (or perhaps it's just us!). Nevertheless we are encouraged by the recent interest in apparent competition and indirect effects in general, and are optimistic that our understanding of the structuring forces in phytophagous insect community ecology will increase significantly in the coming years.

Finally, we have tried to explore some of the consequences of apparent competition in biological control. We note that apparent competition by other names has long been part of the applied entomologist's conceptual armoury. Yet a more explicit consideration of the ecology involved may, we think, help in designing interventions and agricultural landscapes that foster natural biological control. We particular think that a greater fusion of chemical and population ecology is required to get some of the exciting new ideas involving the manipulation of insect behaviour off the drawing board. The self-image of many biological control practitioners is of the good guy promulgating much more environmentally-friendly pest management strategies than his or her agro-chemical colleague. But biological control introductions come with their own risks, and many of these involve indirect effects on native species. It is possible that these detrimental effects may prove minor compared with conventional pest control, but they cannot be ignored and must be assessed and quantified. Again, an explicit consideration of the population ecology of indirect interactions will greatly assist this process. And of course the process will not be one way, our fundamental understanding of ecological processes will continue to be enriched by analyses of the huge (though not often controlled) experiments that are biological control programmes.

6. REFERENCES

Abrams, P. A., 1995, Implications of dynamically variable traits for identifying, classifying, and measuring direct and indirect effects in ecological communities. *Am. Nat.* 146: 112-134.

Abrams, P. A., and Chen, X., 2002, The evolution of traits affecting resource acquisition and predator vulnerability: Character displacement under real and apparent competition. *Am. Nat.* 160: 692-704.

Addicott, J. F., 1978, Competition for mutualists: ants and aphids. *Can. J. Zool.* 56: 2093-2096.

Askew, R. R., and Shaw, M. R., 1974, An account of the Chalcidoidea (Hymenoptera) parasitising leafmining insects of decidious trees in Britain. *Biol. J. Linn. Soc.* 6: 289-335.

Barron, M. C., Barlow, N. D., and Wratten, S. D., 2003, Non-target parasitism of the endemic New Zealand red admiral butterfly (*Bassaris gonerilla*) by the introduced biological control agent *Pteromalus puparum*. *Biol. Cont.* 27: 329-335.

Bersier, L. F., Dixon, P., and Sugihara, G., 1999, Scale-invariant or scale-dependent behavior of the link density property in food webs: A matter of sampling effort? *Am. Nat.* 153: 676-682.

Boettner, G. H., Elkinton, J. S., and Boettner, C. J., 2000, Effects of a biological control introduction on three nontarget native species of saturniid moths. *Cons. Biol.* 14: 1798-1806.

Bonsall, M. B., and Hassell, M. P., 1997, Apparent competition structures ecological assemblages. *Nature* 388: 371-373.

Bribosia, E., Bylemans, D., Migon, M., and Van Impe, G., 2005, In-field production of parasitoids of *Dysaphis plantaginea* by using the rowan aphid *Dysaphis sorbi* as substitute host. *Biocontrol* 50: 601-610.

Chaneton, E. J., and Bonsall, M. B., 2000, Enemy-mediated apparent competition: empirical patterns and the evidence. *Oikos* 88: 380-394.

Cornell, H. V., and Hawkins, B. A., 1993, Accumulation of native parasitoid species on introduced herbivores - a comparison of hosts as natives and hosts as invaders. *Am. Nat.* 141: 847-865.

Cushman, J. H., and Addicott, J. F., 1989, Intra specific and interspecific competition for mutualists - ants as a limited and limiting resource for aphids. *Oecologia* 79: 315-321.

Dicke, M., Sabelis, M. W., Takabayashi, J., Bruin, J., and Posthumus, M. A., 1990, Plant strategies of manipulating predator-prey interactions through allelochemicals - prospects for application in pest-control. *J. Chem. Ecol.* 16: 3091-3118.

Dixon, A. F. G., 1958, The escape responses shown by certain aphids to the presence of the coccinellid *Adalia decempunctata* (L.). *Trans. Roy. Ent. Soc. London* 110: 319-334.

Dixon, A. F. G.,1985, *Aphid Ecology*. Glasgow: Blackie & Son.

Doutt, R., and Nakata, J., 1973, The *Rubus* leafhopper and its egg parasitoid: an endemic biotic system useful in grape-pest management. *Env. Ent.* 2: 381-386.

Elliott, N., Kieckhefer, R., and Kauffman, W., 1996, Effects of an invading coccinellid on native coccinellids in an agricultural landscape. *Oecologia* 105.

Ewel, J. J., O'Dowd, D. J., Bergelson, J., Daehler, C. C., D'Antonio, C. M., Gomez, L. D., *et al.*, 1999, Deliberate introductions of species: Research needs - Benefits can be reaped, but risks are high. *Bioscience* 49: 619-630.

Fischer, M. K., Hoffmann, K. H., and Volkl, W., 2001, Competition for mutualists in an ant-homopteran interaction mediated by hierarchies of ant attendance. *Oikos* 92: 531-541.

Follett, P. A., and Duan, J. J.,2000, *Nontarget Effects of Biological control*: Kluwer Academic Publishers.

Glinwood, R. T., Du, Y. J., and Powell, W., 1999, Responses to aphid sex pheromones by the pea aphid parasitoids *Aphidius ervi* and *Aphidius eadyi*. *Entomol. Exp. Appl.* 92: 227-232.

Glinwood, R. T., Du, Y. J., Smiley, D. W. M., and Powell, W., 1999, Comparative responses of parasitoids to synthetic and plant- extracted nepetalactone component of aphid sex pheromones. *J. Chem. Ecol.* 25: 1481-1488.

Glinwood, R. T., Powell, W., and Tripathi, C. P. M., 1998, Increased parasitization of aphids on trap plants alongside vials releasing synthetic aphid sex pheromone and effective range of the pheromone. *Biocontrol Sci. Technol.* 8: 607-614.

Godfray, H. C. J., 1988, Virginity in haplodiploid populations: a study on fig wasps. *Ecol. Ent.* 13: 283-291.

Guerrieri, E., Poppy, G. M., Powell, W., Rao, R., and Pennacchio, F., 2002, Plant-to-plant communication mediating in-flight orientation of *Aphidius ervi*. *J. Chem. Ecol.* 28: 1703-1715.

Guerrieri, E., Poppy, G. M., Powell, W., Tremblay, E., and Pennacchio, F., 1999, Induction and systemic release of herbivore-induced plant volatiles mediating in-flight orientation of *Aphidius ervi*. *J. Chem. Ecol.* 25: 1247-1261.

Hairston, N. G., Smith, F. E., and Slobodkin, L. B., 1960, Community structure, population control, and competition. *Am. Nat.* 44: 421-425.

Hawkins, B. A., and Marino, P. C., 1997, The colonization of native phytophagous insects in North America by exotic parasitoids. *Oecologia* 112: 566-571.

Henneman, M. L., and Memmott, J., 2001, Infiltration of a Hawaiian community by introduced biological control agents. *Science* 293: 1314-1316.

Holt, R. D., 1977, Predation, apparent competition, and the structure of prey communities. *Theor. Pop. Biol.* 12: 197-229.

Holt, R. D., Grover, J., and Tilman, D., 1994, Simple rules for interspecific dominance in systems with exploitative and apparent competition. *Am. Nat.* 144: 741-771.

Holt, R. D., and Lawton, J. H., 1993, Apparent competition and enemy-free space in insect host-parasitoid communities. *Am. Nat.* 142: 623-645.

Holt, R. D., and Lawton, J. H., 1994, The ecological consequences of shared natural enemies. *Annu. Rev. Ecol. Syst.* 25: 495-520.
Howarth, F. G., 1991, Environmental impacts of classical biological-control. *Ann. Rev. Ent.* 36: 485-509.
Janssen, A., Sabelis, M. W., and Bruin, J., 2002, Evolution of herbivore-induced plant volatiles. *Oikos* 97: 134-138.
Jeffries, M. J., and Lawton, J. H., 1984, enemy free space and the structure of ecological communities. *Biol. J. Linnean Soc.* 23: 269-286.
Kerfoot, W. C., and Sih, A.,1987, *Predation: direct and indirect impacts on aquatic communities.* Hanover, New Hampshire: University Press of New England.
Langer, A., and Hance, T., 2004, Enhancing parasitism of wheat aphids through apparent competition: a tool for biological control. *Agric. Ecosyst. Environ.* 102: 205-212.
Lawton, J. H., 1990, Biological control of plants: a review of generalisations, rules, and principles using insects as agents. In Bassett, C., Whitehouse, L. J. andZabkiewicz, J. A. (Eds.), *Alternatives to the Chemical Control of Weeds. Proceedings of an International Conference* (pp. 3-17): Ministry of Forestry, FRI Bulletin 155.
Lawton, J. H., and Strong, D. R., 1981, Community patterns and competition in folivorous insects. *Am. Nat.* 118: 317-338.
Lewis, O. T., Memmott, J., Lasalle, J., Lyal, C. H. C., Whitefoord, C., and Godfray, H. C. J., 2002, Structure of a diverse tropical forest insect-parasitoid community. *J. Anim. Ecol.* 71: 855-873.
Lotka, A. J.,1925, *Elements of physical biology*. Baltimore: Williams & Wilkins.
Louda, S. M., Kendall, D., Connor, J., and Simberloff, D., 1997, Ecological effects of an insect introduced for the biological control of weeds. *Science* 277: 1088-1090.
Martinez, N. D., 1992, Constant connectance in community food webs. *Am. Nat.* 139: 1208-1218.
McNeill, S., and Lawton, J. H., 1970, Annual production and respiration in animal populations. *Nature* 225: 472.
Menge, B. A., 1995, Indirect effects in marine rocky intertidal interaction webs - patterns and importance. *Ecol. Monogr.* 65: 21-74.
Morris, R. J., Lewis, O. T., and Godfray, H. C. J., 2004, Experimental evidence for apparent competition in a tropical forest food web. *Nature* 428: 310-313.
Morris, R. J., Müller, C. B., and Godfray, H. C. J., 2001, Field experiments testing for apparent competition between primary parasitoids mediated by secondary parasitoids. *J. Anim. Ecol.* 70: 301-309.
Müller, C. B., Adriaanse, I. C. T., Belshaw, R., and Godfray, H. C. J., 1999, The structure of an aphid-parasitoid community. *J. Anim. Ecol.* 68: 346-370.
Müller, C. B., and Godfray, H. C. J., 1997, Apparent competition between two aphid species. *J. Anim. Ecol.* 66: 57-64.
Müller, C. B., and Godfray, H. C. J., 1999a, Indirect interactions in aphid-parasitoid communities. *Res. Popul. Ecol.* 41: 93-106.
Müller, C. B., and Godfray, H. C. J., 1999b, Predators and mutualists influence the exclusion of aphid species from natural communities. *Oecologia* 119: 120-125.
Polis, G. A., and Strong, D. R., 1996, Food web complexity and community dynamics. *Am. Nat.* 147: 813-846.
Pope, T., Croxson, E., Pell, J. K., Godfray, H. C. J., and Muller, C. B., 2002, Apparent competition between two species of aphid via the fungal pathogen *Erynia neoaphidis* and its interaction with the aphid parasitoid *Aphidius ervi*. *Ecol. Ent.* 27: 196-203.
Powell, W., 1998, Semiochemicals to increase parasitism in aphid pest control. *Pestic. Sci.* 54: 291-293.
Powell, W., and Pickett, J. A., 2003, Manipulation of parasitoids for aphid pest management: progress and prospects. *Pest Manag. Sci.* 59: 149-155.
Rand, T. A., and Louda, S. M., 2004, Exotic weed invasion increases the susceptibility of native plants attack by a biocontrol herbivore. *Ecology* 85: 1548-1554.

Redman, A. M., and Scriber, J. M., 2000, Competition between the gypsy moth, *Lymantria dispar*, and the northern tiger swallowtail, *Papilio canadensis*: interactions mediated by host plant chemistry, pathogens, and parasitoids. *Oecologia* 125: 218-228.

Rott, A. S., and Godfray, H. C. J., 2000, The structure of a leafminer-parasitoid community. *J. Anim. Ecol.* 69: 274-289.

Rott, A. S., Müller, C. B., and Godfray, H. C. J., 1998, Indirect population interaction between two aphid species. *Ecol. Lett.* 1: 99-103.

Sakata, H., 1995, Density-dependent predation of the ant *Lasius niger* (Hymenoptera: Formicidae) on two attended aphids *Lachnus tropicalis* and *Myzocallis kuricola* (Homoptera: Aphididae). *Res. Popul. Ecol.* 37: 159-164.

Sakata, H., 1999, Indirect interactions between two aphid species in relation to ant attendance. *Ecol. Res.* 14: 329-340.

Schmidt, M. H., Lauer, A., Purtauf, T., Thies, C., Schaefer, M., and Tscharntke, T., 2003, Relative importance of predators and parasitoids for cereal aphid control. *Proc. R. Soc. Lond. Ser. B-Biol. Sci.* 270: 1905-1909.

Schmitt, R. J., 1987, Indirect interactions between prey: apparent competition, predator aggregation, and habitat segregation. *Ecology* 68: 1887-1897.

Settle, W. H., and Wilson, L. T., 1990, Invasion by the variegated leafhopper and biotic interactions - parasitism, competition, and apparent competition. *Ecology* 71: 1461-1470.

Simberloff, D., and Stiling, P., 1996, How risky is biological control? *Ecology* 77: 1965-1974.

Starý, P., 1993, Alternative host and parasitoid in 1st method in aphid pest-management in glasshouses. *J. Appl. Ent.* 116: 187-191.

Strong, D. R., 1997, Ecology - Fear no weevil? *Science* 277: 1058-1059.

Strong, D. R., Lawton, J. H., and Southwood, S. R.,1984, *Insects on Plants - Community Patterns and Mechanisms* (1 ed.). Oxford: Blackwell Scientific Publications.

Strong, D. R., and Pemberton, R. W., 2000, Ecology - Biological control of invading species - Risk and reform. *Science* 288: 1969-1970.

Thies, C., Roschewitz, I., and Tscharntke, T., 2005, The landscape context of cereal aphid-parasitoid interactions. *Proc. R. Soc. Lond. Ser. B-Biol. Sci.* 272: 203-210.

Thomas, M. B., and Willis, A. J., 1998, Biocontrol - risky but necessary? *TREE* 13: 325-329.

Tilman, D., 1977, Resource competition between planktonic algae: an experimental and theoretical approach. *Ecology* 58: 338-348.

Tompkins, D. M., Dickson, G., and Hudson, P. J., 1999, Parasite-mediated competition between pheasant and grey partridge: a preliminary investigation. *Oecologia* 119: 378-382.

Tompkins, D. M., Draycott, R. A. H., and Hudson, P. J., 2000, Field evidence for apparent competition mediated via the shared parasites of two gamebird species. *Ecol. Lett.* 3: 10-14.

Tompkins, D. M., White, A. R., and Boots, M., 2003, Ecological replacement of native red squirrels by invasive greys driven by disease. *Ecol. Lett.* 6: 189-196.

Turlings, T. C. J., Tumlinson, J. H., and Lewis, W. J., 1990, Exploitation of herbivore-induced plant odors by host-seeking parasitic wasps. *Science* 250: 1251-1253.

van Nouhuys, S., and Hanski, I., 2000, Apparent competition between parasitoids mediated by a shared hyperparasitoid. *Ecol. Lett.* 3: 82-84.

van Veen, F. J. F., Morris, R. J., and Godfray, H. C. J., 2006, Apparent competition, quantitative food webs and the structure of phytophagous insect communities. *Ann. Rev. Ent.* 51: 187-208.

van Veen, F. J. F., Rajkumar, A., Muller, C. B., and Godfray, H. C. J., 2001, Increased reproduction by pea aphids in the presence of secondary parasitoids. *Ecol. Ent.*. 26: 425-429.

Volterra, V.,1926, *Variations and fluctuations in the numbers of individuals in animal species living together*: (Reprinted in 1931. In Chapman, R. N., Animal Ecology, McGraw-Hill, New York, pages 409-448).

Werner, E. E., and Peacor, S. D., 2003, A review of trait-mediated indirect interactions in ecological communities. *Ecology* 84: 1083-1100.

Willis, A. J., and Memmott, J., In press, The effect of alien plants and their biocontrol agents on native biodiversity: food webs as a tool to assess relative impacts. *Biol. Cont.*

Wootton, J. T., 1993, Indirect effects and habitat use in an intertidal community - interaction chains and interaction modifications. *Am. Nat.* 141: 71-89.

Wootton, J. T., 1994, The nature and consequences of indirect effects in ecological communities. *Annu. Rev. Ecol. Syst.* 25: 443-466.

Zimmerman, E. C.,1978, *Microlepidoptera*: University of Hawaii.

ANT-HEMIPTERAN MUTUALISMS: KEYSTONE INTERACTIONS THAT ALTER FOOD WEB DYNAMICS AND INFLUENCE PLANT FITNESS

Consequences of Ant-Hemipteran Mutualisms for Biocontrol

Micky D. Eubanks and John D. Styrsky

Department of Entomology and Plant Pathology, Auburn University

Abstract. Predicting the direct and indirect effects of natural enemies on herbivorous insects in species-rich, highly connected arthropod food webs can be extremely difficult. Community ecologists developed the keystone species concept to help simplify this task. Keystone species are species that have disproportionately large effects on the abundance of many interacting species in a community. Keystone species, however, can be difficult to identify in some communities and the effect of species that seem to play key roles in community dynamics often vary dramatically in both space and time. In some communities, pairwise interactions among species may alter the community-wide effect of a species such that it functions as a keystone species. In this chapter, we term this a 'keystone interaction' and explore the possibility that mutualisms involving ants and honeydew-producing Hemipterans may alter the abundance and distribution of many species in a predictable manner via increased ant predation in the presence of Hemipterans. Mutualisms involving ants and honeydew-producing insects are incredibly widespread in terrestrial ecosystems and may alter the structure of entire arthropod communities. We review cases where these mutualisms result in interference of biological control via intraguild predation of important biological control agents as well as cases where ant mutualisms result in enhanced biological control via intensified ant predation of important plant-damaging herbivores. In addition, we report the results of our own work involving the ecological consequences of fire ant-aphid mutualisms. We conclude that ant-Hemiptera mutualisms rarely disrupt biological, but instead these mutualisms often increase the effectiveness of ants as biological control agents via the removal of insect herbivores that are more important pests than the Hemipterans that benefit from ant mutualisms.

1. INTRODUCTION

Predictable biological control of insect pests is a primary goal of integrated pest management (Pedigo 2002, Harmon and Andow 2004). Predicting herbivore suppression by natural enemies is theoretically straightforward when biological control involves a simple, three trophic level interaction. For instance, textbook examples of biological control typically include a specialized predator or parasitoid, a single herbivorous pest, and a single host plant or crop (Van Driesche and Bellows 1996). In this simplified example, it is relatively easy to accurately estimate the suppressive effect of the natural enemy on the herbivore and the subsequent reduction in plant damage. If biological control of most pests was this straightforward, then consistent, relatively invariable suppression of herbivorous pests would be the norm. Instead, suppression of insect pests by natural enemies can vary dramatically among different crop species and can vary both spatially and

J. Brodeur and G. Boivin (eds.), Trophic and Guild Interactions in Biological Control, 171–189.

temporally within a given crop species (e.g., Rosenheim *et al.* 1995, Teder *et al.* 2000, Menalled *et al.* 2003, Romeis *et al.* 2005). One of the primary reasons that the effects of natural enemies on pests varies is because economically important natural enemies are often not specialist predators or parasitoids that attack only one pest species. Many important natural enemies are generalist predators or parasitoids that attack multiple prey species, making their impact on any single prey species difficult to predict (Murdoch *et al.* 1985, Symondson *et al.* 2002, Koss and Snyder 2005). Furthermore, many generalist natural enemies are either "true" omnivores that feed on both prey and plant food, "trophic" omnivores that feed not only on herbivores but also on other predators (intraguild predators), or both (Rosenheim *et al.* 1995, Coll and Guershon 2002, Eubanks 2005, Eubanks and Styrsky 2005; see also Rosenheim and Harmon, Gillespie and Roitberg, this volume). The pervasiveness of omnivorous natural enemies in agricultural systems produces highly reticulated or highly-connected food webs. These food webs often contain higher order predators that primarily function as intraguild predators, intermediate predators that often consume both prey and plant food and are frequently attacked by intraguild predators, and plant pests that typically feed on multiple plant species and may even feed on other insects (Fig. 1) (Polis and Strong 1996, Rosenheim 1998).

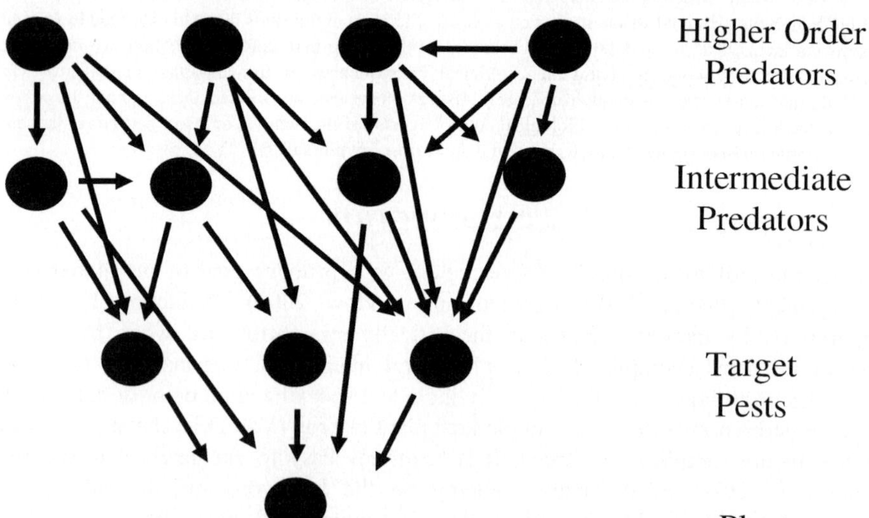

Figure 1. Schematic representation of a highly reticulated food web containing higher order predators that primarily attack other predators, intermediate predators that primarily attack herbivores but that can be intraguild predators or feed directly on plants, and multiple herbivores that attack the focal plant species.

Attempting to predict the impact of one or a few species of natural enemies on one or a few species of herbivores in these highly connected food webs can be extremely daunting. Community ecologists have been attempting to predict the outcome of trophic interactions in complex food webs for decades (MacArthur 1955, Elton 1958) and have developed several concepts to help simplify this difficult task. One of the most prominent and useful of these is the keystone species concept (Paine 1969) whereby researchers attempt to identify species that have disproportionately large effects on food webs by directly or indirectly altering the abundance of many interacting species (Mills *et al.* 1993).

Unfortunately, identifying keystone species in highly connected, species-rich arthropod food webs (e.g., Fig. 1) may be extremely difficult. Furthermore, the effect of keystone species on community structure can vary tremendously with both biotic and abiotic conditions (Power and Tilman 1996, Sanford 1999). For example, interactions among organisms in a community may alter the effect of keystone species, thus altering the interaction strengths linking multiple species. Such interactions could be called 'keystone interactions' and understanding these interactions may allow us to accurately predict when and where keystone species are likely to have strong, community-wide effects. If these interactions occur in agricultural systems, then they may make the efficacy of biological control more predictable. In this chapter we explore the possibility that ant-Hemipteran mutualism function as keystone interactions in many agricultural systems. We have found that ants function as keystone predators in some agricultural systems, but that their effect on other arthropods, especially their suppression of herbivorous insects, depends on the strength of their mutualistic interactions with honeydew-producing insects.

2. THE POWER OF POSITIVE SPECIES INTERACTIONS

Focusing on the effects of a mutualism to predict the outcome of complex trophic interactions is relatively novel because ecologists have historically focused on the importance of negative species interactions as factors regulating the distribution and abundance of organisms (Bronstein 1994, Stachowicz 2001). Insect ecologists, for example, have investigated and debated the relative roles of competition and predation on community structure for over 40 years (e.g., Hairston *et al.* 1960, Lawton and Strong 1981, Strong *et al.* 1984, Denno *et al.* 1995, Chase *et al.* 2002). Recent studies, however, have demonstrated that positive species interactions can dramatically affect the structure and dynamics of ecological communities (Boucher *et al.* 1982, Bronstein 1994, Stachowicz 2001, Bruno *et al.* 2003). For example, facilitative interactions, interactions that benefit one of the participants and cause no harm to either participant, are important in structuring many plant communities (Stachowicz 2001). This is the case when an early plant colonizer of new or unstable habitats provides critical substrate stabilization that, in turn, allows other plant species to colonize the habitat (Clements 1916, Bertness and Leonard 1997, Bruno 2000). Mutualisms, interactions where both species derive benefit, can also play critical roles in structuring communities (Stachowicz 2001, Wimp and Whitham 2001, Bruno *et al.* 2003). Mutualisms between ants and honeydew-producing Hemipterans (aphids, scales, and whiteflies), for example, may have widespread ecological consequences that affect the abundance and distribution of other arthropods, plants, and even microorganisms. Ants benefit by consuming honeydew, a sugary byproduct of herbivory that is excreted by these phloem-feeding insects. Honeydew is comprised of simple sugars mixed with various free amino acids (Mittler 1958, Douglas 1993). Consequently, honeydew is a very attractive food for ants (Way 1963, Hölldobler and Wilson 1990). In exchange for honeydew, ants provide multiple services that benefit honeydew-producing Hemiptera; primarily ants protect their partners from natural enemies (Way 1963, Buckley 1987, Buckley 1990, Stechmann *et al.* 1996, Queiroz and Oliveira 2001) and from competitors (Messina 1981, Mahdi and Whittaker 1993, Karhu 1998).

Not surprisingly, applied ecologists have focused their research almost exclusively on negative species interactions. By definition biological control is the application or manipulation of negative species interactions (parasitism, predation, infection, antagonism, or competition) to suppress a pest population and reduce damage caused by the pest (Van Driesche and Bellows 1996). Manipulating positive species interactions, interactions where one or more species benefit and thus increase in number, in a way that increases the abundance of an herbivorous insect seems counterintuitive to the goals of biological control. In ant-Hemipteran mutualisms, for example, it seems that increases in Hemipteran populations due to the mutualism would result in increased plant damage and reduced yield. As we detail in this chapter, however, surprisingly few studies support the idea that ant-Hemipteran mutualisms ultimately reduce crop yield/plant fitness. In many cases, larger populations of Hemipterans do not affect yield and the mutualism can result in increased predation of more damaging plant pests by ants. Thus, in many agricultural systems ant-Hemipteran mutualisms can result in enhanced biological control and increased plant production.

Understanding positive species interactions may also allow applied ecologists to make more accurate predictions about the effect of natural enemies and the population dynamics of natural enemies and herbivores in species rich, complex communities. Arthropod communities frequently contain many species and these species often directly and indirectly interact with each other (Polis and Strong 1996, Rosenheim 1998). It is inherently difficult to accurately predict how abiotic or biotic disturbances will affect particular species in highly connected, 'reticulate' communities (Polis and Strong 1996, Rosenheim 1998). In these systems, positive interactions may dramatically alter the abundance and distribution of many species in a predictable manner and, consequently, provide great insight into the dynamics of complex ecological systems (Stachowicz 2001, Bruno *et al.* 2003). Since predictable pest control is a major goal of IPM (Pedigo 2002), understanding or manipulating positive species interactions such as mutualisms may enhance sustainable IPM programs.

In this chapter we review the ecological consequences of ant-Hemipteran mutualisms and report our work on fire ant – aphid mutualisms in field and vegetable crops. Although we review research conducted in both managed and natural systems, we focus the chapter on the effects of these mutualisms relative to biological control (herbivore abundance and plant damage). We also explore the extent to which these mutualisms may help applied ecologists explain variation in species interactions in managed systems. Our goal is to highlight work that we think is underappreciated and potentially important to pest management and to highlight related research questions that need to be rigorously investigated.

3. MUTUALISMS EXPLAIN VARIATION IN FOOD WEB INTERACTIONS

Relatively few studies have focused on the community-wide effects of ant-Hemipteran mutualisms. Instead, most studies document the effects of the mutualisms on the survival of the predators of Hemipterans and subsequent changes in the abundance of aphids (e.g., Kaplan and Eubanks 2002). The few studies that have looked for broad changes in arthropod communities due to ant-Hemipteran mutualisms have found them. Protection from aphid predators by wood ants, *Formica aquilonia* (Hymenoptera: Formicidae), for example, results in an approximately 4000% increase in the population size of the honeydew-producing aphid *Symydobius oblongus* on birch trees (Karhu 1998). This mutualism, however, affects the abundance and distribution of other arthropods. Increased ant activity on aphid-infested birch trees results in a 50% reduction in the abundance of defoliating caterpillars, a 77% decrease in the abundance of aphids that do not produce honeydew, and similar reductions in the abundance of parasitoid wasps (Karhu 1998). Likewise, the mutualism between the ant *F. propinqua* and the aphid *Chaitophorus populicola* results in a 57% decrease in arthropod species richness and an 80% decrease in overall arthropod abundance on aphid-infested cottonwood trees (Wimp and Whitham 2001).

We have investigated the effects of ant-Hemipteran mutualisms on the arthropod community associated with cotton in the southeastern United States. Red imported fire ants, *Solenopsis invicta*, referred to simply as fire ants in this chapter, are widespread, invasive ants that are notoriously aggressive and voracious (Vinson 1997). They were originally introduced through the port of Mobile, Alabama, in the early 1900's and now range across the southern United States and have recently invaded parts of California (MacKay and Fagerlund 1997, Vinson 1997, Anonymous 1999). Fire ants can reach extremely high densities in disturbed areas and consume large numbers of other arthropods (Porter and Savignano 1990, Williams 1994, Vinson 1997). Fire ants, like almost all ant species, readily form facultative mutualisms with honeydew-producing insects (Vinson 1997). Fire ants are known to regularly tend aphids (Reilly and Sterling 1983a, 1983b, Vinson and Scarborough 1989), scales (Helms and Vinson 2002), and whiteflies (Morrill 1977).

We have shown that fire ants are very attracted to plants infested with aphids. Fire ants, for example, preferentially foraged on aphid-infested cotton plants ($\overline{x}$ = 103 ants per plant) compared to aphid-free cotton plants ($\overline{x}$ = 5 ants per plant) in a greenhouse experiment and fire ant and cotton aphid abundances were strongly correlated in cotton fields (Kaplan and Eubanks 2002, 2005). Fire ants are also very effective at protecting aphids from predators. Fire ant predation of lady beetle larvae was twice as high on aphid-infested cotton plants as on aphid-free cotton plants with a corresponding 100% increase in aphid survival (Kaplan and Eubanks 2002). Likewise, aphid predators were more than twice as abundant in cotton fields with suppressed densities of fire ants than in control fields with large fire ant populations and, consequently, cotton aphids were three to four times more abundant in fields with large fire ant populations (Kaplan and Eubanks 2002, 2005). This mutualism, however, also affects the abundance of non-aphid predators and other herbivores (Kaplan and Eubanks 2005). In greenhouse cage experiments the survival of herbivores (beet armyworm caterpillars) was dramatically lower in the presence of fire ants and aphids compared to fire ants alone. In a similar greenhouse experiment fire ants and aphids reduced the amount of time that plant bugs (*Lygus lineolaris*) (Hemiptera: Miridae) spent on plant foliage by almost 75%. Using large-scale field manipulations of fire ants along with spatially heterogeneous aphid populations, we found that the ant-aphid mutualism accounted for much of the variation underlying ant impact on cotton herbivores and predators. For example, in 2001 approximately 27% of herbivore taxa and 67% of predator taxa were adversely affected by the ant-aphid mutualism. Similarly, in 2002 approximately 33% of herbivore taxa and 50% of predator taxa were negatively affected. Our results indicate that the relationship between fire ants and aphids serves as a keystone interaction that dramatically alters the structure of cotton arthropod communities (Kaplan and Eubanks 2005).

4. MUTUALISM INCREASES PEST DENSITY AND INCREASES PEST DAMAGE

As mentioned in the introduction, it seems to be a widely held belief that ant-Hemipteran mutualisms ultimately result in increased damage to the plant via increased densities of the Hemipteran. In our review of the literature, however, we found very few studies that documented increased plant damage due to a mutualism (Table 1). It is not clear if the lack of examples in the literature is because this is an uncommon outcome of mutualisms in nature or if it reflects a bias by researchers (e.g., researchers may note ant tending when studying aphid pests, but may not attempt to document the effect of ant tending on aphid population dynamics and plant fitness). In one of the few studies conducted in a managed system, Banks and Macauley (1967) found that colonies of black bean aphids (*Aphis fabae*) tended by black garden ants (*Lasius niger*) were approximately 25% larger than colonies not attended by ants. Differences in aphid abundance affected plant yield; plants with ant-tended aphid colonies produced half as many seeds as plants with untended aphid colonies. Similarly, Sipura (2002) found that the mutualism between the ant *Formica aquilonia* and the aphid *Pterocomma salicis* on the willow *Salix myrsinifolia* resulted in damaging densities of aphids as well as increased levels of an herbivorous beetle that is chemically protected against ant predation. Kay *et al.* (2004) found that treehopper (*Publilia modesta*) (Homoptera: Membracidae) tending by the ant *Formica obscuripes* resulted in larger treehopper populations and more intensive nitrogen reduction in goldenrod plants (*Solidago gigantea*). Renault *et al.* (2005) found that a mutualism between the aphid *Aphis coreopsidis* and a *Camponotus* ant ultimately resulted in reduced production of viable seeds by the plant *Bidens pilosa.* Increased transmission of insect-vectored plant pathogens is another way that mutualisms between ants and honeydew-producing aphids, scales, and whiteflies could negatively affect the host plant. Recent work in our laboratory, for example, suggests that fire ant tending of several different aphid species may result in increased infection levels of tomato plants by the aphid-vectored *Cucumber mosaic virus* (Cooper, Murphy, and Eubanks unpublished data).

Table 1. Studies documenting net negative effect of ant-Hemipteran mutualism on biological control.

Ant mutualist	Hemipteran mutualist	Host plant	System	Plant trait measured	Reference
Lasius niger	*Aphis fabae* (Aphidae)	*Vicia faba* (Faba bean)	Managed: row crop	# viable seeds	Banks and Macaulay 1967
Iridomyrmex sp.	*Sextius virescens* (Membracidae)	*Acacia decurrens*	Natural: woodland	Plant growth, # seeds	Buckley 1983
Crematogaster brevispinosa	*Planococcus citri* Common citrus mealybug	*Schomburgkia tibicinis* (Orchidaceae)	Natural: forest	# flowers, fruits	Rico-Gray and Thien 1989
Camponotus planatus	Unidentified aphid sp.	*Paullinia fuscecens* (Sapindaceae)	Natural: seashore	% flowers with seeds, # seeds per flower	Rico-Gray and Castro 1996
Formica aquilonia	*Pterocomma salicis* (Aphidae)	*Salix myrsinifolia* (Willow sp.)	Natural: lakeshore	herbivory, plant growth	Sipura 2002
Formica obscuripes	*Publilia modesta* (Membracidae)	*Solidago gigantean* (Goldenrod sp.)	Natural: field	Nitrogen content	Kay *et al.* 2004
Camponotus sp.	*Aphis coreopsidis* (Aphidae)	*Bidens pilosa* (Aster sp.)	Natural: field	# seeds	Renault *et al.* 2005

Although some of these studies report fairly dramatic decreases in plant fitness as a consequence of ant-Hemipteran mutualisms, it is surprising that so few studies have documented these effects. Part of the reason may be that aphids and other Hemipterans cause relatively little per capita damage to their host plants, only resulting in significant reductions in growth and reproduction when at extremely high densities for either long periods of time or at critical stages of plant development (Rosenheim *et al.* 1997, Kaplan and Eubanks 2005, Styrsky and Eubanks unpublished data; see section 5 and 6 below). Aphids in particular often have highly variable population dynamics and typically only maintain extremely high densities for a few days or weeks when feeding on annual plants (Kaplan and Eubanks 2005). When honeydew-producing Hemipterans vector plant pathogens, however, the insects may ultimately cause dramatic losses of plant fitness even at low densities. It is in these cases that researchers are likely to find strong, consistent, negative effects of ant-Hemipteran mutualisms on plant fitness (Cooper, Murphy, and Eubanks unpublished data).

5. MUTUALISM DOES NOT AFFECT PLANT DAMAGE

There are few documented examples where ant-Hemipteran mutualisms appear to have limited effects on the population dynamics of the honeydew-producing insects and the mutualism does not alter host plant performance as a result (Table 2). In most of these cases, the herbivorous insects did not appear to suffer significant predation even without the ants, the herbivorous insects fed in a way that did not significantly damage the plants, or the honeydew-producing insect did not reach damaging densities even when tended by ants. For example, although the gamagrass leafhopper *Dalbulus quinquenotatus* (Homoptera: Cicadellidae) benefits from ant tending via reduced predation by spiders, populations of this leafhopper rarely if ever reach damaging densities in the field, probably as a result of incidental predation of leafhoppers by tending ants (Moya-Raygoza and Nault 2000). Likewise, the treehopper *Vanduzca arquata* (Homoptera: Membracidae) does not reach high densities on black locust trees even when tended by the highly aggressive ant *Formica subsericea* (Fritz 1983).

Table 2. Studies documenting lack of net negative or positive effect of ant-Hemipteran mutualism on biological control.

Ant mutualist	Hemipteran mutualist	Host plant	System	Plant trait measured	Reference
Formica subsericea	*Vanduzca arquata* (Membracidae)	*Robinia pseudoacacia* (Black locust)	Natural: old field	herbivory, plant growth	Fritz 1983
Brachymyrmex obscurior	*Dalbulus quinquenotatus* (Cicadellidae)	*Tripsacum pilosum* (Gamagrass)	Natural: field	# leaves, plant growth	Moya-Raygoza and Nault 2000

6. MUTUALISM INCREASES HEMIPTERAN DENSITY BUT DECREASES OVERALL PEST DAMAGE

Studies documenting a net positive effect of ant-Hemipteran mutualisms to biological control are more numerous (n = 14) than those reporting a net negative effect or no net effect (Table 3). In these studies, ant-Hemipteran mutualisms result in increased ant predation or harassment of other herbivores that are more damaging to the host plant than are the ant-tended Hemipterans. As a consequence, ant-Hemipteran mutualisms increased ant suppression of herbivory by these other herbivores, resulting in increased plant fitness in some cases (Table 3). *Formica* ants tending the Membracid *Publilia concava* on goldenrod (*Solidago altissima*), for example, attack and deter feeding by the adults and larvae of two Chrysomelid species, *Trihabda vigata* and *T. borealis*. As a consequence, goldenrod plants naturally infested with *P. concava* were less defoliated and produced more seeds than their nearest neighbors that did not host a Membracid colony. Another *Formica* ant species, *F. yessensis*, similarly protected the oak *Quercus dentata* as a consequence of tending the aphid *Tuberculatus quercicola* (Ito and Higashi 1991). Aphid-tending ants decreased the abundance of leaf-feeding Lepidopteran larvae resulting in reduced leaf loss on oak trees hosting the aphid. Additionally, although acorn production did not differ between trees that did and did not host the ant-tended aphid, the percentage of acorns damaged by acorn-boring Lepidopteran larvae was reduced on the trees that did host the ant-tended aphid, presumably because of increased ant predation.

Many of the studies that report a net positive effect of ant-Hemipteran mutualisms to biological control assess the benefit to the host plant in terms of reduced herbivory or increased vegetative growth (Table 3). Although suppression of herbivory and increased plant growth may be correlated with increased plant fitness, any reported benefit to biological control would be better assessed by measuring seed production, herbivore damage to seeds or fruit, or some other trait related more closely to plant fitness or yield. Of the fourteen published studies that report enhanced biological control in the presence of an ant-Hemipteran mutualism, only eight considered the effect of the mutualism on fitness traits such as seed production and damage to seeds or fruit, including the two studies described above. Further, five of these eight studies (and five of the fourteen studies total) are correlational, relying on natural patterns of Hemipteran presence and absence on plants to estimate the benefit of ant-Hemipteran mutualisms to biological control in terms of plant fitness or yield.

Of the nine studies that are experimental, all but two (Floate and Whitham 1994, Sipura 2002) manipulated the presence and absence of the ant-mutualist rather than Hemipteran-mutualist. Though perhaps logistically more difficult in some systems, manipulating Hemipteran abundance provides a more realistic test of the hypothesis that ant-Hemipteran mutualisms can benefit biological control because it is the ants that typically respond to the presence of honeydew-producing Hemipterans rather the Hemipterans responding to the presence of ants. Further, manipulating ants rather than Hemipterans makes it difficult to assess whether the effects of ants as predators is a truly a consequence of the ant-Hemipteran mutualism. Floate and Whitham (1994) manipulated the presence and absence of the aphid *Chaitophorus populicola* on cottonwood trees and demonstrated that, via their recruitment of *Formica* ants, aphids indirectly increased ant predation of eggs and larvae of a Chrysomelid beetle, resulting in a two-fold reduction in beetle herbivory.

Of the fourteen studies that report a net positive effect of ant-Hemipteran mutualisms to biological control, nine were conducted in natural systems and focused on plants species of little economic importance. The other five studies were conducted in managed systems and focused on tree crops (citrus, cocoa, sapodilla, and coconut). Ant-Hemipteran mutualisms have long been exploited to enhance biological control of more damaging, non-Hemipteran herbivores in some orchard systems. As early as the fourth century A.D., growers in China cultured the citrus ant (*Oecophylla smarigdina*) in orange orchards to establish ant-tended colonies of a mealybug species with the goal of increasing ant protection of developing fruit from more damaging pests (Huang and Yang 1987). Whether positive or negative, the consequences of ant-Hemipteran mutualisms may be more pronounced in orchard systems because of the greater stability of these habitats relative to more ephemeral crops and because trees in these systems often host Hemipterans such as scales that are well protected from natural enemies thus allowing long-term establishment of colonies.

Whether ant-Hemipteran mutualisms can enhance biological control in more ephemeral managed systems such as in row and vegetable crops is practically unknown. We have been investigating this possibility in the Southeastern cotton agroecosystem since 2002. Cotton plants (*Gossypium hirsutum*) serve as host plants for cotton aphids (*Aphis gossypii*), a honeydew-producing herbivore typically considered an economic pest. Red imported fire ants ('fire ants' from here on) are primarily ground foragers, but they are attracted to cotton aphid honeydew and readily ascend cotton plants to tend cotton aphid colonies. Because fire ants are extremely aggressive and incredibly abundant in cotton fields in the Southeast, their activity on cotton plants in the presence of cotton aphids disrupts biological control of the aphids but may result in increased ant suppression of herbivory by other more economically important pests. Cotton aphids generally only threaten yield at extremely high densities and at specific times during the growing season, whereas Lepidopteran larvae such as the cotton bollworm (*Helicoverpa zea*) (Noctuidae) and the beet armyworm (*Spodoptera exigua*) (Noctuidae), and other cotton pests such as the boll weevil (*Anthonomus grandis*) (Curculionidae) represent a greater risk to yield at lower densities.

Table 3. Studies documenting net positive effect of ant-Hemipteran mutualism on biological control.

Ant mutualist	Hemipteran mutualist	Host plant	System	Plant trait measured	Reference
Crematogaster africana	*Cataenococcus loranthi Stictococcus sjostedti*	*Tapinanthus bangwensis* (Mistletoe)	Managed: cocoa plantation	herbivory, plant growth	Room 1972
Azteca sp.	*Coccus viridis* (Soft green scale)	*Citrus* sp.	Managed: orchard	Herbivory	Jutsum *et al.* 1981
Formica sp.	*Publilia concava* (Membracidae)	*Solidago altissima* (Goldenrod sp.)	Natural: old field	plant growth, # seeds	Messina 1981
Formica rufa	*Periphyllus testudinaceus* (Aphidae)	*Acer pseudoplatanus* (Sycamore)	Natural: woodland	Herbivory, plant growth, # seeds	Whittaker and Warrington 1985
Formica yessensis	*Tuberculatus quercicola* (Aphidae)	*Quercus dentata* (Oak sp.)	Natural: coastal chaparral	# leaves, flowers, acorns	Ito and Higashi 1991
Dolichoderus thoracicus (Black cocoa ant)	*Cataenococcus hispidus* (mealybug)	*Theobroma cacao* (Cocoa)	Managed: plantation	Herbivore damage to fruits	Khoo and Ho 1992
Anopeplis custodiens	*Cerataphis variabilis* (Aphididae)	*Cocos nucifera* (Coconut)	Managed: plantation	Nutlet damage, nut set	Löhr 1992
Formica sp.	*Chaitophorus populicola* (Aphidae)	*Populus fremontii, P. angustifolia* (Cottonwood)	Natural: woodland	Herbivory	Floate and Whitham 1994
Camponotus brutus	*Hilda undata* (Tettigometridae)	*Ficus vallischoudae* (Fig)	Natural: woodland	Damage to unripe fruit	Dejean *et al.* 1997

Table 3 continued.

Ant mutualist	Hemipteran mutualist	Host plant	System	Plant trait measured	Reference
Formica aquilonia	*Symydobius oblongus* (Aphidae)	*Betula pubecens* (White Birch)	Natural: woodland	Herbivory	Karhu 1998
Dolichoderus thoracicus (Black cocoa ant)	*Planococcus lilacinus* (mealybug)	*Manilkara zapota* Sapodilla	Managed: plantation	Damage to fruits	Van Mele and Cuc 2001
Formica aquilonia	*Pterocomma salicis* (Aphidae)	*Salix phylicifolia* (Willow sp.)	Natural: lakeshore	Herbivory, plant growth	Sipura 2002
Lasius japonicus, *Tetramorium tsushimae*	*Aphis craccivora*, *Acyrthosiphon pisum*, *Megoura crassicauda*	*Vicia angustifolia* (Vetch sp.)	Natural: field	# seeds	Suzuki *et al.* 2004
Complex of five species	*Guayaquila xiphias* Membracidae	*Didymopanax vinosum* Araliaceae	Natural: cerrado savannah	Herbivory	Oliveira and Del-Claro 2005

We tested the hypothesis that the fire ant-cotton aphid mutualism can benefit cotton yield by increasing ant suppression of caterpillar pests in a two-year field cage experiment in which we manipulated cotton aphid density on plants in fields naturally infested with fire ants. Cotton plants in each cage were seeded with cotton aphids collected from the surrounding field. Beet armyworm caterpillars were applied to plants weekly at a rate of ten caterpillars per plant in 2003 and ten to thirty caterpillars per plant in 2004. We sampled each plant weekly during the period of flowering and boll production to record the number of aphids on six leaves, the total number of fire ants and caterpillars, the percentage leaf area consumed by caterpillars, and the number of cotton squares (flower buds), flowers, and bolls (cotton fruit). The field cages (2003: n = 30; 2004: n = 36) excluded all other herbivores and natural enemies.

Averaged over all sampling dates, fire ant abundance on plants increased significantly with cotton aphid abundance in both 2003 ($F_{1,28} = 3.69$, one-tailed $p = 0.03$, $R^2 = 0.12$; Fig. 2a) and 2004 ($F_{1,34} = 69.62$, one-tailed $p < 0.0001$, $R^2 = 0.67$; Fig. 2b) as a result of fire ants tending cotton aphids for their honeydew.

The increase in fire ant activity on plants with more aphids did not result in a detectable decrease in beet armyworm caterpillar abundance in 2003 ($F_{1,28} = 0.45$, one-tailed $p = 0.25$, $R^2 = 0.02$;Fig. 3a), but increased fire ant abundance on plants at least marginally significantly decreased caterpillar damage to leaves ($F_{1,28} = 2.40$, one-tailed $p = 0.06$, $R^2 = 0.08$; Fig. 3b). In 2004, increased fire ant abundance on plants significantly decreased both the abundance of beet armyworm caterpillars ($F_{1,34} = 14.73$, one-tailed $p = 0.0003$, $R^2 = 0.30$; Fig. 3c) and caterpillar damage to leaves ($F_{1,34} = 19.31$, one-tailed $p < 0.0001$, $R^2 = 0.36$; Fig. 3d).

The overall consequence to the plant of the aphid-induced increase in fire ant predation of caterpillars was positive in both years. Although cotton aphid abundance did not affect the number of cotton squares ($F_{1,28} = 0.23$, one-tailed p = 0.32, $R^2 = 0.01$; Fig. 4a) in 2003, increased aphid abundance did result in a greater number of cotton bolls ($F_{1,28} = 2.65$, one-tailed p = 0.05, $R^2 = 0.09$; Fig. 4b). In 2004, increased cotton aphid abundance increased both the number of squares ($F_{1,34} = 10.21$, one-tailed p = 0.002, $R^2 = 0.23$; Fig. 4c) and bolls ($F_{1,34} = 3.86$, one-tailed p = 0.03, $R^2 = 0.10$; Fig. 4d). These results suggest that the fire ant-cotton aphid mutualism can benefit plant yield by enhancing biological control of caterpillar pests by fire ants, at least at low to moderate densities of aphids and before cotton bolls open and the lint is exposed.

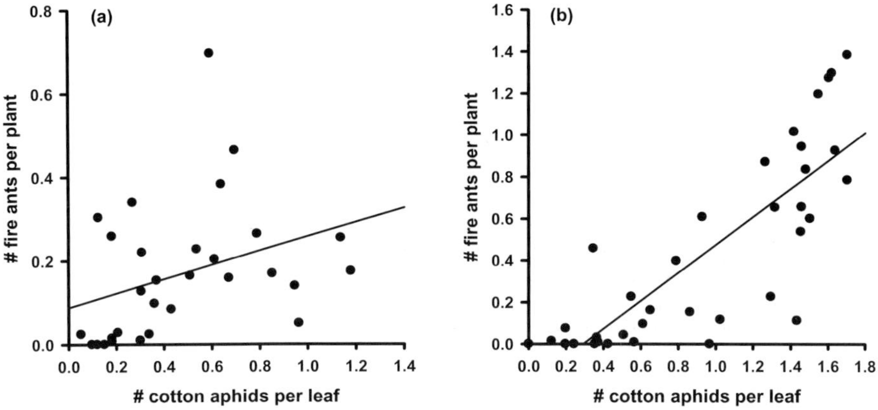

Figure 2. The effect of cotton aphid abundance on the recruitment of red imported fire ants onto cotton plants in field cages in (a) 2003 (n = 30 cages) and (b) 2004 (n = 36 cages). Each point represents mean abundance per plant averaged over four weeks and seven weeks in 2003 and 2004, respectively. Abundance data were log10(n+1)-transformed to meet assumptions of normality.

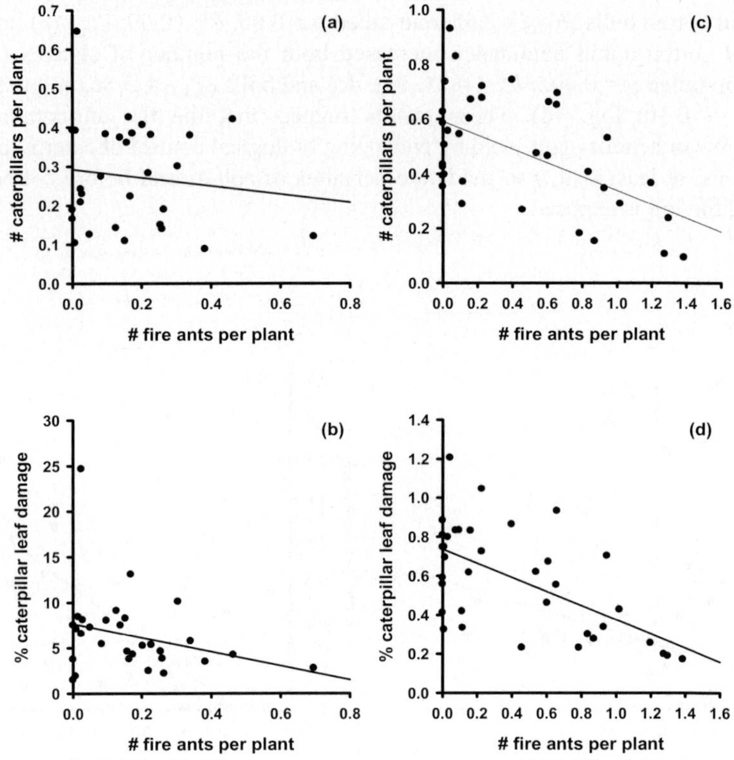

Figure 3. The effect of red imported fire ant abundance on cotton plants on the abundance of beet armyworm caterpillars and percent leaf herbivory by caterpillars in 2003 (a and b, respectively) and in 2004 (c and d, respectively). Each point represents mean abundance per plant averaged over four weeks and seven weeks in 2003 and 2004, respectively. Abundance data and 2004 herbivory data were log10(n + 1) transformed to meet assumptions of normality.

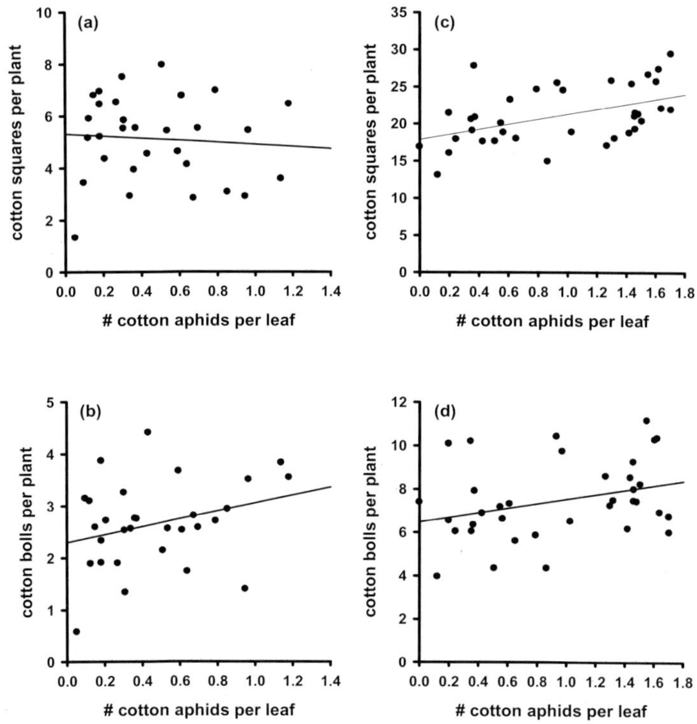

Figure 4. The indirect effect of cotton aphid abundance (via increased fire ant predation of caterpillars) on the number of squares and bolls in 2003 (a and b, respectively) and in 2004 (c and d, respectively). Each point represents mean abundance per plant averaged over four weeks and seven weeks in 2003 and 2004, respectively. Cotton aphid abundance data were log10(n + 1) transformed to meet assumptions of normality.

7. CONCLUSIONS

In both our own work and in our review of the literature we found that ant – Hemipteran mutualisms act as 'keystone interactions' in many terrestrial arthropod communities. The mutualism results in dramatic increases in the abundance of ants on plants and dramatic increases in their subsequent suppression of predaceous and herbivorous arthropods. In most cases, suppression of herbivores by Hemiptera-tending ants is beneficial to the plant. This is because most Hemiptera do little damage to their host plants relative to the damage caused by herbivores such as caterpillars, stinkbugs, etc. The results of this study suggest that ant – Hemipteran mutualisms could be manipulated (i.e., encouraged) in many agricultural systems to increase the efficacy of ants as biological control agents and that understanding the community-wide effects of these mutualisms may make biological control in species rich, highly connected arthropod communities more predictable.

8. REFERENCES

Anonymous, 1999, New fire ant endangers California agriculture. Calif Grow. 23(8): 18.

Bach, C.E., 1991, Direct and indirect interactions between ants (*Pheidole megacephala*), scales (*Coccus viridis*) and plants (*Pluchea indica*). Oecologia 87: 233-239.

Banks, C. J., and Macaulay, E. D. M., 1967, Effects of *Aphis fabae* Scop. and its attendant ants and insect predators on yields of field beans (*Vicia faba* L.). Ann Appl Biol. 60: 445-453.

Bertness, M. D., and G. H. Leonard, 1997, The role of positive interactions in communities: lessons from intertidal habitats. Ecology 78: 1976-1989.

Boucher, D. H., James, and K. H. Keeler, 1982, The ecology of mutualism. Ann Rev Ecol Syst. 13: 315-347.

Bronstein, J. L., 1994*a*, Conditional outcomes in mutualistic interactions. Trends Ecol Evol. 9: 214-217.

Bronstein, J. L., 1994*b*, Our current understanding of mutualism. Quart Rev Biol. 69: 31-51.

Bruno, J.F., 2000, Facilitation of cobble beach plant communities through habitat modification by *Spartina alterniflora*. Ecology 81: 1179-1192.

Bruno J.F., Stachowicz J.J., Bertness M.D., 2003, Inclusion of facilitation into ecological theory. Trends Ecol Evol. 18: 119–25.

Buckley, R., 1983, Interaction between ants and membracid bugs decreases growth and seed set of host plant bearing extrafloral nectaries. Oecologia 58: 132-136.

Buckley, R., 1987, Interactions involving plants, Homoptera, and ants. Ann Rev Ecol Syst. 18: 111-135.

Buckley, R., 1990, Ants protect tropical Homoptera against nocturnal spider predation. Biotropica 22: 207-209.

Chang, G. C., 1996, Comparison of single versus multiple species of generalist predators for biological control. Environ Entomol. 25: 207-213.

Clements, F. E., 1916, Plant succession: an analysis of the development of vegetation. Carnegie Inst Wash Pub. 242: 1-512.

Coll, M. and M. Guershon, 2002, Omnivory in terrestrial arthropods: Mixing plant and prey diets. Ann Rev Entomol 47: 267-297.

Chase, J. M., P. A. Abrams, J. P. Grover, S. Diehl, P. Chesson, R. D. Holt, S. A. Richards, R. M. Nisbet, and T. J. Case, 2002, The interaction between predation and competition: A review and synthesis. Ecology Letters 5: 302-315.

DeJean, A., T. Bourgoin, and M. Gibernau, 1997, Ant species that protect figs against other ants: result of territoriality induced by a mutualistic homopteran. Ecoscience 4: 446-453.

Denno, R. F.,M. S. McClure, and J. R. Ott, 1995, Interspecific interactions in phytophagous insects: competition revisited and resurrected. Ann Rev Entomol. 40: 297-331.

Douglas, A. E., 1993, The nutritional quality of phloem sap utilized by natural aphid populations. Ecol Entomol. 18: 31-38.

Elton, C. S., 1958, The ecology of invasion by animals and plants. Methuen, London.

Eubanks, M. D., 2001, Estimates of the direct and indirect effects of red imported fire ants on biological control. Biol Control 21: 35-43.

Eubanks, M.D., S.A. Blackwell, C.J. Parrish, Z.D. DeLamar, and H. Hull-Sanders, 2002, Intraguild predation of beneficial arthropods by red imported fire ants in cotton. Environ Entomol. 31: 1168-1174.

Eubanks, M.D. and J.D. Styrsky, 2005, Effects of plant feeding on the performance of omnivorous 'predators', in: Plant-Provided Food and Herbivore-Carnivore Interactions, F. Wäckers, P. van Rijn and J. Bruin, eds. Cambridge University Press, London, pp. 148-177.

Eubanks, M.D., 2005, Predaceous herbivores and herbivorous predators: The biology of omnivores and the ecology of omnivore – prey interactions, in: Ecology of Predator – Prey Interactions, P. Barbosa and I. Castellanos, eds. Oxford University Press, Oxford, pp. 3-17.

Floate, K. D., and T. G. Whitham, 1994, Aphid-ant interaction reduces chrysomelid herbivory in a cottonwood hybrid zone. Oecologia 97: 215-221.

Fritz, R. S., 1983, Ant protection of a host plant's defoliator: consequence of an ant-membracid mutualism. Ecology 64: 789-797.

Hairston, N. G., F. E. Smith, and L. B. Slobodkin, 1960, Community structure, population control and competition. Am Nat. 94: 421–425.

Harmon, J.P. and D.A. Andow, 2004, Indirect effects between shared prey: Predictions for biological control. Biocontrol 49: 605-626.

Helms, K. R., and S. B. Vinson, 2002, Widespread association of the invasive ant *Solenopsis invicta* with an invasive mealybug. Ecology 83: 2425-2438.

Hölldobler, B., and E. O. Wilson, 1990, The Ants. Harvard University Press, Cambridge.

Holway, D. A., L. Lach, A. V. Suarez, N. D. Tsutsui, and T. J. Case, 2002, The causes and consequences of ant invasions. Ann Rev Ecol Syst. 33: 181-233.

Huang, H.T. and P. Yang, 1987, The ancient cultured citrus ant. BioScience 37: 665-671.

Ito, F. and S. Higashi, 1991, An indirect mutualism between oaks and wood ants via aphids. J Anim Ecol. 60: 463-470.

Jutsum, A.R., J.M. Cherrett, M. Fisher, 1981, Interactions between the fauna of citrus trees in Trinidad and the ants *Atta cephalotes* and *Azteca* sp. J. App Ecol. 18: 187-195.

Kaplan, I. and M.D. Eubanks, 2002, Disruption of cotton aphid (Hompoptera: Aphididae)-natural enemy dynamics by red imported fire ants (Hymenoptera: Formicidae). Environ Entomol. 31: 1175-1183.

Kaplan, I. and M.D. Eubanks, 2005, Aphids alter the community-wide impact of fire ants. Ecology 86: 1640-1649.

Karhu, K.J., 1998, Effects of ant exclusion during outbreaks of a defoliator and a sap-sucker on birch. Ecol Entomol. 23: 185-194.

Kay, A.D., S.E. Scott, J.D. Schade, and S.E. Hobbie, 2004, Stoichiometric relations in an ant-treehopper mutualism. Ecol Letters 7: 1024-1028.

Khoo, K. C. and C.T. Ho, 1992, The influence of *Dolichoderus thoracicus* (Hymenoptera: Formicidae) on losses due to *Helopeltis heivora* (Heteroptera: Miridae), black pod disease, and mammalian pests in cocoa in Malaysia. Bull Entomol Res. 82: 485-491.

Koss, A.M. and W.E. Snyder, 2005, Alternative prey disrupt biocontrol by a guild of generalist predators. Biol Control 32: 243-251.

Laine K.J. and P. Niemela, 1980, The influence of ants on the survival of mountain birches during an *Oporinia autumnata* (Lep., Geometridae) outbreak. Oecologia 47: 39-42.

Lawton, J. H., and D. R. Strong, 1981, Community patterns and competition in folivorous insects. Am Nat. 118: 317-338.

Löhr, B., 1992, The pugnacious ant, *Anopeplis custodiens* (Hymenoptera: Formicidae), and its beneficial effect on coconut production in Tanzania. Bull Entomol Res. 82: 213-218.

MacArthur, R. H., 1955, Fluctuations of animal populations and a measure of community stability. Ecology 36: 533-536.

MacKay, W.P. and R. Fagerlund, 1997, Range expansion of the red imported fire ant, *Solenopsis invicta* Buren (Hymenoptera: Formicidae), into New Mexico and extreme western Texas. Proc Entomol Soc Wash. 99: 757-758.

Mahdi, T., and J. B. Whittaker, 1993, Do birch trees (*Betula pendula*) grow better if foraged by wood ants? J Anim Ecol. 62: 101-116.

Menalled, F.D., A.C. Costamagna, P.C. Marino, and D.A. Landis, 2003, Temporal variation in the response of parasitoids to agricultural landscape structure. Agric Ecosyst Environ. 96: 29-35.

Messina, F. J, 1981, Plant protection as a consequence of an ant membracid mutualism: interactions on goldenrod (*Solidago* sp.). Ecology 62: 1433-1440.

Mills, L. S., M. E. Soule, and D. F. Doak, 1993, The keystone-species concept in ecology and conservation. BioScience 43: 219-224.

Mittler, T. E, 1958, Studies on the feeding and nutrition of *Tuberolachnus salignus*. II. The nitrogen and sugar composition of ingested phloem sap and excreted honeydew. J Exp Biol. 35: 74-84.

Moya-Raygoza, G. and R. Nault, 2000, Obligatory mutualism between *Dalbulus quinquenotatus* (Homoptera: Cicadellidae) and attendant ants. Ann Entomol Soc Am. 93: 929-940.

Murdoch, W. W., J. Chesson, and P. L. Chesson, 1985, Biological control in theory and practice. Am Nat. 125: 344-366.

Nickerson, J. C., C. A. Rolph Kay, L. L. Buschman, and W. H. Whitcomb, 1977, The presence of *Spissistilus festinus* as a factor affecting egg predation by ants in soybeans. Flor Entomol. 60: 193-199.

Oliveira, P.S. and K. Del-Claro, 2005, Multitrophic interactions in a neotropical savannah: ant-hemipteran systems, associated insect herbivores and a host plant, in: Biotic Interactions in the Tropics: Their Role in the Maintenance of Species Diversity, D. Burslem, M. Pinard, and S. Hartley, eds. Cambridge University Press, Cambridge, pp. 414-438.

Paine, R. T. 1969, A note on trophic complexity and community stability. Am Nat. 103: 91-93.

Pedigo, L. P., 2002, Entomology and Pest Management, 4th ed., Prentice Hall, Upper Saddle River.

Polis, G. A., 1991, Complex trophic interactions in deserts: an empirical critique of food-web theory. Am Nat. 138: 123-155.

Polis, G. A., Myers, C. A., and Holt, R. D., 1989, The ecology and evolution of intraguild predation: potential competitors that eat each other. Ann Rev Ecol Syst. 20: 297-330.

Polis, G. A., and Strong, D. R., 1996, Food web complexity and community dynamics. Am Nat. 147: 813-846.

Porter, S. D., and D. A. Savignano, 1990, Invasion of polygyne fire ants decimates native ants and disrupts arthropod community. Ecology 71: 2095-2106.

Power, M. E., and D. Tilman, 1996, Challenges in the quest for keystones. Bioscience 46: 609-621.

Queiroz, J. M., and P. S. Oliveira, 2001, Tending ants protect honeydew-producing whiteflies (Homoptera: Aleyrodidae). Environ Entomol. 30: 295-297.

Reilly, J.J. and Sterling, W.L., 1983a, Interspecific association between the red imported fire ant (Hymenoptera: Formicidae), aphids, and some predaceous insects in a cotton agroecosystem. Environ Entomol. 12: 541-545.

Reilly, J.J. and Sterling, W.L., 1983b, Dispersion patterns of the red imported fire ant (Hymenoptera: Formicidae), aphids, and some predaceous insects in east Texas cotton fields. Environ Entomol. 12: 380-385.

Renault, C.K., L.M. Buffa, and M.A. Delfino, 2005, An aphid-ant interaction: effects on different trophic levels. Ecol Res. 20: 71-74.

Rico-Gray, V. and G. Castro, 1996, Effect of an ant-aphid interaction on the reproductive fitness of *Paullinia fuscecens* (Sapindaceae). Southwest Nat. 41: 434-440.

Rico-Gray, V. and L.B. Thein, 1989, Ant-mealybug interaction decreases reproductive fitness of *Schomburgkia tibicinis* (Orchidaceae) in Mexico. J Trop Ecol. 5: 109-112.

Romeis, A., D. Babendreier, F.L. Wackers, and T.G. Shanower, Habitat and plant specificity of *Trichogramma* egg parasitoids – underlying mechanisms and implications. Bas App Ecol 6: 215-236.

Room, P.M., 1972, The fauna of the mistletoe *Tapinanthus bangwensis* (Engl. and K. Krause) growing on cocoa in Ghana: relationships between fauna and mistletoe. J Anim Ecol. 41: 611-621.

Rosenheim, J. A., 1998, Higher order predators and the regulation of insect herbivore populations. Annu Rev Entomol. 43: 421-447.

Rosenheim, J. A., 2001, Source-sink dynamics for a generalist insect predator in habitats with strong higher-order predation. Ecol Monogr 71: 93-116.

Rosenheim, J. A., Wilhoit, L. R., and Armer, C. A., 1993, Influence of intraguild predation among generalist insect predators on the suppression of an herbivore population. Oecologia 96: 439-449.

Rosenheim, J. A., Kaya, H. K., Ehler, L. E., Marois, J. J., and Jaffee, B. A., 1995, Intraguild predation among biological-control agents: Theory and evidence. Biol Control 5: 303-335.

Sanford, E., 1999, Regulation of keystone predation by small changes in ocean temperature. Science 283: 2095-2097.

Sipura, M., 2002, Contrasting effects of ants on the herbivory and growth of two willow species. Ecology 83: 2680-2690.

Stachowicz, J. J., 2001, Mutualisms, positive interactions, and the structure of ecological communities. BioScience 51: 235-246.

Stechman, D. H., W. Volkl, and P. Stary, 1996, Ant attendance as a critical factor in the biological control of the banana aphid, *Pentalonia nigronervossa* Coq. (Hom. Aphididae) in Oceania. J App Entomol. 120: 119-123.

Strong, D. R., Jr., J. H. Lawton, and T. R. E. Southwood, 1984, Insects on plants: community patterns and mechanisms. Harvard University Press, Cambridge.

Symondson, W.O.C., K.D. Sunderland, and M.H. Greenstone, 2002, Can generalist predators be effective biocontrol agents. Ann Rev Entomol. 47: 561-594.

Suzuki, N., K. Ogura and N. Katayama, 2004, The efficiency of herbivore exclusion by ants attracted to aphids on the vetch *Vicia angustifolia* L. (Leguminosae). Ecol Res. 19: 275-282.

Teder, T., M. Tanhuanpaa, K. Ruohomaki, P. Kaitaniemi, and J. Henriksson, 2000, Temporal and spatial variation of larval parasitism in non-outbreaking poupulations of a folivorous moth. Oecologia 123: 516-524.

Van Driesche, R. G. and T.S. Bellows, 1996, Biological control. Chapman and Hall, New York.

Van Mele, P. and N.T.T. Cuc, 2001, Farmers' perceptions and practices in use of *Dolichoderus thoracicus* (Smith) (Hymenoptera: Formicidae) for biological control of pests of *Sapodilla*. Biol Control 20: 23-29.

Vinson, S. B., 1997, Invasion of the red imported fire ant (Hymenoptera: Formicidae): spread, biology, and impact. Am Entomol. 43: 23-39.

Vinson, S.B. and Scarborough, T.A., 1989, Impact of the imported fire ant on laboratory populations of cotton aphid (*Aphis gossypii*) predators. Flor Entomol. 72: 107-111.

Way, M. J., 1963, Mutualism between ants and honeydew-producing Homoptera. Ann Rev Entomol. 8: 307-344.

Way, M. J. and K. C. Khoo, 1992, Role of ants in pest management. Ann Rev Entomol. 37: 479-503.

Whittaker, J.B. and S. Warrington, 1985, An experimental field study of different levels of insect herbivory induced by *Formica rufa* predation on sycamore (*Acer pseudoplatanus*). III. Effects on tree growth. J App Ecol. 22: 797-811.

Williams, D.F., 1994, Exotic ants: Biology, Impact, and Control of Introduced Species. Westview Press, Boulder.

Wimp, G. M. and T. G. Whitham, 2001, Biodiversity consequences of predation and host plant hybridization on an aphid-ant mutualism. Ecology 82: 440-452.

INTERSPECIFIC COMPETITION AMONG NATURAL ENEMIES AND SINGLE VERSUS MULTIPLE INTRODUCTIONS IN BIOLOGICAL CONTROL

Nick Mills

Department of Environmental Science, Policy and Management, Mulford Hall, University of California, Berkeley, CA 94720-3114

Abstract. One of the most challenging questions in biological control has been whether the single best natural enemy will provide greater suppression of the abundance of an arthropod pest than a combination of natural enemies due to the effects of competition for a shared resource (the pest). From a theoretical perspective, simple predator-prey models clearly indicate that for multiple natural enemies sharing a single pest, only a single natural enemy will persist in the system, the species that can reduce the pest to the lowest equilibrium density. In addition, the biological control record of natural enemy introductions against invasive insect pests suggests that establishment rates are higher for projects with single versus multiple introductions, and that competitive exclusion can result from a sequence of parasitoid introductions against a pest. From natural field populations, however, it is clear that insect herbivores frequently support a diverse assemblage of both parasitoid and predator species suggesting that resource partitioning can mediate the effects of competition. So how important are competitive interactions in the context of biological control? Here, I consider the asymmetries of competition among natural enemies of arthropod pests, the incidence of competitive exclusion, the mechanisms of coexistence and their consequences for the success of biological control, and the need for a more experimental approach to the study of competition among natural enemies.

1. INTRODUCTION

Biological control represents the action of living natural enemies in suppressing the abundance or activity of pests. At its most dramatic in the context of introductions of natural enemies from the region of origin of invasive pests, biological control also includes the periodic release of individuals for immediate or season-long suppressions of pests, habitat enhancement for conservation of natural enemies, and the natural biological control provided by natural enemies in both production and wildland ecosystems (Van Driesche and Hoddle, 2003). As a naturally-occurring ecosystem service, biological control has been loosely valued at U.S. $400 billion per year (Costanza *et al.*, 1997), and has been considered as one of the most important in the context of human welfare and food security (Kremen and Chaplin, 2006).

Arthropod individuals represent discrete packets of limiting resources for the development of natural enemies, and provide ample opportunity for competition

J. Brodeur and G. Boivin (eds.), Trophic and Guild Interactions in Biological Control, 191–220.

among the natural enemy species that form an assemblage in association with any particular host (Mills, 1999). Although a host individual represents only a small part of the resources needed by a predator to complete its development, a single host is sufficient to allow the complete development of parasites, including parasitoids, parasitic nematodes and microbial pathogens. Thus competition between natural enemies can be intense at the individual level and can include players that span different kingdoms (Hochberg and Lawton, 1990).

Following Mills (1999), interspecific competition is defined as the negative effects that one species has upon another by consuming or controlling access to a resource that is limited in availability or through susceptibility to a shared natural enemy. Interspecific competition can be mediated either through exploitative (scramble) competition, in which competitors interact through the consumption of a diminishing supply of an essential resource, through interference (contest) competition, in which the activity of one species reduces the access of a competitor to a limited but essential resource, or through apparent competition, in which competitors share a common natural enemy. At an individual level, competition has frequently been shown to have strong negative impacts on the performance of competing species of natural enemies, and has been particularly well documented for interactions among predators (e.g., Obrycki *et al.*, 1998), among parasitoids (e.g., Wang and Messing, 2004), and between parasitoids and pathogens (e.g., Nguyen *et al.*, 2005). In the case of interactions among parasitoids, Smith (1929) further distinguished between extrinsic competition among adult parasitoids in their search for hosts, and intrinsic competition among parasitoid larvae in acquiring resources from a single host individual.

Biological control introductions have generated some classic examples of competitive exclusion among parasitoid species (Bess *et al.*, 1961; Murdoch *et al.*, 1996), and more recently of habitat displacement of native coccinellids following the introductions of exotic species, such as *Coccinella septempunctata* and *Harmonia axyridis* into the North American continent (Alyokhin and Sewell, 2004; Evans, 2004). In contrast, natural communities provide us with ample evidence for the coexistence of natural enemies through mechanisms such as niche partitioning (Tagaki and Hirose, 1994), behavioural segregation through aggregation (Bonsall *et al.*, 2004), spatial productivity gradients (Amarasekare, 2000), or shared natural enemies (van Nouhuys and Hanski, 2000).

In this chapter I will begin by considering the three key mechanisms of interspecific competition among natural enemies, and subsequently the theory and experimental evidence for the impact of competition among natural enemies in the context of biological control. In addition, the focus of this chapter will be on mechanisms other than intraguild predation (IGP), as IGP is reviewed in detail by Rosenheim and Harmon (this volume) and Janssen *et al.* (this volume).

2. MECHANISMS OF COMPETITION AMONG NATURAL ENEMIES

2.1. Exploitative Competition

To illustrate the three different mechanisms of interspecific competition among natural enemies, I adopt Levins' (1973) interaction networks in which positive influences are represented by arrows and negative influences by clubs (Fig. 1). Exploitative competition (Fig. 1a), by definition, occurs among all natural enemies that share a common and limiting host, and is characterized by an absence of direct interaction between competitors, but a negative impact that is mediated through a limiting availability of common hosts (Grover, 1997). The competing natural enemies both gain from the host, and at the same time have a negative impact on the availability of the host for other natural enemies. Thus exploitative competition occurs between individuals of the same species as well as between individuals of different species.

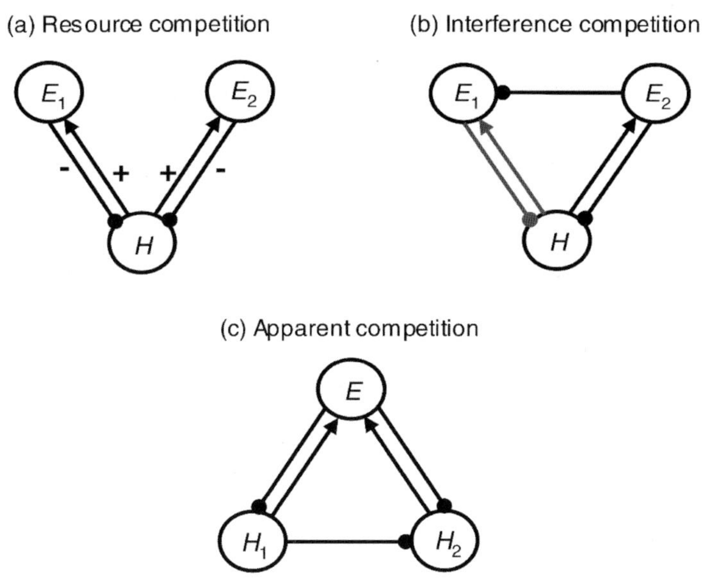

Figure 1. Interaction networks for (a) exploitative, (b) interference, and (c) apparent competition. The circles show hosts (H) and natural enemies (E), and the linking lines show benefits (+, arrows), losses (–, clubs), and strong (black) and weak (grey) interaction strengths.

For interactions between predators it has proved difficult to separate the effects of exploitative competition from those of IGP, although for both acarine predators (Walzer *et al.*, 2002; Onzo *et al.*, 2003) and coccinellids (Obrycki *et al.*, 1998; Yasuda *et al.*, 2004) it appears that IGP has a stronger influence in the context of short-term laboratory experiments. In addition, in the case of coccinellids, there is some evidence that a disparity in body size may increase the likelihood of IGP (Yasuda *et al.*, 2004). In contrast, for parasitoids and pathogens the distinction is more obvious and has been studied in greater detail. The outcome of intrinsic (within-host) interactions can often be symmetrical, particularly when development times of the natural enemies are equivalent. For example, several studies of competition between aphid parasitoids show no clear intrinsic advantage between the first and second species to parasitize the host. The outcome of multiple parasitism then depends on the relative timing of attack, with the species that oviposits first tending to be the victor (Völkl and Stadler, 1991; Persad and Hoy, 2003). Similarly, the outcome of competition between the two whitefly parasitoids *Amitus fuscipennis* (Hymenoptera: Platygasteridae) and *Encarsia formosa* (Hymenoptera: Aphelinidae) favours the species that oviposits first with no consistent advantage of either species (de Vis *et al.*, 2003). In many cases, intrinsic interactions between parasitoids and pathogens are also symmetrical. For example, Hochberg (1991) showed that the presence of a granulovirus reduced the combined weight of the brood of *Cotesia glomerata* (Hymenoptera: Braconidae) by 28%, whereas the parasitoid reduced the reproductive output of the virus by 29%. In addition, the outcome of the exploitative competition depended on the relative timing of the completion of parasitoid development in relation to host death from viral infection. Similar results have been obtained for other parasitoid-pathogen interactions involving microsporidia, fungi and viruses (Brooks, 1993 [for a review]; Down *et al.*, 2005; Kim *et al.*, 2005; Nguyen *et al.*, 2005).

Under other circumstances, intrinsic interactions among natural enemies can be highly asymmetrical and predictable, most notably when the development times of the two natural enemies are distinctly different. Larval ectoparasitoids often develop more rapidly on their hosts than larval endoparasitoids, and in cases of multiple parasitism, are frequently the superior competitors through exploitative competition. Force (1970) showed that the larval ectoparasitoids *Torymus baccharidis* and *T. koebeli* (Hymenoptera: Torymidae) were competitively superior to the egg-prepupal endoparasitoid *Platygaster californica* (Hymenoptera: Platygasteridae), and Zaviezo and Mills (2001) document the superiority of the larval ectoparasitoid *Hyssopus pallidus* (Eulophidae) over the egg-prepupal endoparasitoid *Ascogaster quadridentata* (Braconidae). Similarly, among scale insect hosts, ectoparasitoids are also competitively superior to endoparasitoids (Flanders, 1971), as shown for various *Aphytis* (Hymenoptera: Aphelinidae) species, as ectoparasitoids, in comparison to *Coccophagoides* (Aphelinidae), *Encarsia* (Aphelinidae) and *Pteropterix* species, as endoparasitoids (Huffaker and Kennett, 1966; Gerson, 1968; Steinberg *et al.*, 1987; Yu *et al.*, 1990).

2.2. Interference Competition

Interference competition results from the direct interaction of competitors or from the modification of a resource by one competitor such that it becomes unusable by another (shown by the - interaction of E_2 on E_1 and the diminished interaction between E_1 and H in Fig. 1b). In the simplest case, interference among natural enemy species represents a contest between competitors where the victim is excluded from access to the host through direct fighting or conditioning of the host as a resource. Thus success in contest competition is often a specific characteristic of the biology of a natural enemy species although it can also be influenced by the relative timing of attack (see Boivin and Brodeur this volume). In addition, as with exploitative competition, interference competition can occurs intraspecifically as well as interspecifically. However, when interference occurs among predator species (Rosenheim, 1998) or between a facultative hyperparasitoid and a primary parasitoid (Amarasekare, 2000; Hunter and Woolley, 2001) the contest frequently leads to IGP, a more complex case, in which the victor gains resources directly from its competitor (see Rosenheim and Harmon this volume, Janssen *et al.* this volume).

There are numerous examples of direct fighting or antagonism among parasitoid species that parasitize a common host (see Boivin and Brodeur this volume). The fighting or antagonism can be extrinsic, between adult females as they search for hosts to parasitize or even between adult females and juveniles, or it can be intrinsic, between larvae that hatch from eggs in or on the same host individual. An interesting example of extrinsic interference between searching female parasitoids has been documented from direct observations of parasitoids searching in the field for bark beetle larvae on ash logs in Switzerland (Mills, 1991). Surprisingly, the smaller females of *Cheiropachus quadrum* (Pteromalidae) and *Eurytoma morio* (Eurytomidae) were more aggressive than the larger females of *Coeloides filiformis* (Braconidae) and were able to displace the latter from suitable oviposition sites on the surface of the bark. Even more aggressive interference has been observed under laboratory conditions among the bethylid parasitoids of coffee berry borer that guard their hosts following oviposition (Perez-Lachaud *et al.*, 2002; Batchelor *et al.*, 2005). *Cephalonomia stephanoderis* is the most aggressive species and was able to win contests against *C. hyalipennis* as the prior owner of the host, and against *Prorops nasuta* as either owner or intruder, with the losing female being killed in 69% of the contests. Ovicide has also been documented as a form of extrinsic interference between ovipositing female parasitoids and juvenile competitors. Although ovicide has more frequently been documented in cases of intraspecific competition (Strand and Godfray, 1989; Netting and Hunter, 2000), Pedata *et al.* (2002) provide evidence that female *Encarsia formosa* killed 20% of the eggs of *E. pergandiella* laid 72 h earlier in whitefly nymphs.

The most frequent form of interference among parasitoids, however, is intrinsic, occurring among larvae that share an individual host. The relative rarity of interspecific discrimination among parasitoid species leads to frequent examples of multiple parasitism, the occurrence of immature stages of more than one parasitoid species in or on the same individual host. As individual hosts can only support the development of a single solitary parasitoid (see Boivin and Brodeur this volume for exceptions) there is intense competition between parasitoid immatures that takes the form of physical attack or physiological suppression. Of these two forms of larval competition, physical attack in which mandibulate first instar parasitoid larvae are able to bite and kill competitors, is by far the most frequent. In some cases the outcome of the contest is highly asymmetrical and predictable, as in all examples of cleptoparasitism (Mills, 1994a, 2003). In other cases the order of oviposition or the interval between ovipositions influences the outcome (Strand, 1986; Mackauer, 1990). For example, larvae of *Cotesia plutellae* can consistently kill larvae of *Diadegma semiclausum* (Ichneumonidae) unless the latter precedes the former by two days or more (Shi *et al.*, 2004). In some cases, solitary parasitoids are consistently superior to gregarious species (which are often thought to be less aggressive), as for the solitary *Cotesia rubecula* and gregarious *C. glomerata* (Laing and Corrigan, 1987). However, there are several recent examples that show gregarious parasitoids to be aggressive and superior competitors when sharing a host with a solitary species (Boivin and van Baaren, 2000; Marktl *et al.*, 2002; Pexton and Mayhew, 2004). Although most frequently reported among hymenopteran parasitoids, physical attack has also been documented from competition among dipteran parasitoids (Schmid-Hempel and Schmid-Hempel, 1996) and coleopteran parasitoids (Royer *et al.*, 1999).

Physiological suppression as a form of intrinsic interference provides a means of controlling access to the host through mechanisms that are poorly understood (Vinson and Hegazi, 1998), and in general, seems to be less common than physical attack in interspecific interactions among parasitoids (Fisher, 1971; Godfray, 1994). However, both Hågvar (1988) and Mackauer (1990) suggest, from observations of parasitism of aphids, that eggs of an *Aphidius* (Braconidae) species may be physiologically suppressed through oviposition by an *Ephedrus* (Aphidiidae) species. Similarly, physiological suppression has been shown to be an effective mechanism for the fruit fly parasitoid *Fopius arisanus* (Braconidae) to eliminate its competitors, with 83% of the eggs of *Diachasmimorpha kraussii* (Braconidae), 80% of the eggs of *Psyttalia concolor* (Braconidae) (Wang and Messing, 2002), and 90% of the eggs of *D. longicaudata* (Wang *et al.*, 2003) killed in the presence of *F. arisanus* larvae in multiparasitized hosts.

2.3. Apparent Competition

Although predation has long been known to mediate the coexistence of competing prey species (Chase *et al.*, 2002), Holt (1977) was the first to show that prey species can also compete for natural enemy free space. In contrast to exploitative competition, in which two natural enemy species compete directly for a shared host, apparent competition results from the indirect interaction of two host species that share a common natural enemy (Fig. 1c). For example, a more abundant host species can have a strong indirect negative impact on a less abundant host species by supporting a large natural enemy population that can use both hosts. In this way the less abundant host experiences a greater level of predation that it would in the absence of its competitor.

Documentation of apparent competition among natural enemies appears to be infrequent in the literature. However, van Nouhuys and Hanski (2000) demonstrated the potential for apparent competition among parasitoids of the Glanville fritillary butterfly in Finland through a shared cocoon hyperparasitoid *Gelis agilis* (Ichneumonidae). In this study a one-time addition of cocoon clusters of *Cotesia glomerata*, a parasitoid of *Pieris* butterflies, was made to field sites supporting larval nests of the Glanville fritillary. Following this addition, the abundance of cocoons of *C. melitaearum*, a parasitoid of the Glanville fritillary, declined, in some cases to extinction. While hyperparasitism of *C. glomerata* and *C. melitaearum* was not monitored in this study, the most likely mechanism for the decline of *C. melitaearum* is apparent competition via *G. agilis*. A second study provides evidence of the potential for apparent competition between aphid parasitoids mediated by a shared hyperparasitoid (Morris *et al.*, 2001). In this study, pots of beans with *Acyrthosiphon pisum* parasitized by *Aphidius ervi* and of nettles with *Microlophium carnosum* parasitized by *A. microlophii* were exposed to hyperparasitism in the field. A short-term, but significant, effect of the presence of *A. microlophii* on the frequency of hyperparasitism of *A. ervi* by a suite of eight hyperparasitoid species was observed, again indicating a potential for apparent competition.

Both aphidiine mummies and microgastrine cocoons are exposed on leaf surfaces and are well known to be susceptible to high level of hyperparasitism (Sullivan and Völkl, 1999), and these are perhaps the most likely situations in which to expect apparent competition to occur at the third trophic level. Another situation in which apparent competition might be expected to occur is among the multiple species of coccinellids and chrysopids that attack aphid colonies. While the focus in these communities has been on IGP (Brodeur and Rosenheim, 2000; Lucas, 2005), these predators are often common, occur at differential levels of abundance, and show differential susceptibility to commonly-occurring shared parasitoids.

3. COMPETITION AND BIOLOGICAL CONTROL: THEORY

3.1. Competition for a shared and limiting resource

The familiar Lotka-Volterra competition model is a continuous-time model that describes the competitive interaction between two species with logistic population growth. The classical properties of this model are either competitive exclusion, coexistence, or a priority effect (see Mills, 1999). Competitive exclusion occurs when the ratio of the carrying capacities of the two competitors (winner to loser) exceeds the interspecific competitive effect of the weaker competitor. Coexistence occurs if competition generates a stronger intraspecific than interspecific influence on population growth for both species. Then finally, a priority effect, such that either species can exclude the other when present first, occurs if competition generates a stronger interspecific than intraspecific influence on population growth for both species. The Lotka-Volterra competition model makes three important assumptions, first, that the limiting resource is constant (and thus the model fails to distinguish between interference and exploitative competition), second, that the zero-growth isoclines for each competitor are linear (whereas most population processes are inherently nonlinear), and third, that the environment is homogenous with a uniform distribution of the limiting resource.

To address the first assumption, Tilman (1982, 1990) extended the Lotka-Volterra competition model to consider the importance of a dynamic resource population and exploitative competition. This led to the R^* rule, which states that in the absence of mediating factors exploitative competitors cannot coexist, and that the competitively superior species is the one that is more efficient in exploiting the resource and able to drive resource densities to the lowest level (see Grover, 1997 for a recent review). This leads to competitive exclusion, as confirmed by Murdoch *et al.* (2003) in the context of exploitative competition between parasitoids, and by Murdoch *et al.* (1996) to account for the competitive exclusion of *Aphytis lingnanensis* by *A. melinus* in the biological control of California red scale (see below). Although competitive exclusion is the typical outcome of exploitative competition, Briggs *et al.* (1993) provide evidence of a limited range of conditions for parasitoid coexistence associated with temporal niche partitioning in the attack of different host stages. Similarly, both Amarasekare (2000) and Bonsall *et al.* (2002) show some limited potential for life history and ecological trade-offs to mediate competition among parasitoids and promote coexistence.

The distinction between interference and exploitative competition is often overlooked, but has been addressed by Vance (1984) and Amarasekare (2002). Vance (1984) extended the Lotka-Volterra competition model with its implicit constant resource to show that interference can promote coexistence if each species interferes more with its own resource acquisition than that of its competitor, and that the interference is sufficient to outweigh the greater efficiency of the superior exploitative competitor. When a dynamic resource population is included, however, a simple trade-off in superiority between exploitative and interference competition is no longer sufficient to support coexistence, although it can lead to priority effects in the exclusion of competitors (Amarasekare, 2002). It is only when interference leads to some positive benefit, as in the gain of resources through IGP, that coexistence becomes more likely. As noted earlier, interference competition among natural enemies, particularly among predators, often leads to IGP, and the interplay between exploitative competition and IGP is discussed in detail by Holt *et al.* (1994), Grover and Holt (1998) and Rosenheim and Harmon (this volume).

The second assumption of linear zero-growth isoclines was questioned by Ayala *et al.* (1973) when they found that the Lotka-Volterra model could not accurately predict the equilibrium densities observed from experimental populations of *Drosophila* (Diptera: Drosophilidae) species. These authors point out that the zero-growth isoclines for each competitor are more likely to be concave than linear, which does not affect the predicted set of outcomes from interspecific competition, but does influence the equilibrium abundance of the competitors under conditions of coexistence. The competition coefficients of the Lotka-Volterra model imply that the per capita effect on each population of adding an individual of the competing species, relative to the effect of adding an individual of the same species, remains constant at all densities. More generally, however, per capita competition effects are not constant, but vary with the relative abundance of competitor populations leading to nonlinearity in the equilibrium densities (Abrams, 1987). Other common sources of nonlinearity, that generate concave zero-growth isoclines, arise from the functional response for resource acquisition (Vandermeer *et al.*, 2002), and intraspecific density dependence (Ayala *et al.*, 1973).

Shorrocks *et al.* (1979) addressed the third assumption of a homogeneous environment by extending the Lotka-Volterra model to include a refuge from competition. A refuge allows for a proportion of individuals of each species to have non overlapping distributions in a fragmented environment, and to occupy patches exclusively of their competitors. Similarly, Hochberg and Hawkins (1993) showed how refuges could promote coexistence among parasitoids that share a common host. By reducing interspecific relative to intraspecific competition, the inclusion of refuges allows for greater coexistence between competitors. The refuge effect was further developed by Ives and May (1985) and Ives (1988) using aggregated distributions of competitors to generate refuges, confirming and generalizing the earlier observations that aggregation and refuges generally promote coexistence between competitors both for exploitative and interference competition. Similarly, May and Hassell (1981) developed a discrete-time model for competition between two parasitoids with an aggregated distribution of attack among individuals of a common host, and Hochberg *et al.* (1990) developed a model for the competition between a parasitoid and a pathogen. Aggregated attacks by the parasitoid typically led to coexistence between two parasitoids (May and Hassell, 1981), but to a broader range of outcomes, from exclusion through priority effects to coexistence, for competition between a parasitoid and a pathogen (Hochberg *et al.*, 1990).

Kakehashi *et al.* (1984) pointed out that coexistence among parasitoids was supported only if the aggregation of attacks was independent for each parasitoid (independent niches), with exclusion of the inferior competitor as the outcome for a common distribution of aggregation for both parasitoids (identical niches). Interestingly, coexistence in the May and Hassell (1981) model required the two parasitoids to limit their own population growth more than that of their competitor (intraspecific greater than interspecific), a condition qualitatively similar to the properties of the Lotka-Volterra model. Klopfer and Ives (1997) subsequently explored the effect of other forms of aggregation on parasitoid coexistence, finding that although direct host density-dependent and inverse host density-dependent aggregation can also facilitate parasitoid coexistence, it is host density-independent aggregation, as used by May and Hassell (1981), that has the greatest positive influence on coexistence.

From a series of spatially-explicit models of competition, including both patch occupancy models (Hastings, 1980; Nee and May, 1992) and lattice models (Hassell *et al.*, 1994; Comins and Hassell, 1996) it has become clear that coexistence can also result form a trade-off between competition and colonization. Two natural enemies can then coexist if the inferior within-patch competitor (fugitive species) is the superior colonizer. As in the case of aggregation, explicit spatial structure provides a refuge for a fugitive species through the mechanism of niche partitioning. In addition to spatial structure, environmental disturbance can promote the coexistence of fugitive species. Using a patch occupancy model Barradas and Cohen (1994) show that disturbance of intermediate frequency can promote stability even in the absence of a colonization advantage to compensate for competitive inferiority. However, the promotion of coexistence by intermediate disturbance is a complex process that can operate at a variety of scales through a variety of mechanisms, and deserves closer attention in the future (Roxburgh *et al.*, 2004; Shea *et al.*, 2004).

To further explore the influence of niche partitioning and within-host competition on the suppression of host densities, Pedersen and Mills (2004) combined the niche partitioning model of Kakehashi *et al.* (1984) with the parasitoid species richness model of Hochberg (1996) to generate a model that can be used to explore all known forms of antagonistic interactions among parasitoids (Mills, 2003). The model has the form:

$$N_{g+1} = \lambda N_g h_N (sf_{PN} + tf_{PN} f_{QN} + uf_{QN} + v) \tag{1}$$

$$P_{g+1} = c\lambda N_g (s[1 - f_{PN}] + t[1 - f_{PN}]f_{QP}) \tag{2}$$

$$Q_{g+1} = c\lambda N_g (u[1 - f_{QN}] + t[1 - f_{QN}]f_{PN} + t[1 - f_{PN}][1 - f_{QP}]) \tag{3}$$

where N, P and Q are the abundance of host, primary and interactive parasitoid respectively, λ is the per capita rate of increase of the host population, h_N is a density dependence function acting on the host population, f_{YX} is the escape response representing the escape of X from parasitism by Y, c is the number of parasitoids produced per host parasitized, and s, t, u, and v are regions of niche overlap between the host and parasitoids representing the proportions of hosts uniquely accessible to P, to both P and Q, to Q only, and to neither parasitoid respectively. A Ricker function was used for host density dependence, and the Getz and Mills (1996) function that includes egg limitation and aggregated attacks for the escape responses. This model provides the opportunity to examine the influence of a host refuge (v), parasitoid niche overlap (t) and parasitoid interactions (Q is a later-attacking interactive species that can either compete for or gain resources from the earlier-attacking primary parasitoid *P*) on the suppression of host abundance.

To illustrate this point, consider a pest *N* with a moderate rate of population increase ($\lambda = 2$) to a carrying capacity ($K = 10{,}000$) that exceeds an economic injury level, and two possible scenarios for parasitoid introductions. In the first scenario, a solitary endoparasitoid *P* with low search efficiency ($a_{PN} = 0.25$) is introduced, but fails to provide sufficient control due to the presence of a host refuge from parasitism ($v = 0.5$). A second solitary parasitoid *Q* is introduced that can break into the host refuge from parasitism by *P*, but is interactive with *P* through exploitative competition and either facultative hyperparasitism (greater search rate for primary than secondary hosts, $a_{QN} = 0.75$, $a_{QP} = 0.25$). For the second scenario, endoparasitoid *P* has a greater search efficiency ($a_{PN} = 0.25$), but still fails to generate sufficient control, and so a refuge-breaking parasitoid *Q* is introduced that interacts with *P* through cleptoparasitism (lower search rate for primary than secondary hosts, $a_{QN} = 0.25$, $a_{QP} = 0.75$). Would the introduction of either form of *Q* be detrimental to the contribution of *P* to control of the pest, or are there any circumstances in which two parasitoids together could provide better control than *P* alone? One might imagine that the addition of a mild hyperparasitoid could perhaps lead to improved control, but that such strong cleptoparasitism would be particularly detrimental.

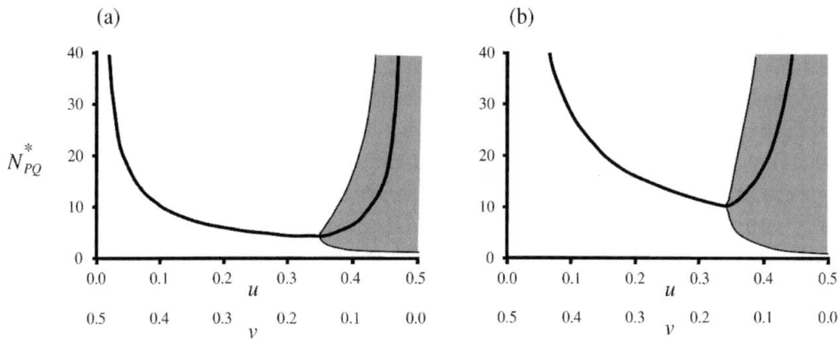

*Figure 2. A bifurcation diagram from Pedersen and Mills (2004) of the equilibrium pest density (N*PQ with both parasitoids present) after the introduction of (a) a facultative hyperparasitoid Q (aQN = 0.75, aQP = 0.25) that breaks the pest refuge from parasitism by a primary parasitoid P with lower search efficiency (aPN = 0.25), and (b) a cleptoparasitoid Q (aQN = 0.25, aQP = 0.75) that breaks the pest refuge from a primary parasitoid P with high search efficiency (aPN = 0.75). Along the x-axis, the host refuge from parasitism (v) decreases from 0.5 to 0 while the proportion of pests available to attack only by the interactive parasitoid Q (u) increases from 0 to 0.5. The grey region indicates loss of stability and increasing pest population fluctuations as the refuge from parasitism is lost.*

In this context it might be surprising to find that both forms of *Q*, as a second introduction, lead to a substantial reduction in pest abundance provided that a sufficient host refuge remains to stabilize the interaction. From the first scenario (Fig. 2a), in the presence of the primary parasitoid *P* alone the pest equilibrium density (N^*_P) is reduced to 303.37, but subsequent introduction of the facultative hyperparasitoid *Q* leads to substantially greater control. For example, the pest equilibrium density in the presence of both parasitoids (N^*_{PQ}) drops to below 10 with *Q* able to break into 20% of the pest refuge from parasitism by *P* (i.e., $u = 0.1$ and v declines from 0.5 to 0.4). While the degree of pest suppression is positively linked to the refuge breaking ability of *Q*, the stability of the three species equilibrium is lost as the refuge from parasitism by both parasitoids becomes too small ($v < 0.15$). From the second scenario (Fig. 2b), a more efficient *P* can reduce pest equilibrium density to 218.29, but the refuge-breaking ability of even a strong cleptoparasitoid, as a subsequent introduction *Q*, can again lead to substantial improvement in control. Under this second scenario, the reductions in pest equilibrium density that result from *Q*'s incursion into the pest refuge from parasitism by *P* are not quite as good as under the first scenario, but still far greater than can be achieved through increasing the search efficiency of the single primary parasitoid *P*. As suggested by Hochberg and Hawkins (1994), refuge breaking by a parasitoid is a particularly effective way to achieve greater pest suppression, and can outweigh the negative interactions between parasitoid species that occur where their niches overlap (Pedersen and Mills, 2004).

3.2. Competition for enemy free space

In addition to competing for shared resources, natural enemies can also compete for enemy free space. The outcome of apparent competition is analogous to that of exploitative competition, in that it also leads to the exclusion of one of the two competitors. Apparent competition leads to the P^* rule, in which the victor is the host species that is more efficiently exploited by the natural enemy and supports the greatest natural enemy densities (Holt and Lawton, 1993). Of course, hosts may compete through the combined effects of both exploitative and apparent competition. From a theoretical perspective (Holt *et al.*, 1994) this leads to the corresponding R^{**} and P^{**} rules that under an assumption of equivalent predation, competitive exclusion is also frequent and the winning host simultaneously persists at the lowest resource densities and supports the greatest natural enemy densities. However, if the shared natural enemy shows some preference between the two host species, the outcome of the competition is more complex and depends on a trade-off between susceptibility to predation (superiority in apparent competition) and efficiency in extraction of resources (superiority in exploitative competition). Apparent competition, similarly to resource mediated competition, is more frequently asymmetrical than symmetrical in its impact on competing species, which can result either from differential host quality or enemy preference (Chaneton and Bonsall, 2000), or simply from the effects of egg limitation in case of a shared parasitoid (Heimpel *et al.*, 2003).

4. COMPETITION AND BIOLOGICAL CONTROL: EVIDENCE

4.1. Competitive Displacement

In a recent review, Reitz and Trumble (2002) documented 42 cases of interspecific competition leading to competitive displacement, and of these 14 involved natural enemies used in the biological control of insect pests. The most widely known and most extensively documented example of competitive displacement from the biological control record concerns the biological control of California red scale (DeBach and Sundby, 1963; Rosen and DeBach, 1979; Luck and Podoler, 1985; Murdoch *et al.*, 1996). The introduction to California in 1947 of the parasitoid *Aphytis lingnanensis* from India led to satisfactory control of red scale along the coast, eliminating the earlier accidentally introduced *A. chrysomphali.* But *A. lingnanensis* failed to provide sufficient control of California red scale in the inland valleys, and so an additional parasitoid *A. melinus* was imported from China in 1957. In coastal citrus groves where scale had been adequately suppressed, *A. lingnanensis* coexisted with *A. melinus*, but for groves in the inland valleys *A. lingnanensis* had been completely excluded by 1972 and scale densities were more effectively suppressed.

The displacement of *A. lingnanensis* by *A. melinus* occurred surprisingly rapidly, taking from 1-3 years or 2-9 scale generations, and over extensive areas. Interestingly, from laboratory studies *A. lingnanensis* appeared to be the better competitor, as it is intrinsically superior in larval competition and has an extrinsically greater search rate (Rosen and DeBach, 1979). Although both species were subsequently found to avoid multiple parasitism, *A. melinus* was also found to be able to produce female progeny from younger stages of scale than was the case for *A. lingnanensis*, and to be able to produce two progeny from largest-size scale in comparison to one for *A. lingnanensis* (Luck and Podoler, 1985). Using a stage structured host-parasitoid model, Murdoch *et al.* (1996) show that the former characteristic is sufficient to account for the displacement of *A. lingnanensis* by *A. melinus*, and that both characteristics help to account for the rapidity of the displacement and the degree of scale suppression observed following the release of *A. melinus*.

There are a number of other examples of competitive displacement among parasitoid species. Displacement of one introduced parasitoid by another, as in the *Aphytis* species on California red scale, include the displacement of *Fopius vandenboschi* by *F. arisanus* in Hawaii (Bess *et al.*, 1961), of *Anagyrus antoninae* (Encyrtidae) by *Neodusmetia sangwani* (Encyrtidae) in Texas (Schuster and Dean, 1976), of *Aphidius smithi* by *A. ervi* in North America (Mackauer and Khambhampati, 1986), of *Cotesia glomerata* by *C. rubecula* in western North America (Wilkinson, 1966; Biever, 1992), of *Aphytis holoxanthus* by *Pteropterix smithi* in Israel (Steinberg *et al.*, 1987), of *Bathyplectes curculionis* (Ichneumonidae) by *B. anurus* in eastern North America (Harcourt, 1990), and of *Trichogrammatoidea* (Trichogrammatidae) species by *Copidosoma floridanum*

(Encyrtidae) in New Zealand (Howarth, 1991). In other cases, introduced parasitoids have led to the displacement of indigenous parasitoids, such as the displacement of *Lysiphlebus* (Braconidae) species by *L. testaceipes* in the Mediterranean (Tremblay, 1984), of *Pseudhomalopoda prima* (Encyrtidae) in Florida and of *A. costalimai* in Brazil by *Aphytis holoxanthus* (Bennett, 1993), and of *Praon pequodorum* (Braconidae) by *A. ervi* in North America (Schellhorn *et al.*, 2002). In many cases, the superior competitor under field conditions cannot be predicted from the outcome of intrinsic competition at the level of an individual host, as extrinsic factors acting at the population level, such as search efficiency, availability of adult food, host density effects, and tolerance of environmental conditions, frequently reverse the competitive superiority (Steinberg *et al.*, 1987)

Recent introductions of coccinellids into the North American environment have also led to competition with native coccinellids, causing a reduction in the abundance and habitat usage of the native species. Although *Coccinella septempunctata*, *Harmonia axyridis* and *Propylea quatordecimpunctata* have all been deliberately released as biological control agents on various occasions, it is believed that they inadvertently colonized the continent around seaports where they are likely to have been contaminants of ship cargoes (Day *et al.*, 1994). Despite the fact that coccinellid populations are notoriously variable in abundance from year to year (Evans, 2000), several sets of observations from field crops in South Dakota (Elliot *et al.*, 1996), from apple orchards in West Virginia (Brown and Miller, 1998), from potato fields in Maine (Alyokhin and Sewell 2004) and from alfalfa fields in both Manitoba (Turnock *et al.*, 2002) and Utah (Evans, 2004) show a consistent decline in the abundance of native coccinellid species following the rise to dominance of one or other of the introduced species. The longest-term data set is that of Alyokhin and Sewell (2004) who present data from 1971-2001 on visual counts of adult coccinellids from weekly sampling of 25 plants in each of 5 - 30 plots each year. Before 1980 two native coccinellids *Coccinella transversoguttata* and *Hippodamia tredecimpunctata* dominated, but the relative abundance of the introduced *C. septempunctata* increased from 6% in 1980 to 100% in 1994. Subsequently, both *H. axyridis* and *P. quatordecimpunctata* have become a consistent component of the coccinellid assemblage in these potato plots and the relative abundance of *C. septempunctata* has consequently declined. *C. transversoguttata* was not seen at all during the last three years of the surveys, while *H. tredecimpunctata* remains a small and variable component of the assemblage. The authors note that there was a significant decline in the abundance of aphids in the potato plots both in 1980 when *C. septempunctata* appeared, and again after both *H. axyridis* and *P. quatordecimpunctata* arrived in 1993-1995. These declines suggest that the exotic coccinellids may have been better exploiters of the aphids present on this exotic crop, and that exploitative competition may have played a role in the habitat displacement of native species by the exotic competitors (Evans, 2004), although it is possible that the exotic species may also be more effective intraguild predators (Yasuda *et al.*, 2004).

In all cases of competitive displacement, whether of introduced species by others or of indigenous species by introduced species, and whether it is parasitoids or predators, the displacement has been associated with increased suppression of the pest population. In many cases the mechanism for the displacement is suggested to be exploitative competition or an extrinsic superiority in the exploitation of the host or prey even though the displaced species may be intrinsically superior in larval competition. In addition, in terms of non-target environmental impacts, it is interesting to note that for the displaced indigenous natural enemies this represents a retreat back to their natural habitat from a more recent expansion of their distribution and range to include exotic hosts on exotic crops (Bennett, 1993; Evans, 2000, 2004). In their natural habitat they are likely to continue to be extrinsically superior in the use of indigenous hosts on indigenous plants in contrast to the exotic invaders.

4.2. Competitive Coexistence

In contrast to the strong theoretical prediction of competitive exclusion from exploitative and interference competition, the general field experience is that multiple natural enemies coexist, at least on a seasonal basis, on shared hosts. There are many examples of at least short-term coexistence on a local scale of parasitoids, predators and pathogens (e.g., Phoofolo *et al.*, 2001; Choate and Rieske, 2005; Nakashima and Akashi, 2005), although the dominance of a particular group can vary with host feeding niche (Hawkins *et al.*, 1997), and the additivity of the mortality can vary with the specific nature of the players and interactions (Snyder *et al.*, 2005). As an example, it is well known that the majority of host insects support a suite of coexisting parasitoid species. Hawkins and Lawton (1987) reported that the mean number of primary and secondary parasitoid species per host species in Britain is 4.0 for 76 species of Homoptera and 9.4 for 87 species of Lepidoptera. Thus we must conclude that there are a number of factors that mitigate competition in real systems through a variety of mechanisms.

Coexistence between natural enemies due, at least in part, to a greater role of self limitation in comparison to interspecific competition has been implicated as a mechanism for the coexistence of the two coccinellids *Adalia bipunctata* and *Coleomegilla maculata* in corn fields in Minnesota (Schellhorn and Andow, 1999). Observations suggested that the species-specific oviposition behaviour of the two coccinellids, *A. bipunctata* at the tops of the plants and *C. maculata* near the bottom, resulted in niche partitioning and a level of cannibalism of eggs by foraging adults that was at least equal to, if not greater than, interspecific predation of eggs.

Although temporal segregation has frequently been cited as a possible form of resource partitioning within parasitoid assemblages (Plantard *et al.*, 1996; Aukema *et al.*, 2004; Herz and Heitland, 2005) it seems less likely to mitigate exploitative competition, as the attack of different life stages of a host can readily lead to preemptive competition favouring the earlier-attacking species. However, the temperature-dependent seasonal switch in dominance of *Aphytis maculicornis* and *Coccophagoides utilis* as parasitoids of olive scale (Rochat and Gutierrez, 2001) provides an interesting example of temporal partitioning from the biological control literature.

Another suggested mechanism for coexistence is host density specialization, as occurs between introduced parasitoids of arrowhead scale where *Aphytis yanonensis* is more effective at higher host densities due to its shorter generation time and higher reproductive potential, in comparison with *Coccobius fulvus* (Hymenoptera: Aphelinidae) that is more effective in maintaining scale populations at a low density (Takagi and Hirose, 1994). It is not clear, however, that competition can be effectively avoided through either temporal or host density partitioning in host-parasitoid models (Murdoch *et al.*, 2003).

Although self limitation, temporal and host density partitioning may account for occasional examples of coexistence among natural enemies, the most frequent mechanisms are likely to be spatial partitioning and intermediate disturbance. From a 19 year study of parasitism of cabbage root fly involving both a specialist (*Trybliographa rapae*) (Hymenoptera: Figitidae) and a generalist (*Aleochara bilineata*) (Coleoptera: Staphylinidae) parasitoid, Bonsall *et al.* (2004) suggest that coexistence was made possible by the aggregated distribution of parasitism of the specialist in contrast to the random distribution of parasitism of the generalist, creating refuges for each competitor through the differential distributions of parasitoid attack. Similarly, coexistence can result from differential attack of hosts on different parts of a host plant, such as bark thickness in relation to parasitoid ovipositor length for parasitism of cerambycid larvae (Paine *et al.*, 2000), different heights on cotton plants for parasitoids of silverleaf whitefly (Bográn *et al.*, 2002), and leaves versus stems of citrus linked to substrate preference for parasitoids of California red scale (Borer *et al.*, 2004). A differential response of aphidophagous predators to host plants, perhaps based on colour, also provides evidence of spatial niche partitioning (Lorenzetti *et al.*, 1997). In addition, trade-offs in ecological characteristics have been noted, such that an intrinsically inferior parasitoid in terms of larval fighting can either be extrinsically superior in terms of search efficiency (Pschorn-Walcher, 1987), or longevity (Bonsall *et al.*, 2002). Similarly, Amarasekare (2000) suggests that coexistence of two egg parasitoids of the harlequin bug may result from a tradeoff between exploitative and interference competition, although in this case interference takes the form of IGP. There have been few studies of intermediate disturbance in the context of biological control, but Szentkiralyi and Kozar (1991) have looked at the intensity of pesticide application on the diversity of arthropods in apple orchards in Hungary. In this study although the species richness of apple pests was limited by disturbance, that of the natural enemies was influenced more by the diversity of the surrounding vegetation than by disturbance. In contrast, Brown and Schmitt (2001) found evidence in support of a greater diversity of insect parasitoids in apples disturbed by insecticide treatments in comparison to untreated apples.

The consequences of natural enemy coexistence in the context of biological control are generally considered beneficial or inconsequential. For example, the combination of two parasitoids established on olive scale in California is considered complementary (Rochat and Gutierrez, 2001), and either complementary (Takagi and Hirose, 1994) or of no consequence (Matsumoto *et al.*, 2003) for arrowhead scale in Japan. In addition, experimental studies with different combinations of whitefly parasitoids have shown that despite the potential for interference competition, almost all combinations resulted in equally good short-term suppression of whitefly abundance (Bográn *et al.*, 2002; Hunter *et al.*, 2002). Similarly, in the context of augmentative biological control, despite the potential for strong competitive interactions at an individual level, interactions at a population level appear to have been of little importance to the goal of host suppression for various combinations of parasitoids and coccinellids for whitefly control in the field (Heinz and Nelson, 1996), predators of whiteflies in glasshouses (Lucas and Alomar, 2002), predators of aphids (Chang, 1996), and predators of spider mites (Schausberger and Walzer, 2001). Other field studies of coexisting parasitoids in a variety of agroecosystems have shown an abundance of hosts, a low incidence of multiple parasitism, and no evidence for competition among parasitoid species (Miller, 1980; Flanders and Oatman, 1987; Takasu *et al.*, 1998). In contrast, Force (1974) provides evidence from the parasitoid assemblage of a gall-forming midge that an intrinsically inferior competitor *Platygaster californica* could suppress host densities more effectively alone than in competition with larval ectoparasitoids. Although Briggs and Latto (2001) were unable to verify this observation from a short term (two month) field cage study, they did find that *P. californica* alone produced the most parasitoid offspring as the larval ectoparasitoids killed a large fraction of hosts without offspring production, which suggests that *P. californica* may indeed be more effective in the longer term. Thus the time scale of experiments looking at competition among natural enemies can be of particular importance (see Janssen *et al.* this volume). A similar observation was obtained from a detailed study of the coexistence of two egg parasitoids *Trissolcus murgantiae* (Hymenoptera: Scelionidae) and *Ooencyrtus johnsonii* (Encyrtidae) of the harlequin bug (Amarasekare, 2000, 2003). Only the more efficient host exploiter *T. murgantiae* was able to persist at the lowest host densities, while coexistence with the intrinsically superior *O. johnsonii* (through IGP) was possible at higher host densities. Coexistence reduced the efficiency of *T. murgantiae*, suggesting that as in the case of *P. californica*, the more efficient host exploiter alone is likely to be able to reduce host densities to a greater extent than in the presence of intrinsically superior competitors.

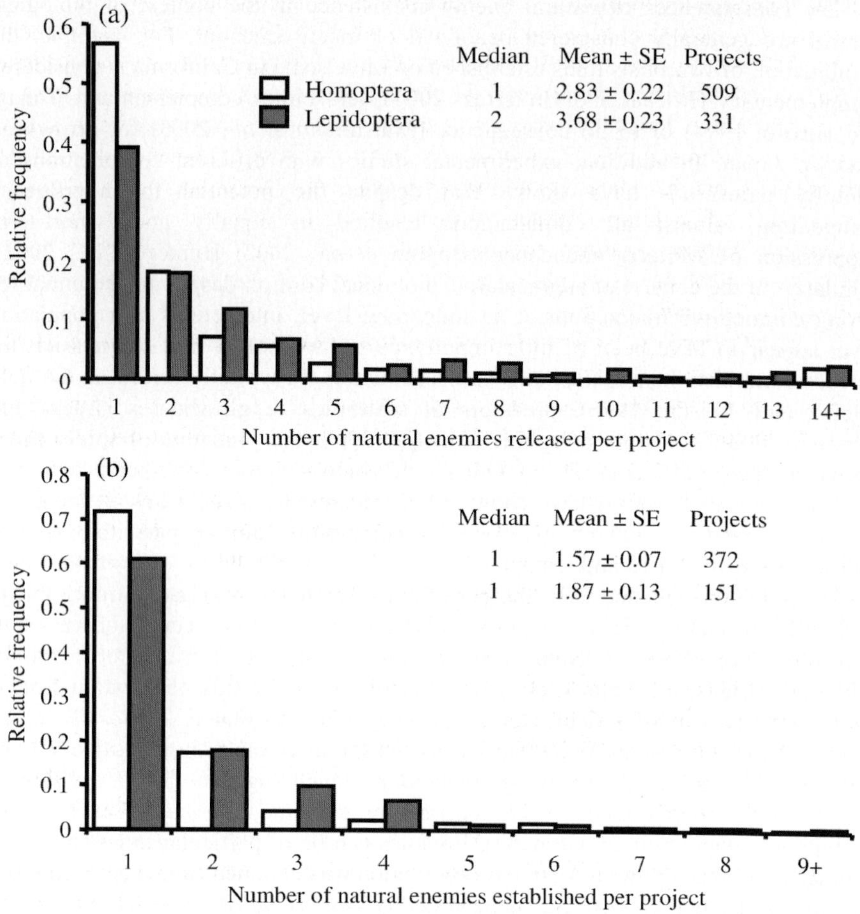

Figure 3. A comparison of the relative frequency of the number of natural enemies (a) introduced and (b) established against Homoptera and Lepidoptera in classical biological control projects worldwide (from the BIOCAT database, Greathead and Greathead, 1992).

4.3. Competition and Multiple Introductions

There has been a long debate regarding the merit of single versus multiple introductions of natural enemies in classical biological control (see Denno and Finke this volume). Ehler and Hall (1982) were the first to suggest that the rate of establishment of exotic natural enemies introduced against invasive pests was inversely related to the number of natural enemy species already present. This observation was challenged by Keller (1984) who suggested that competition is unlikely to be of importance during colonization, when pests remain abundant, and that the inverse relationship may have been driven by factors other than competition. Subsequently, Denoth *et al.* (2002) questioned whether the number of natural enemy species released against a pest influences the rate of establishment, and whether single versus multiple natural enemy introductions leads to greater success in biological control. From 108 projects against insect pests, they found that natural enemy establishment was significantly greater for single rather than multiple introductions, but that there was no improvement in the rate of success in relation to the number of natural enemies introduced. From this they concluded that competition between natural enemies could be an important constraint in the biological control of insect pests, in that it may limit the probability of natural enemy establishment.

From the record of classical biological control (Greathead and Greathead, 1992), the greatest difference in overall success rates has been observed between introductions against Homoptera and Lepidoptera (Mills, 1994b, 2006; Kimberling, 2004). If competition is an important constraint on the success of introduced natural enemies, then it might be expected to be most readily apparent through a comparison of the results of introductions against these two groups of pests. Here, I use the BIOCAT database of worldwide natural enemy introductions up to 1990 (Greathead and Greathead, 1992) to test whether single versus multiple introductions have had a differential effect on establishment and control in projects targeting Homoptera and Lepidoptera.

From 509 projects against homopteran pests and 331 projects against lepidopteran pests, the number of natural enemies introduced in a single project has varied from 1 to 34 for Lepidoptera and from 1 to 53 for Homoptera. A comparison of the relative frequency of introductions of increasing numbers of natural enemies in biological control projects for these two groups of pests indicates that although single species introductions have been slightly more frequent for homopteran than lepidopteran pests (Fig. 3a), the distributions are not significantly different ($\chi^2 = 133$, $df = 121$, $P = 0.22$). Of 372 natural enemy establishments against homopteran pests and 151 establishments against lepidopteran pests, the number of natural enemies established in a single project has varied from 1 to 14 for Lepidoptera and 1 to 12 for Homoptera. The corresponding relative frequencies of natural enemies established show a clear predominance of single species establishments for both pest groups (Fig. 3b). Interestingly, the drop off in relative frequency with an increasing number of natural enemies established differs significantly for the two groups of pests ($\chi^2 = 60$, $df = 42$, $P = 0.04$), dropping more rapidly for homopteran than for lepidopteran pests.

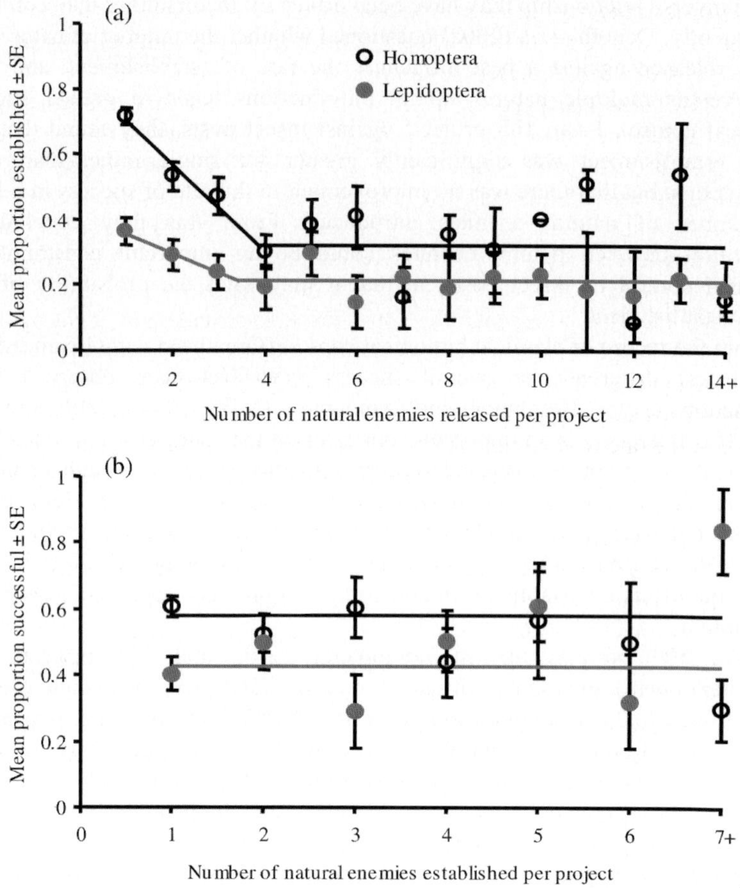

Figure 4. A comparison of the proportion of natural enemies that (a) became established in relation to the numbers introduced, and (b) became successful in relation to the numbers established, against Homoptera and Lepidoptera in classical biological control projects worldwide (from the BIOCAT database, Greathead and Greathead, 1992). Establishment in relation to natural enemies introduced (regression for 1-4 species introduced only), Homoptera $y = 0.85 - 0.14x$, $R^2 = 0.96$, *Lepidoptera* $y = 0.42 - 0.06x$, $R^2 = 0.96$; *(mean ± SE for 4-14+ species introduced), Homoptera 0.32 ± 0.03, Lepidoptera 0.21 ± 0.03. Success in relation to natural enemies established (mean ± SE), Homoptera 0.58 ± 0.02, Lepidoptera 0.43 ± 0.04.*

For smaller numbers of natural enemies introduced, up to a threshold of four, their rate of establishment (proportion established) declined linearly in relation to the number introduced both for Homoptera and Lepidoptera (Fig. 4a). For Homoptera, establishment rate declined from 0.72 to 0.30 for one to four species introduced. The corresponding decline for Lepidoptera was 0.36 to 0.19, indicating the lower overall establishment of natural enemies for this group of pests. Beyond a threshold of four natural enemies introduced, the rate of establishment showed some variation, but was essentially constant, although significantly different (Mann-Whitney, $P = 0.001$) for both groups of pests at 0.32 for Homoptera and 0.21 for Lepidoptera. These data are consistent with the notion that competition has a significant influence on the rate of establishment of natural enemies in biological control projects, at least for projects with less than five natural enemies introduced. In contrast, the rate of success of the established natural enemies (proportion that contributed to the complete, substantial or partial control of the pest) was independent of the number of natural enemies established (Fig. 4b), and was again significantly greater (Mann-Whitney, $P = 0.001$) for Homoptera (0.58) than for Lepidoptera (0.43). These data therefore do not provide any evidence for an influence of competitive interactions on the impact of the established natural enemies on a pest or the success of a biological control project.

It is perhaps surprising that the same pattern as noted by Denoth *et al.* (2002), of a decline in the rate of establishment with the number of natural enemies introduced, persisted for both Homoptera and Lepidoptera using a much larger data set of biological control introductions. All the more so, as no such relationship has been found for natural enemy introductions in the biological control of weeds, even though the range in numbers of natural enemies released and established in such projects is similar (Denoth *et al.*, 2002). Although pest densities are generally considered sufficiently abundant to obviate competition between natural enemies during the phase of establishment in biological control projects (Keller, 1984), the substantial decline in rate of establishment with an increasing level of multiple introduction against insect pests is intriguing. The reduced rate of establishment for multiple introductions against both Homoptera and Lepidoptera is certainly consistent with notion of competitive exclusion, but cannot be used to infer causation, as the same pattern may equally well be caused by other factors, such as host specificity and project continuity. For example, one important difference between introductions against weeds and insect pests is that the natural enemies used against weeds have necessarily been more host specific due to a greater requirement for host range testing. Those insect pest projects in which the single best-adapted natural enemy was introduced first are likely to have readily led to establishment and success. However, for those projects in which a quick success was not achieved, if continued, would have led to a sequence of successive introductions involving less specialized natural enemies which would have been less likely to become established. Thus the role of competition in influencing the establishment of natural enemies in the biological control of insect pests remains an open question, and one that needs to be tested experimentally in order to resolve the patterns observed from the historical record.

5. CONCLUSIONS

Parasitoids and predators that share a common host are frequently involved in interspecific interactions that appear to be indicative of intense competition. In terms of intrinsic competition, it is clear that natural enemies compete through both exploitative and interference competition, but that apparent competition is less likely to play an important role at the third trophic level. Multiple parasitism by different species of parasitoids provides a dramatic illustration of the intense struggle between solitary parasitoid larvae in their fight to secure the limiting resources of a single host individual, and intrinsic competition at the individual level presents clear gains and losses for the competing species. The outcome of multiple parasitism via interference competition is frequently predictable due to the strong asymmetry of the interaction, as in the case of cleptoparasitoids, ectoparasitoids and those species with more rapid development rates. It is tempting to use these types of observations to provide guidelines for the most effective use of natural enemies in the biological control of arthropod pests (e.g., Mills, 1990), but evidence from both field observations (DeBach and Sundby, 1963; Steinberg *et al.*, 1987) and host-parasitoid models (Murdoch *et al.*, 2003; Pedersen and Mills, 2004) indicate the overriding influence of extrinsic factors in determining the outcome of competitive interactions among natural enemies.

However, in general, far less is known regarding the extrinsic factors that mediate exploitative competition at the population level. There is ample evidence from the biological control record that competitive displacement of both parasitoids and predators is a real, though not particularly frequent, outcome of multiple introductions of natural enemies. In each case, one of the most important consequences of competitive displacement is improved suppression of host densities, and there can be no question that the displacement results from competition for a limiting resource. The mechanism involved appears to be superior efficiency in the exploitation of the resource, which in the case of the competitive displacement among parasitoids of California red scale results from an ability to produce female parasitoids from earlier stage scale (Murdoch *et al.*, 1996).

Nonetheless, the numerous examples of coexistence among natural enemies in both natural and production ecosystems provide overwhelming evidence that in the majority of cases the potential for interspecific competition is avoided. Although there are some examples from the biological control literature of ecological (host density) and life history trade-offs, and temporal niche partitioning, as mechanisms for avoidance of competition, both theoretical and observational studies suggest that environmental heterogeneity and the spatial partitioning of niches is likely to be the most frequent mechanism. Another interesting yet neglected mechanism for coexistence is intermediate disturbance, which deserves greater attention in future natural enemy studies in the context of biological control.

In addition, natural enemies provide some unique advantages for experimental manipulative studies of competitive interactions and their role in population and community ecology. As small short-lived species, natural enemies are relatively easy to manipulate and their population growth rates can be directly estimated in short-term experiments using field cages or enclosures. To examine the impact of competitive interactions on host suppression, Sih *et al.* (1998) recommend use of a 2 x 2 factorial design, and a multiplicative risk model (Soluk and Collins, 1988) to estimate expected mortality rates in a two-way ANOVA on log transformed data. A greater emphasis on experimental studies of the population effects and emergent properties of competition among natural enemies is clearly needed, as this is the only way in which the compatibility of multiple natural enemies for biological control programs can be adequately assessed, and the question of the role of competition in the establishment of multiple natural enemies resolved.

6. ACKNOWLEDGEMENTS

It is a pleasure to thank Jacques Brodeur and Guy Boivin for inviting me to contribute this chapter as a presentation at an IOBC/Biocontrol Network symposium entitled 'Trophic and guild interactions in biological control' that was held in Magog-Orford, Canada on 8-11 May 2005.

7. REFERENCES

Abrams, P., 1987, The nonlinearity of competitive effects in models of competition for essential resources, Theor. Popul. Biol. 32: 50-65.

Alyokhin, A., and Sewell, G., 2004, Changes in a lady beetle community following the establishment of three alien species, Biol. Invasions 6: 463-471.

Amarasekare, P., 2000, Coexistence of competing parasitoids on a patchily distributed host: Local vs. spatial mechanisms, Ecology 81: 1286-1296.

Amarasekare, P., 2002, Interference competition and species coexistence, Proc. R. Roc. Lond. B 269: 2541-2550.

Amarasekare, P., 2003, Diversity-stability relationships in multitrophic systems: an empirical exploration, J. Anim. Ecol. 72: 713-724.

Aukema, B. H., Richards, G. B., Krauth, S. J., and Raffa, K. F., 2004, Species assemblage arriving at and emerging from trees colonized by *Ips pini* in the Great Lakes region: Partitioning by time since colonization, season, and host species, Ann. Entomol. Soc. Am. 97: 117-129.

Ayala, F. J., Gilpin, M. E., and Ehrenfeld, J. G., 1973, Competition between species, theoretical models and experimental tests, Theor. Popul. Biol. 4: 331-356.

Barradas, I., and Cohen, J. E., 1995, Disturbances allow coexistence of competing species, J. Math. Biol. 32: 663-676.

Batchelor, T. P., Hardy, I. C. W., Barrera, J. F., and Perez-Lachaud, G., 2005, Insect gladiators II: Competitive interactions within and between bethylid parasitoid species of the coffee berry borer, *Hypothenemus hampei* (Coleoptera: Scolytidae), Biol. Control 33: 194-202.

Bennett, F. D., 1993, Do introduced parasitoids displace native ones?, Fla. Entomol. 76: 54-63.

Bess, H. A., van den Bosch, R., and Haramoto, F. H., 1961, Fruit fly parasites and their activities in Hawaii, Proc. Hawaiian Entomol. Soc. 17: 367-378.

Biever, K. D., 1992, Distribution and occurrence of *Cotesia rubecula* (Hymenoptera: Braconidae) a parasite of *Artogeia rapae* in Washington and Oregon, J. Econ. Entomol. 85: 739-742.

Bográn, C. E., Heinz, K. M., and Ciomperlik, M. A., 2002, Interspecific competition among insect parasitoids: Field experiments with whiteflies as hosts in cotton, Ecology 83: 653-668.

Boivin, G., and van Baaren, J., 2000, The role of larval aggression and mobility in the transition between solitary and gregarious development in parasitoid wasps, Ecol. Lett. 3: 469-474.

Bonsall, M. B., and Hassell, M. P., 1999, Parasitoid-mediated effects: Apparent competition and the persistence of host-parasitoid assemblages, Res. Popul. Ecol. 41: 59-68.

Bonsall, M. B., Hassell, M. P., and Asefa, G., 2002, Ecological trade-offs, resource partitioning, and coexistence in a host-parasitoid assemblage, Ecology 83: 925-934.

Bonsall, M. B., Hassell, M. P., Reader, P. M., and Jones, T. H., 2004, Coexistence of natural enemies in a multitrophic host-parasitoid system, Ecol. Entomol. 29: 639-647.

Borer, E. T., Murdoch, W. W., and Swarbrick, S. L., 2004, Parasitoid coexistence: Linking spatial field patterns with mechanism, Ecology 85: 667-678.

Briggs, C. J., and Latto, J., 2001, Interactions between the egg and larval parasitoids of a gall-forming midge and their impact on the host, Ecol. Entomol. 26: 109-116.

Briggs, C. J., Nisbet, R. M., and Murdoch, W. W., 1993, Coexistence of competing parasitoid species on a host with a variable life cycle, Theor. Popul. Biol. 44: 341-373.

Brodeur, J., and Rosenheim, J. A., 2000, Intraguild interactions in aphid parasitoids, Entomol. Exp. Appl. 97: 93-108.

Brooks, W. M., 1993, Host-parasitoid-pathogen interactions, in: Parasites and Pathogens of Insects, Volume 2, N. E. Beckage, S. N. Thompson, and B. A. Federici, eds., Academic Press, New York, pp. 231-272.

Brown, M. W., and Miller, S. S., 1998, Coccinellidae (Coleoptera) in apple orchards of eastern West Virginia and the impact of invasion by *Harmonia axyridis*, Entomol. News 109: 143-151.

Brown, M. W., and Schmitt, J. J., 2001, Seasonal and diurnal dynamics of beneficial insect populations in apple orchards under different management intensity, Environ. Entomol. 30: 415-424.

Chaneton, E. J., and Bonsall, M. B., 2000, Enemy-mediated apparent competition: Empirical patterns and the evidence, Oikos 88: 380-394.

Chang, G. C., 1996, Comparison of single versus multiple species of generalist predators for biological control, Environ. Entomol. 25: 207-212.

Chase, J. M., Abrams, P. A., Grover, J. P., Diehl, S., Chesson, P., Holt, R. D., Richards, S. A., Nisbet, R. M., and Case, T. J., 2002, The interaction between predation and competition: a review and synthesis, Ecol. Lett. 5: 294-315.

Choate, B. A., and Rieske, L. K., 2005, Life history and age-specific mortality of eastern tent caterpillar (Lepidoptera : Lasiocampidae), Ann. Entomol. Soc. Am. 98: 496-502.

Comins, H. N., and Hassell, M. P., 1996, Persistence of multispecies host-parasitoid interactions in spatially distributed models with local dispersal, J. Theor. Biol. 183: 19-28.

Costanza, R., d'Arge, R., de Groot, R., Farber, S., Grasso, M., Hannon, B., Limburg, K., Naeem, S., O'Neill, R. V., Paruelo, J., Raskin, R. G., Sutton, P., and van den Belt, M., 1997, The value of the world's ecosystem services and natural capital, Nature 387: 253-260.

Day, W. H., Prokrym, D. R., Ellis, D. R., and Chianese, R. J., 1994, The known distribution of the predator *Propylea quatuordecimpunctata* (Coleoptera: Coccinellidae) in the United States, and thoughts on the origin of the species of five other exotic lady beetles in eastern North America, Entomol. News 105: 244-256.

De Vis, R. M. J., Mendez, H., and van Lenteren, J. C., 2003, Comparison of foraging behavior, interspecific host discrimination, and competition of *Encarsia formosa* and *Amitus fuscipennis*, J. Insect Behav. 16: 117-152.

DeBach, P., and Sundby, R. A., 1963, Competitive displacement between ecological homologues, Hilgardia 34: 105-166.

Denoth, M., Frid, L., and Myers, J. H., 2002, Multiple agents in biological control: improving the odds?, Biol. Control 24: 20-30.

Down, R. E., Smethurst, F., Bell, H. A., and Edwards, J. P., 2005, Interactions between the solitary endoparasitoid, *Meteorus gyrator* (Hymenoptera : Braconidae) and its host, *Lacanobia oleracea* ((Lepidoptera : Noctuidae), infected with the entomopathogenic microsporidium, *Vairimorpha necatrix* (Microspora : Microsporidia), Bull. Entomol. Res., 95: 133-144.

Ehler, L. E., and Hall, R. W., 1982, Evidence of competitive exclusion of introduced natural enemies in biological control, Environ. Entomol. 11: 1-4.

Elliott, N., Kieckhefer, R., and Kauffman, W., 1996, Effects of an invading coccinellid on native coccinellids in an agricultural landscape, Oecologia 105: 537-544.

Evans, E. W., 2000, Morphology of invasion: body size patterns associated with the establishment of *Coccinella septempunctata* in western North America, Eur. J. Entomol. 97: 469-474.

Evans, E. W., 2004, Habitat displacement of North American ladybirds by an introduced species, Ecology 85: 637-647.

Fisher, R. C., 1971, Aspects of the physiology of endoparasitic Hymenoptera, Biol. Rev. 46: 243-278.

Flanders, S. E., 1971, Multiple parasitism of armored coccids (Homoptera) by host regulative aphelinids (Hymenoptera): ectoparasites vs endoparasites, Can. Entomol. 103: 857-872.

Flanders, R. V., and Oatman, E. R., 1987, Competitive interactions endophagous parasitoids of potato tuberworm larvae in southern California, Hilgardia 55: 2-34.

Force, D. C., 1970, Competition among four hymenopterous parasites of an endemic insect host, Ann. Entomol. Soc. Amer. 63: 1675-1688.

Force, D. C., 1974, Ecology of insect host parasitoid communities, Science 184: 624-632.

Gerson, U., 1968, The comparative biologies of two hymenopterous parasites of the chaff scale, *Parlatoria pergandii*, Entomophaga 13: 163-173.

Getz, W. M., and Mills, N. J., 1996, Host-parasitoid coexistence and egg-limited encounter rates, Am. Nat. 148: 333-347.

Godfray, H. C. J., 1994, Parasitoids: Behavioural and Evolutionary Ecology, Princeton University Press, Princeton.

Greathead, D.J., and Greathead, A., 1992, Biological control of insect pests by insect parasitoids and predators: the BIOCAT database, Biocontrol News Inform. 13: 61N-68N.

Grover, J. P., 1997, Resource Competition, Chapman and Hall, London.

Grover, J. P., and Holt, R. D., 1998, Disentangling resource and apparent competition: realistic models, J. Theor. Biol. 191: 353-376.

Hågvar, E. B., 1988, Multiparasitism of the green peach aphid, *Myzus persicae*: competition in the egg stage between *Aphidius matricariae* and *Ephedrus cerasicola*, Entomol. Exp. Appl. 47: 275-282.

Harcourt, D. G., 1990, Displacement of *Bathyplectes curculionis* (Thoms.) (Hymenoptera: Ichneumonidae) by *Bathyplectes anurus* (Thoms.) in eastern Ontario [Canada] populations of the alfalfa weevil, *Hypera postica* Gyll. (Coleoptera: Curculionidae), Can. Entomol. 122: 641-646.

Hassell, M. P., Comins, H. N., and May, R. M., 1994, Species coexistence and self-organizing spatial dynamics, Nature 370: 290-292.

Hastings, A., 1980, Disturbance, coexistence history, and competition for space, Theor. Popul. Biol. 18: 363-373.

Hawkins, B. A., and Lawton, J. H., 1987, Species richness for parasitoids of British phytophagous insects, Nature 326: 788-790.

Hawkins, B. A., Cornell, H. V., and Hochberg, M. E., 1997, Predators, parasitoids, and pathogens as mortality agents in phytophagous insect populations, Ecology 78: 2145-2152.

Heimpel, G. E., Neuhauser, C., and Hoogendoorn, M., 2003, Effects of parasitoid fecundity and host resistance on indirect interactions among hosts sharing a parasitoid, Ecol. Lett. 6: 556-566.

Heinz, K. M., and Nelson, J. M., 1996, Interspecific interactions among natural enemies of *Bemisia* in an inundative biological control program, Biol. Control, 6: 384-393.

Herz, A., and Heitland, W., 2005, Species diversity and niche separation of cocoon parasitoids in different forest types with endemic populations of their host, the common pine sawfly Diprion pini (Hymenoptera : Diprionidae), Eur. J. Entomol. 102: 217-224.

Hochberg, M. E., 1991, Intra-host interactions between braconid endoparasitoid *Apanteles glomeratus* and a baculovirus for larvae of *Pieris brassicae*, J. Anim. Ecol. 60: 51-64.

Hochberg, M. E., 1996, Consequences for host population levels of increasing natural enemy species richnesss in biological control, Am. Nat. 147: 307-318.

Hochberg, M. E., and Hawkins, B. A., 1993, Predicting parasitoid species richness, Am. Nat. 142: 671-693.

Hochberg, M. E., and Hawkins, B. A., 1994, The implications of population dynamics theory to parasitoid diversity and biological control, in: Parasitoid Community Ecology, B. A. Hawkins, and W. Sheehan, eds., Oxford University Press, Oxford, pp. 451-471.

Hochberg, M. E., and Lawton, J. H., 1990, Competition between kingdoms, Trends Ecol. Evolut. 5: 367-371.

Hochberg, M. E., Hassell, M. P., and May, R. M., 1990, The dynamics of host-parasitoid-pathogen interactions, Am. Nat. 135: 74-94.

Holt, R. D., 1977, Predation, apparent competition, and the structure of prey communities, Theor. Popul. Biol. 12: 197-229.

Holt, R. D., and Lawton, J. H., 1993, Apparent competition and enemy-free space in insect host-parasitoid communities, Am. Nat. 142: 623-645.

Holt, R. D., and Lawton, J. H., 1994, The ecological consequences of shared natural enemies, Annu. Rev. Ecol. Syst. 25: 495-520.

Holt, R. D., Grover, H., and Tilman, D., 1994, Simple rules for interspecific dominance in systems with exploitative and apparent competition, Am. Nat. 144: 741-771.

Howarth, F. G., 1991, Environmental impacts of classical biological control, Annu. Rev. Entomol. 36: 485-510.

Huffaker, C. B., and Kennett, C. E., 1966, Studies of two parasites of olive scale, *Parlatoria oleae* (Colvée). IV. Biological control of *Parlatoria oleae* (Colvée) through the compensatory action of two introduced parasites, Hilgardia 37: 283-335.

Hunter, M. S., and Woolley, J. B., 2001, Evolution and behavioral ecology of heteronomous aphelinid parasitoids, Annu. Rev. Entomol. 46: 251-290.

Hunter, M. S., Collier, T. R., and Kelly, S. E., 2002, Does an autoparasitoid disrupt host suppression provided by a primary parasitoid?, Ecology 83: 1459-1469.

Ives, A. R., 1988, Aggregation and the coexistence of competitors, Ann. Zool. Fenn. 25: 75-88.

Ives, A. R., and May, R. M., 1985, Competition within and between species in a patchy environment: relations between microscopic and macroscopic models, J. Theor. Biol. 115: 65-92.

Kakehashi, N., Suzuki, Y., and Iwasa, Y., 1984, Niche overlap of parasitoids in host-parasitoid systems: its consequence to single versus multiple introduction controversy in biological control, J. Appl. Ecol. 21: 115-131.

Kim, J. J., Kim, K. C., and Roberts, D. W., 2005, Impact of the entomopathogenic fungus *Verticillium lecanii* on development of an aphid parasitoid, *Aphidius colemani*, J. Invert. Pathol. 88: 254-256.

Kimberling, D. N., 2004, Lessons from history: predicting successes and risks of intentional introductions for arthropod biological control, Biol. Invasions 6: 301-318.

Klopfer, E. D., and Ives, A. R., 1997, Aggregation and the coexistence of competing parasitoid species, Theor. Popul. Biol. 52: 167-178.

Kremen, C., and Chaplin, R., 2006, Insects as providers of ecosystem services, in: Insect Conservation Biology, A. Stewart, O. Lewis, and T. New, eds., CABI Publishing, Wallingford, (in press).

Laing, J. E., and Corrigan, J. E., 1987, Intrinsic competition between the gregarious parasite Cotesia glomeratus and the solitary parasite Cotesia rubecula (Hymenoptera: Braconidfae) for their host Artogeia rapae (Lepidoptera: Pieridae), Entomophaga 32: 493-502.

Levins, R., 1973, The qualitative analysis of partially specified systems, Ann. NY Acad Sci 231: 123-138.

Lorenzetti, F., Arnason, J. T., Philogene, B. J. R., and Hamilton, R. I., 1997, Evidence for spatial niche partitioning in predaceous Aphidophaga: Use of plant colour as a cue, Entomophaga 42: 49-56.

Lucas, E., 2005, Intraguild predation among aphidophagous predators, Eur. J. Entomol. 102: 351-363.

Lucas, E., and Alomar, O., 2002, Impact of the presence of *Dicyphus tamaninii* Wagner (Heteroptera: Miridae) on whitefly (Homoptera: Aleyrodidae) predation by *Macrolophus caliginosus* (Wagner) (Heteroptera: Miridae), Biol. Control 25: 123-128.

Luck R. F., and Podoler, H., 1985, Competitive exclusion of *Aphytis lingnanensis* by *Aphytis melinus*: potential role of host size, Ecology 66: 904-913.

Lynch, L. D., Ives, A. R., Waage, J. K., Hochberg, M. E., and Thomas, M. B., 2002, The risks of biocontrol: Transient impacts and minimum nontarget densities, Ecol. Appl. 12: 1872-1882.

Mackauer, M., 1990, Host discrimination and larval competition in solitary endoparasitoids, in: Critical Issues in Biological Control, M. Mackauer, L. E. Ehler, and J. Roland, eds., Intercept, Andover, pp. 41-62.

Mackauer, M., and Kambhampati, S., 1986, Parasitoids of the pea aphid in North America, in: Ecology of Aphidophaga, I. Hodek, ed., Junk, Dordrecht, pp. 347-356.

Marktl, R. C., Stauffer, C., and Schopf, A., 2002, Interspecific competition between the braconid endoparasitoids *Glyptapanteles porthetriae* and *Glyptapanteles liparidis* in *Lymantria dispar* larvae, Entomol. Exp. Appl. 105: 97-109.

Matsumoto, T., Itioka, T., Nishida, T., and Kaneko, S., 2003, Is one parasitoid enough? A test comparing one with a pair of parasitoid species in the biological control of arrowhead scales, Popul. Ecol. 45: 61-66.

May, R. M., and Hassell, M. P., 1981, The dynamics of multiparasitoid-host interactions, Am. Nat. 117: 234-261.

Miller, J. C., 1980, Niche relationships among parasitic insects occurring in a temporary habitat, Ecology 61: 270-275.

Mills, N. J., 1990, Biological control, a century of pest management, Bull. Entomol. Res. 80: 359-362.

Mills, N. J., 1991, Searching strategies of the parasitoids of the ash bark beetle (*Leperisinus varius* F.) and its relevance to biological control, Ecol. Entomol. 16: 461-470.

Mills, N. J., 1994a, The structure and complexity of parasitoid communities in relation to biological control, in: Parasitoid Community Ecology, B. A. Hawkins and W. Sheehan, eds., Oxford University Press, Oxford, pp. 397-417.

Mills, N. J., 1994b, Biological control: some emerging trends, in: Individuals, Populations and Patterns in Ecology, S. R. Leather, A. D. Watt, N. J. Mills, and K. F. A. Walters, Eds., Intercept, Andover, pp. 213-222.

Mills, N. J., 1999, Interspecific competition in insects, in: Ecological Entomology, Second Edition, C. B. Huffaker and A. P. Gutierrez, eds., Wiley, New York, pp. 355-387.

Mills, N. J., 2003, Parasitoid interactions and biological control, in: 1st International Symposium on Biological Control of Arthropods, R. Van Driesche, ed., USDA Forest Service FHTET-03-05, pp. 108-113.

Mills, N. J., 2006, Accounting for differential success in the biological control of homopteran and lepidopteran pests, N. Z. J. Ecol. (in press).

Morris, R. J., Mueller, C. B., and Godfray, H. C. J., 2001, Field experiments testing for apparent competition between primary parasitoids mediated by secondary parasitoids, J. Anim. Ecol. 70: 301-309.

Murdoch, W. W., Briggs, C. J., and Nisbet, R. M., 1996, Competitive displacement and biological control in parasitoids: A model, Am. Nat. 148: 807-826.

Murdoch, W. W., Briggs, C. J., and Nisbet, R. M., 2003, Consumer-Resource Dynamics, Princeton University Press, Princeton.

Nakashima, Y., and Akashi, M., 2005, Temporal and within-plant distribution of the parasitoid and predator complexes associated with *Acyrthosiphon pisum* and *A. kondoi* (Homoptera : Aphididae) on alfalfa in Japan, Appl. Entomol. Zool. 40: 137-144.

Nee, S., and May, R. M., 1992, Dynamics of metapopulations, habitat destruction and competitive coexistence, J. Anim. Ecol. 61: 37-40.

Netting, J. F., and Hunter, M. S., 2000, Ovicide in the whitefly parasitoid, *Encarsia formosa*, Anim. Behav. 60: 217-226.

Nguyen, D. H., Nakai, M., Takatsuka, J., Okuno, S., Ishii, T., and Kunimi, Y., 2005, Interaction between a nucleopolyhedrovirus and the braconid parasitoid *Meteorus pulchricornis* (Hymenoptera : Braconidae) in the larvae of *Spodoptera litura* (Lepidoptera: Noctuidae), Appl. Entomol. Zool. 40: 325-334.

Obrycki, J. J., Giles, K. L., and Ormord, A. M., 1998, Interactions between an introduced and indigenous coccinellid species at different prey densities, Oecologia 117: 279-285.

Onzo, A., Hanna, R., and Sabelis, M. W., 2003, Interactions in an acarine predator guild: Impact on *Typhlodromalus aripo* abundance and biological control of cassava green mite in Benin, West Africa, Exp. Appl. Acarol. 31: 225-241.

Paine, T. D., Paine, E. O., Hanks, L. M., and Millar, J. G., 2000, Resource partitioning among parasitoids (Hymenoptera: Braconidae) of *Phoracantha semipunctata* in their native range, Biol. Control 19: 223-231.

Pedata, P. A., Giorgini, M., and Guerrieri, E., 2002, Interspecific host discrimination and within-host competition between *Encarsia formosa* and *E. pergandiella* (Hymenoptera: Aphelinidae), two endoparasitoids of whiteflies (Hemiptera: Aleyrodidae), Bull. Entomol. Res. 92: 521-528.

Pedersen, B. S., and Mills, N. J., 2004, Single vs. multiple introduction in biological control: the roles of parasitoid efficiency, *antagonism* and niche overlap, J. Appl. Ecol. 41: 973-984.

Perez-Lachaud, G., Hardy, I. C. W., and Lachaud, J-P., 2002, Insect gladiators: Competitive interactions between three species of bethylid wasps attacking the coffee berry borer, *Hypothenemus hampei* (Coleoptera: Scolytidae), Biol. Control 25: 231-238.

Persad, A. B., and Hoy, M. A., 2003, Intra- and interspecific interactions between *Lysiphlebus testaceipes* and *Lipolexis scutellaris* (Hymenoptera: Aphidiidae) reared on *Toxoptera citricida* (Homoptera: Aphididae), J. Econ. Entomol. 96: 564-569.

Pexton, J. J., and Mayhew, P. J., 2004, Competitive interactions between parasitoid larvae and the evolution of gregarious development, Oecologia 141: 179-190.

Phoofolo, M. W., Obrycki, J. J., and Lewis, L. C., 2001, Quantitative assessment of biotic mortality factors of the European corn borer (Lepidoptera: Crambidae) in field corn, J. Econ. Entomol. 94: 617-622.

Plantard, O., Rasplus, J-Y., and Hochberg, M. E., 1996, Resource partitioning in the parasitoid assemblage of the oak galler *Neuroterus quercusbaccarum* L. (Hymenoptera: Cynipidae), Acta Oecol. 17: 1-15.

Pschorn-Walcher, H., 1987, Interspecific competition between the principal larval parasitoids of the pine sawfly *Neodiprion sertifer* Geoff. (Hymenoptera: Diprionidae), Oecologia 73: 621-625.

Reitz, S. R., and Trumble, J. T., 2002, Competitive displacement among insects and arachnids, Annu. Rev. Entomol. 47: 435-465.

Rochat, J., and Gutierrez, A. P., 2001, Weather-mediated regulation of olive scale by two parasitoids, J. Anim. Ecol. 70: 476-490.

Rosen, D., and DeBach, P., 1979, Species of *Aphytis* of the World (Hymenoptera: Aphelinidae), Israel University Press, Jerusalem.

Rosenheim, J. A., 1998, Higher-order predators and the regulation of insect herbivore populations, Annu. Rev. Entomol. 43: 421-447.

Roxburgh, S. H., Shea, K., and Wilson, J. B., 2004, The intermediate disturbance hypothesis: Patch dynamics and mechanisms of species coexistence, Ecology 85: 359-371.

Royer, L., Fournet, S., Brunel, E., and Boivin, G., 1999, Intra- and interspecific host discrimination by host seeking larvae of coleopteran parasitoids, Oecologia 118: 59-68.

Schausberger, P., and Walzer, A., 2001, Combined versus single species release of predaceous mites: Predator-predator interactions and pest suppression, Biol. Control 20: 269-278.

Schellhorn, N. A., and Andow, D. A., 1999, Cannibalism and interspecific predation: Role of oviposition behavior, Ecol. Appl. 9: 418-428.

Schellhorn, N. A., Kuhman, T. R., Olson, A. C., and Ives, A. R., 2002, Competition between native and introduced parasitoids of aphids: Nontarget effects and biological control, Ecology 83: 2745-2757.

Schmid-Hempel, R., and Schmid-Hempel, P., 1996, Larval development of two parasitic flies (Conopidae) in the common host *Bombus pascuorum*, Ecol. Entomol. 21: 63-70.

Schuster, M. F., and Dean, H. A., 1976, Competitive displacement of *Anagyrus antoninae* (Hym.: Encyrtidae), by its ecological homologue *Neodusmetia sangwani* (Hym.: Encyrtidae), Entomophaga 21: 127-30.

Shea, K., Roxburgh, S. H., and Rauschert, E. S. J., 2004, Moving from pattern to process: coexistence mechanisms under intermediate disturbance regimes, Ecol. Lett. 7: 491-508.

Shi, Z.-H., Li, Q.-B., and Li, X., 2004, Interspecific competition between *Diadegma semiclausum* Hellen (Hym., Ichneumonidae) and *Cotesia plutellae* (Kurdjumov) (Hym., Braconidae) in parasitizing *Plutella xylostella* (L.) (Lep., Plutellidea), J. Appl. Entomol. 128: 437-444.

Shorrocks, B., Atkinson, W., and Charlesworth, P., 1979, Competition of a divided and ephemeral resource, J. Anim. Ecol. 48: 899-908.

Sih, A., Englund, G., and Wooster, D., 1998, Emergent impacts of multiple predators on prey, Trends Evol. Ecol. 13: 350-355.

Smith, H., 1929, Multiple parasitism: its relation to the biological control of insect pests, Bull. Entomol. Res. 20: 141-149.

Snyder, W.E., Chang, G. C., and Prasad, R. P., 2005, Conservation biological control: biodiversity influences the effectiveness of predators, in: Ecology of Predator-Prey Interactions, P. Barbosa, and I. Castellanos, eds., Oxford University Press, Oxford, pp. 324-343.

Soluk, D. A., and Collins, N. C., 1988, Synergistic interactions between fish and stoneflies: facilitation and interference among stream predators, Oikos 52: 94-100.

Steinberg, S., Podoler, H., and Rosen, D., 1987, Competition between two parasites of the Florida red scale in Israel, Ecol. Entomol. 12: 299-310.

Strand, M. R., 1986, The physiological interactions of parasitoids with their hosts and their influence on reproductive strategies, in: Insect Parasitoids, J. Waage, and D. Greathead, eds., Academic Press, London, pp. 97-136.

Strand, M. R., and Godfray, H. C. J., 1989, Superparasitism and ovicide in parasitic Hymenoptera theory and a case study of the ectoparasitoid *Bracon hebetor*, Behav. Ecol. Sociobiol. 24: 421-432.

Sullivan, D. J., and Völkl, W., 1999, Hyperparasitism: multitrophic ecology and behavior, Annu. Rev. Entomol. 44: 291-315.

Szentkiralyi F., and Kozar, F., 1991, How many species are there in apple insect communities? Testing the resource diversity and intermediate disturbance hypotheses, Ecol. Entomol. 16: 491-504.

Takagi, M., and Hirose, Y., 1994, Building parasitoid communities: the complementary role of two introduced parasitoid species in a case of successful biological control, in: Parasitoid Community Ecology, B. A. Hawkins, and W. Sheehan, eds., Oxford University Press, Oxford, pp. 437-448.

Takasu, K., Hirose, Y., and Takagi, M., 1998, Occasional interspecific competition and within plant microhabitat preference in egg parasitoid of the bean bug, *Riptortus clavatus* (Hemiptera: Alydidae) in soybean, Appl. Entomol. Zool. 33: 391-399.

Tilman, D., 1982, Resource Competition and Community Structure, Princeton University Press, Princeton.

Tilman, D., 1990, Mechanisms of plant competition for nutrients: the elements of a predictive theory of competition, in: Perspectives on Plant Competition, J. B. Grace, and D. Tilman, eds., Academic Press, San Diego, pp. 117-142.

Tremblay, E., 1984, The parasitoid complex (Hymenoptera Ichneumonoidea) of *Toxoptera aurantii* (Homoptera: Aphidoidea) in the Mediterranean area, Entomophaga 29: 203-210.

Turnock, W. J., Wise, I. L., and Matheson, F. O., 2002, Abundance of some native coccinellines (Coleoptera: Coccinellidae) before and after the appearance of *Coccinella septempunctata*, Can. Entomol. 135: 391-404.

Van Driesche, R. G., and Hoddle, M. S., 2003, Biological control of insect pests, in: Encyclopedia of Insects, V. H. Resh, and R. T. Cardé, eds., Academic Press, San Diego, pp. 103-115.

van Nouhuys, S., and Hanski, I., 2000, Apparent competition between parasitoids mediated by a shared hyperparasitoid, Ecol. Lett. 3: 82-84.

Vance, R. R., 1984, Interference competition and the coexistence of two competitors on a single limiting resource, Ecology 65: 1349-1357.

Vandermeer, J., Evans, M. A., Foster, P., Höök, T., Reiskind, M., and Wund, M., 2002, Increased competition may promote species coexistence, Proc. Nat. Acad. Sci. 99: 8731-8736.

Vinson, S. B., and Hegazi, E. M., 1998, A possible mechanism for the physiological suppression of conspecific eggs and larvae following superparasitism by solitary endoparasitoids, J. Insect Physiol. 44: 703-712.

Völkl, W., and Stadler. B., 1991, Inter-specific larval competition between *Lysiphlebus testaceipes* and *Aphidius colemani* (Hymenoptera: Aphidiidae), J. Appl. Entomol. 111: 63–71.

Walzer, A., Blumel, S., and Schausberger, P., 2002, Population dynamics of interacting predatory mites, *Phytoseiulus persimilis* and *Neoseiulus californicus*, held on detached bean leaves, Exp. Appl. Acarol. 25: 731-743.

Wang, X. G., and Messing, R. H., 2002, Newly imported larval parasitoids pose minimal competitive risk to extant egg-larval parasitoid of tephritid fruit flies in Hawaii, Bull. Entomol. Res. 92: 423-429.

Wang, X. G., and Messing, R. H., 2004, Two different life-history strategies determine the competitive outcome between *Dirhinus giffardii* (Chalcididae) and *Pachycrepoideus vindemmiae* (Pteromalidae), ectoparasitoids of cyclorrhaphous Diptera, Bull. Entomol. Res. 94: 473-480.

Wang, X. G., Messing, R. H., and Bautista, R. C., 2003, Competitive superiority of early acting species: A case study of opiine fruit fly parasitoids, Biocontrol Sci. Techn. 13: 391-402.

Yasuda, H., Evans, E. W., Kajita, Y., Urakawa, K., and Takizawa, T., 2004, Asymmetric larval interactions between introduced and indigenous ladybirds in North America, Oecologia 141: 722-731.

Yu, D. S., Luck, R. F., and Murdoch, W. W., 1990, Competition, resource partitioning, and coexistence of an endoparasitoid *Encarsia perniciosi* and an ectoparasitoid *Aphytis melinus* of the California red scale, Ecol. Entomol. 15: 469-480.

Zaviezo, T., and Mills, N. J., 2001, The response of *Hyssopus pallidus* to hosts previously parasitized by *Ascogaster quadridentata*: heterospecific discrimination and host quality, Ecol. Entomol. 26: 91-99.

EXPERIMENTAL APPROACHES TO UNDERSTANDING THE RELATIONSHIP BETWEEN PREDATOR BIODIVERSITY AND BIOLOGICAL CONTROL

Cory S. Straub and William E. Snyder

Department of Entomology, Washington State University, Pullman, WA 99164

Abstract. Conservation biological control (CBC) involves the manipulation of the environment to favour the natural enemies of pests. Alternative agricultural practices, such as organic farming, are more biodiversity-friendly than conventional agricultural practices and generally lead to greater predator species richness and abundance. This is desirable from a conservation perspective, but it is unclear how greater predator diversity affects biological control. Unfortunately, the predator ecology literature provides little guidance: increasing the number of predator species has been shown to enhance, diminish, and not affect prey suppression. In this chapter we explore how the experimental approach used in biodiversity and ecosystem functioning (BEF) studies, which focus on the ecological consequences of species loss, may be used to study how increasing predator diversity affects biological control. The notable features of this approach are: 1) realistic levels of species richness (i.e., > 2 predator species), 2) the use of substitutive, rather than additive, experimental designs, and 3) experimentally distinguishing the effect of species richness from the effects of species abundance, composition, and identity. This experimental approach can be used to identify which components of predator biodiversity—species richness, abundance, composition, and identity—should be targeted by CBC practitioners to maximize pest suppression. Further, it can be used to assess whether predator biodiversity conservation and biological control are truly compatible goals. Ultimately, we hope that this chapter will serve to motivate future research into this important problem.

1. INTRODUCTION

Conservation biological control (CBC) involves the manipulation of the environment to favour the natural enemies of pests (Barbosa 1998). This may be achieved by removing deleterious factors, such as broad spectrum pesticides, and by providing limiting resources that natural enemies need to survive and flourish in agroecosystems. CBC is appealing because it can be implemented directly by growers and because it is a potentially sustainable pest management practice (Barbosa 1998).

While the primary goal of CBC is to control pests by enhancing naturally occurring predator populations, CBC serves the additional function of conserving predator species, which are generally more vulnerable to extinction than lower-trophic level species (Kruess and Tscharntke 1994, Duffy 2002). The conservation component of CBC is particularly relevant given that modern agriculture is

J. Brodeur and G. Boivin (eds.), Trophic and Guild Interactions in Biological Control, 221–239.

predicted to threaten biodiversity on a scale that rivals global warming (Tilman *et al.* 2001). Alternative agricultural practices, such as organic farming, are widely believed to be more biodiversity friendly and may thus be an important strategy for mitigating this biodiversity loss (Krebs *et al.* 1999, DEFRA 2002, EU 2002). Numerous studies have compared biodiversity on conventional and organic farms to test the assertion that biodiversity is conserved by organic agriculture. A recent review of these studies (Hole *et al.* 2005) and a meta-analysis (Bengtsson *et al.* 2005) found that biodiversity is enhanced by organic management practices. For predatory arthropods in particular, the meta-analysis showed that both species richness and abundance increases with organic management (Bengtsson *et al.* 2005). Ground beetles (Carabidae) also showed greater species richness and abundance on organic compared with conventional farms. Carabids were excluded from the predatory arthropod category but are often predacious (Ball and Bousquet 2001), suggesting that the benefit of organic management to predator biodiversity is even more pronounced. One explanation for why organic agriculture promotes greater predator diversity is that there is less reliance on persistent, broad-spectrum pesticides. As these pesticides continue to fall out of favour because of the risks of non-target effects (EPA 2003), it seems likely that predator diversity will also be on the rise in conventional agroecosystems.

Greater predator biodiversity on farms is clearly desirable from a conservation perspective, but its consequences for pest suppression are less clear (Snyder *et al.* 2005). This is because the conserved predators are often generalists. While generalist predators can be effective biological control agents (Chang and Kareiva 1999, Symondson *et al.* 2002), they may not reliably improve pest suppression because they also eat non-pest prey, including other predators (Polis *et al.* 1989, Rosenheim *et al.* 1995). Such food-web complexity may make the top-down control of prey populations unlikely (Strong 1992, Polis and Strong 1996). Indeed, adding intraguild predators to experimental communities has been shown to disrupt herbivore suppression (Rosenheim *et al.* 1993, Snyder and Ives 2001, see also Rosenheim and Harmon, this volume, who present a meta-analysis of the role of intraguild predators in biological control) and dampen trophic cascades (Snyder and Wise 2001, Finke and Denno 2004). However, agroecologists have often suggested that greater species richness should generally improve pest suppression (Pimentel 1961, Altieri 1994, Kruess and Tscharnkte 2000). There is some experimental evidence for this view as well (Cardinale *et al.* 2003, Aquilino *et al.* 2005, Björkman and Liman 2005). In addition to the positive and negative effects of predator diversity, several studies have found no effect of increasing predator species richness on herbivore suppression (Evans 1991, Chang 1996, Rodriguez and Hawkins 2000, Schmitz and Sokol-Hessner 2002, Vance-Chalcraft *et al.* 2004, Aukema and Raffa 2004, Straub and Snyder 2006). Thus, the predator ecology literature provides conflicting views on the value of predator biodiversity to biological control.

The uncertain relationship between predator biodiversity and biological control presents an important problem for agriculture and the environment. Organic growers have fewer chemical pest control options and more predator species than conventional growers. Thus, whether increasing predator biodiversity strengthens, weakens, or does not affect biological control will at least partially determine the extent to which growers find it profitable to adopt organic agricultural practices. This, in turn, will determine the extent to which agriculture continues to threaten biodiversity. Finke and Denno (2004) recently brought attention to this problem by questioning whether biodiversity conservation and biological control are compatible goals. In this essay we explore how CBC researchers can address this problem by adapting an experimental approach developed by ecosystem ecologists.

2. BIODIVERSITY AND ECOSYSTEM FUNCTIONING (BEF) AND MULTIPLE-PREDATOR STUDIES

Two lines of research, biodiversity and ecosystem functioning (BEF) and multiple-predator studies, are relevant to our understanding of the relationship between predator biodiversity and biological control. It has recently been appreciated that an exchange of ideas among these historically distinct sub-disciplines can lead to valuable insights (Wilby and Thomas 2002a, Wilby and Thomas 2002b, Ives *et al.* 2005). We believe this to be particularly true in the case of CBC, where the uncertain consequence of increasing predator diversity has emerged as a problem of paramount importance. Here we examine the approaches, objectives, and experimental designs of BEF and multiple-predator studies, pointing out their similarities and differences.

2.1. Approaches and objectives of BEF and multiple-predator studies

Interest in the relationship between biodiversity and ecosystem functioning has been motivated by alarming declines in global biodiversity (Kinzig *et al.* 2002, Loreau *et al.* 2002). Ecosystem functioning refers to ecological processes such as primary production, decomposition, pollination, and natural pest suppression. Many of these ecosystem functions directly impact human welfare, and the concern is that ecosystems will lose the ability provide these valuable services as species are lost. Early studies focused on plant communities (e.g., Tilman *et al.* 1997) and sought to distinguish between alternative hypotheses describing the relationship between biodiversity and ecosystem functioning. These hypotheses are presented graphically in Fig. 1a, and are extended to the relationship between predator biodiversity and biological control in Fig. 1b. The final point on the x-axis represents communities with the full complement of naturally occurring diversity, the first point represents communities with no species, and the trajectory of the line connecting these points represents how ecosystem functioning (or biological control) is hypothesized to change as diversity is lost. To test these hypotheses, BEF studies start with high levels of diversity and experimentally subtract species to mimic extinction. We refer to this as a top-down approach. By contrast, multiple-predator studies have used what we will call a bottom-up approach. These studies often start with a focal predator-prey interaction and add predator species to see how (or if) the focal interaction is modified. Because more species are included in BEF studies than in multiple-predator studies, the species interactions that give rise to the observed patterns are more tractable in the latter. This represents a difference in the objectives of BEF and multiple-predator studies—the former have emphasized pattern (Fig. 1) and the latter process.

While BEF studies generally include numerous species and have emphasized pattern over process, they have evolved from more process-oriented studies. These earlier studies typically focused on competition within and between plant species. Such plant competition studies are particularly common in the intercropping literature, where the objective has been to identify species combinations that maximize primary productivity (Vandermeer 1989). These studies look to see if combining plant species leads to overyielding, or greater primary productivity in species mixtures than is expected from the performance of each of the constituent species in monoculture (e.g., Petchey 2003). When this emergent effect is observed, the interacting species are described as complementary. Species complementarity can occur through resource partitioning, when plant species use resources in different ways and combine to use more resources than any one species could by itself, or through facilitation, when one plant species increases the consumption rate and growth of a second species (Loreau *et al.* 2001). As plant ecologists began to explore more complex communities in BEF studies, they were able to make inferences about the underlying species interactions based on knowledge gained from the earlier plant competition studies. For example, many BEF studies have measured primary productivity, an important ecosystem property, and used overyielding to infer complementarity among species in diverse assemblages.

The goal of multiple-predator studies has also been to identify the emergent effects of combining species. Here the response variable of interest is prey population size (rather than biomass). Emergent effects are observed when the prey population size differs from that predicted by summing the impacts of each predator species when in monoculture (Sih *et al.* 1998). Emergent predator impacts may be positive, such that multiple predators suppress prey to a greater extent than predicted from their individual impacts, or negative, such that prey suppression is less than that predicted by summing individual predator impacts. Positive emergent impacts are called synergism, facilitation, or prey risk enhancement, and negative emergent impacts are called antagonism or prey risk reduction (Sih *et al.* 1998). When the impact of multiple predators does not differ from expectations based on individual predator impacts, there is no evidence for predator-predator interaction and the combined impact is additive (Sih *et al.* 1998). Most ecologists would likely agree that predator-predator interactions and their consequences for prey suppression are well understood. Thus, like the plant ecologists who pioneered the field of BEF, it seems that predator ecologists are now well prepared to investigate more complicated consumer communities. As we discuss in the next section, predator ecologists will want to carefully choose their experimental design.

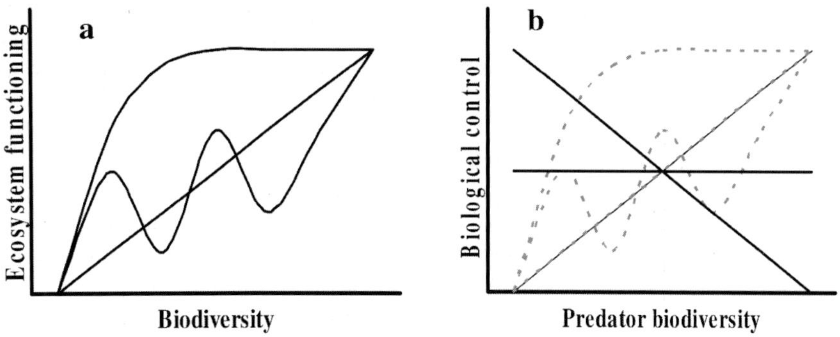

Figure 1. a) Hypothesized relationships between biodiversity and ecosystem functioning. From left to right, biodiversity ranges from 0 species to the natural level of biodiversity. The redundancy hypothesis (the line that increases and then levels off) states that ecosystem functioning increases with biodiversity until a point at which additional species have no additional effect. This hypothesis implies that species occupy several functional roles, but that most species are functionally identical and thus redundant. The linear hypothesis (the straight line) states that ecosystem functioning increases with biodiversity in a linear fashion. This hypothesis implies that species are functionally unique and have positive, but equivalent, effects. The idiosyncratic hypothesis (the wavy line) states that ecosystem functioning changes with biodiversity in an idiosyncratic fashion. This hypothesis implies that species are unique and have positive and negative effects that vary in magnitude. b) Hypotheses about the relationship between biodiversity and ecosystem functioning can be extended to the relationship between predator biodiversity and biological control. The predator ecology literature suggests two additional relationships: no relationship, and a negative (but not necessarily linear) relationship between predator biodiversity and biological control.

2.2. Experimental designs used by BEF and multiple-predator studies

One important difference between BEF and multiple-predator studies is that workers in these sub-disciplines have typically employed different experimental designs. BEF researchers have generally used substitutive, or replacement series designs, while multiple-predator studies have typically used additive series designs. Here we argue that the substitutive design may be more appropriate for assessing the value of predator diversity *per se,* but acknowledge that this approach also has its limitations.

2.2.1. Substitutive versus additive designs

In substitutive experimental designs, species richness is manipulated but the total abundance (or initial biomass) of individuals is held constant across richness treatments. To hold the total abundance of individuals constant, intraspecific densities are systematically reduced at increasing levels of richness (Connolly 1998, Jolliffe 2000). Thus, as species richness increases, the opportunity for intraspecific interactions decreases, while the opportunity for interspecific interactions increases. This approach allows a direct comparison of communities dominated by intra- versus interspecific interactions. In essence, the substitutive design tests the simple hypothesis that competition (and interference) is greater among conspecifics than among heterospecifics. This is expected if different species utilize different resources. If this hypothesis is supported, then conserving species richness (i.e., adding new species to a community) will lead to greater total resource exploitation than conserving abundance (i.e., adding more individuals of the species that already exist in the community). In the case of plant communities this is manifested as greater primary productivity (an ecosystem property); in the case of predator communities it is manifested as greater herbivore suppression (an ecosystem process and an important service in the context of agriculture).

While BEF researchers directly compare communities dominated by intra- and interspecific interactions, ecologists interested in multiple-predator effects typically focus on how interspecific interactions among predators affect prey populations. Thus, they hold intraspecific densities constant across levels of species richness in an attempt to isolate the effect of interspecific interactions. As a consequence, the total density of predators increases with species richness. The advantage of this additive experimental design is that it can be analyzed by a factorial ANOVA, and significant interaction terms can be used to identify the emergent effects of predators on their prey. However, these emergent effects may be caused by increasing species richness, by increasing the total abundance of predators, or by both factors. Importantly, synergism and antagonism are not restricted to interspecific interactions--adding more of the same species can lead to these emergent effects, just as adding different species can. Sih *et al.* (1998) note that this problem may be particularly acute in the case of antagonism. This is because increasing the density of single predator species generally leads to a decline in its per-capita effect on prey, a phenomenon known as predator interference. If there is no difference in the strength of intra- and interspecific interference, it is incorrect to conclude that antagonism is an emergent effect of increasing species richness (Sih *et al.* 1998). Thus, substitutive designs, which hold the total abundance of predators constant and directly compare the strength of intra- and interspecific interference, may be more appropriate for assessing the effect of increasing predator species richness. While we are only aware of a few multiple-predator studies that have used the substitutive design (Evans 1991, Chang 1996, White and Eigenbrode 2000, Schmitz and Sokol-Hessner 2002, Vance-Chalcraft *et al.* 2004, Aquilino *et al.* 2005, Björkman and Liman 2005, Straub and Snyder 2006), it is interesting to note that none of these studies have found that predator interference increases with species richness.

From our perspective, a clear objective of CBC programs should be to conserve predator abundance, because there is little doubt that increasing the total abundance of predators will generally improve biological control. However, predator interference (both intra- and interspecific) will almost always place limits on the benefits of increasing abundance. If adding new predator species leads to less interference, and a greater per-capita impact, than adding more of the same species, then there may also be value in conserving predator species richness. It is only through careful experimentation that CBC researchers can determine if, in addition to predator abundance, predator species richness itself should be targeted for conservation. Because additive experimental designs compound the effects of intra- and interspecific interference, while substitutive designs directly compare them, we believe the latter to be more appropriate for assessing whether greater predator species richness is good, bad, or neutral for biological control.

2.2.2. Limitations of the substitutive design

As with any experimental approach, the substitutive design also has its limitations. The primary problem is that, to hold the total abundance of individuals constant, the density of each species is systematically reduced at increasing levels of species richness. This can lead to confusion in the interpretation of the experimental results (Connolly 1998, Jolliffe 2000). For example, if herbivore suppression is strengthened by increasing predator species richness in a substitutive design (e.g., White and Eigenbrode 2000, Aquilino *et al.* 2005, Björkman and Liman 2005), the underlying mechanism may be a release from intraspecific interference--if some predators have a greater per-capita impact in the presence of fewer conspecifics, then the richer communities will exert stronger top-down suppression despite having the same total number of individuals. However, stronger herbivore suppression in the more diverse communities may also be driven by facilitation among predator species, that is, it may be caused by interspecific interactions. And of course, both mechanisms may be operating simultaneously. Thus, substitutive designs do not readily reveal the mechanisms driving differences between levels of species richness. This reiterates our earlier statement that BEF studies have emphasized pattern over process.

A potentially fruitful approach might be to follow the substitutive experiment with an additive experiment. Having answered our first research question—in our example, increasing species richness improves biological control—we can use an additive design to test for antagonism, additivity, or facilitation. Both antagonism and additivity would suggest that, even if interspecific interference also occurs (in the case of antagonism), it was a release from intraspecific interference that led to greater biocontrol in the substitutive experiment. This result suggests that strong negative interactions among conspecifics places a limit on their ability to suppress pests, and new species must be added to suppress herbivore populations to a lower level. By contrast, facilitation would indicate that at least some of the predator species became more effective *because* of greater species richness. This latter outcome provides the most compelling reason to conserve predator species richness. Thus, if interpreted appropriately, additive experimental designs may complement substitutive designs by improving our understanding of how predator biodiversity affects biological control (Ives *et al.* 2005).

A second potential criticism of substitutive designs is that they do not accurately mimic changes in predator biodiversity because, while substitutive designs uncouple predator species richness and total predator abundance, these factors are often positively correlated in the field (Cardinale *et al.* 2003, Bengtsson *et al.* 2005, Finke and Denno 2005, Hole *et al.* 2005). Additive experimental designs avoid this problem by increasing both predator species richness and total predator abundance in the high diversity treatment. However, additive designs may also fail to accurately mimic changes in predator biodiversity because, while predator and prey abundance are often positively correlated in the field (Cardinale *et al.* 2003, Koss *et al.* 2005), additive designs increase predator abundance in the high diversity treatment but hold the total prey density constant across levels of diversity. Thus, by increasing the total predator abundance in the high diversity treatment, additive designs successfully mimic the natural correlation between predator richness and abundance but may fail to capture the natural correlation between predator and prey abundance. Conversely, by holding predator abundance constant across levels of diversity, the substitutive design fails to capture the natural correlation between predator richness and total predator abundance but may successfully mimic the natural correlation between predator and prey abundance (Straub and Snyder 2006). One solution to this apparent trade-off might be to increase both predator and prey abundance in high diversity treatments. By combining the strengths of both substitutive and additive designs, this synthetic experimental approach may model changes in predator biodiversity more accurately.

3. DISENTANGLING THE COMPONENTS OF PREDATOR BIODIVERSITY

To assess the compatibility of predator conservation and biological control, studies that isolate the effect of predator species richness on biocontrol are clearly needed. To improve CBC, studies will also have to evaluate the importance of predator abundance, composition and identity. These components of predator biodiversity--species richness, abundance, composition and identity--are often correlated in nature, and careful experimental designs are needed to tease apart their independent effects. Fortunately, BEF studies have made considerable progress on this front. As we hope to demonstrate in this section, an exciting opportunity for experimental work in CBC awaits.

3.1. Species richness and abundance

Both predator species richness and abundance are increasing in response to organic management (Bengtsson *et al.* 2005). Each of these components of predator biodiversity can affect the strength of biological control. Because BEF studies have almost always used species richness as their metric of biodiversity, they have largely ignored the effects of abundance. Multiple-predator studies, by virtue of their heavy reliance on additive experimental designs, have usually manipulated both abundance and richness simultaneously. As mentioned above, this approach captures the natural correlation between increasing species richness and total predator abundance (Cardinale *et al.* 2003, Bengtsson *et al.* 2005, Finke and Denno 2005, Hole *et al.* 2005), but is unable to identify which component of predator biodiversity, if not both, should be conserved to maximize pest suppression.

Interestingly, by overlooking abundance effects, BEF researchers may have oversimplified an important concept in both ecosystem and predator ecology. This is the concept of *functional redundancy*. While any two species will often differ in many ways, they can be identical with respect to any one function, such as the suppression of a prey species (Sih *et al.* 1998, Chalcraft and Resetarits 2003). A premise of much BEF work is that adding functionally identical (or substitutable) species to the community will have no effect on ecosystem functioning (Fig. 1). In theory, this is not entirely accurate because it ignores abundance effects. Functionally identical species will often have additive effects at low densities but at increasing densities their combined effect will become increasingly sub-additive. This is because at high densities the functional role is more completely filled—in the case of predators and their prey, the prey population is eventually depleted (although total prey extinction is not certain and the proportion of prey that survives may be a function of the strength of predator interference). Thus, functionally identical species become functionally redundant at some critical abundance. This same logic applies to individuals of a single species: adding individuals to a population (e.g., predator population) will affect ecosystem functioning (e.g., herbivore suppression) until some critical abundance at which conspecifics become functionally redundant.

To improve the practice of CBC, we need to better understand how predator abundance, richness, and function are interrelated. Experiments that manipulate predator abundance and richness *independently* can make progress towards this goal. An experiment with the following treatments would be particularly valuable: 1) Low density, Low richness (typical of conventional agroecosystems), 2) High density, Low richness, and 3) High density, High richness (typical of organic agroecosystems). Such an experiment could be used to asses the independent effects of abundance and richness. Further, it could directly test the hypothesis that adding different predator species to an agroecosystem will lead to better biocontrol than adding more of the same predator species. This outcome would be expected if conspecifics become functionally redundant before heterospecifics, a prediction that makes sense because different species are more likely to be functionally diverse.

3.2. Species richness and composition

To evaluate the relationship between biodiversity and ecosystem functioning, plant studies have often examined how ecosystem functioning changes along a gradient of species richness (Loreau *et al.* 2002). At each level of species richness, communities with multiple species compositions are included. These species compositions are generated by random draws from a large species pool. In the statistical analysis, variance caused by species richness is assigned to the treatment term, while variance caused by species composition is assigned to the error term. Thus, this experimental approach attempts to isolate the effect of species richness by minimizing the influence of species composition (Huston and McBride 2002).

Few predator studies have attempted to distinguish between the effects of species richness and species composition (but see Finke and Denno 2005, Wilby *et al.* 2005). While predator ecologists may not have recognized this as an interesting or important distinction, there are probably additional reasons for this disparity between BEF and multiple-predator research. First, predator species richness is generally less than that of lower trophic-level consumers (Duffy 2003), a logistical constraint that limits the opportunity to experimentally address the problem. Predator ecologists often limit the available species pool even further by concentrating on strong interactors, or the subset of the predator community that most obviously affects prey population dynamics. Finally, predator ecologists have largely focused on the emergent effects of multiple predators on their prey. In these factorial experiments, the high richness treatment is composed of a single species composition. Thus, these studies all confound the effects of species richness with species composition.

A casual inspection of the multiple-predator literature reveals that species composition is extremely important. Examples of synergism (Soluk and Collins 1988, Losey and Denno 1998), antagonism (Rosenheim *et al.* 1993, Ferguson and Stiling 1996, Finke and Denno 2004, Snyder and Ives 2001), and additivity (Hurd and Eisenberg 1990, Snyder and Ives 2003) all exist, and this variation in predator-predator interactions is almost certainly caused by variation in species composition (although other factors may be important too). Obviously, these studies cannot be used to assess the value of species richness by themselves, because they each include only a single species composition. And while a meta-analysis of these studies would be a valuable exercise in its own right, it too would fall short of the goal of assessing the value of species richness. This is because, as we noted in the previous section, the additive designs most commonly employed by these studies confound the effects of species richness with density. Thus, experiments will have to be carefully designed to disentangle the effects of predator species richness and composition.

Recently, two well-designed predator studies have experimentally distinguished between the effects of predator species richness and composition (Finke and Denno 2005, Wilby *et al.* 2005). In each of these studies, multiple predator compositions were nested within each level of species richness, and the independent effects of species richness and composition were analysed. In both cases, the effect of species composition on herbivore suppression was significant. Further, Finke and Denno (2005) found an interaction such that increasing species richness led to weakened herbivore suppression in compositions that included intraguild predators, but not in compositions that excluded intraguild predators. This study takes an important first step towards understanding why species compositions vary in their effectiveness.

While there is a growing need to assess whether predator species conservation and biological control are compatible goals, the primary objective of many CBC research programs will be to improve the biological control of pests. This will likely require the targeted conservation of specific species (Snyder *et al.* 2005). Thus, CBC researchers will want to know which species compositions are the most effective. Experiments that vary species composition to better isolate the effect of species richness also allow the researcher to screen for particularly effective compositions (e.g. Finke and Denno 2005, Wilby *et al.* 2005). In follow-up studies, particular species compositions may be selected to confirm, with greater replication, which combinations should be targeted for conservation. Unfortunately, there will often be a trade-off such that both species richness and composition cannot be powerfully tested in the same experiment: to test for the effect of species richness, many species compositions must be included; to test for the effect of species composition, species compositions should be replicated, which will often mean that fewer compositions can be included. As researchers in the field know all too well, BEF experiments reflect a complex mix of trade-offs, sacrifice and compromise (Allison 1999).

3.3. Species richness and identity

As we have noted, many BEF studies create communities with varying levels of richness by randomly drawing species (with replacement) from a large species pool. As a consequence of this experimental approach, the probability that any one species occurs in a community increases with species richness. If a species with an extreme trait value that is important to the ecosystem function of interest occurs more often in richer communities, these communities may show higher (or lower, as the case may be) levels of functioning. Such sampling effects are caused by changes in species identity, not by changes in diversity *per se* (because species-poor communities may also include these key species). Further, they depend on the assumption that community assemblage is a random process, which is debatable (Hooper *et al.* 2005). Thus, whether sampling effects are an experimental artifact or a legitimate mechanism by which biodiversity affects ecosystem functioning has been debated (Loreau *et al.* 2001). This debate has important implications for biodiversity conservation because it can potentially affect whether resource managers focus their conservation efforts on one or few species, or on species richness more broadly.

Interestingly, a similar debate can be found in the classical biological control literature. On average, 3.4 natural enemy species have been released against each pest (Waage and Mills 1992). However, whether multiple species actually provide better control than a single species is uncertain (Myers 1985, Myers *et al.* 1989, Ehler 1990, Denno and Finke this volume, Janssen *et al.* this volume, Rosenheim and Harmon this volume). Myers (1985) proposed two models to explain how multiple introductions can lead to successful biological control. In the lottery model, multiple introductions increase the success of biological control by increasing the probability that a single, effective agent is introduced. In the cumulative stress model, multiple species act in an additive fashion and are cumulative in their suppressive effects on the target pest. The lottery and cumulative stress models of classical biological control are directly analogous to the sampling and species complementarity effects of BEF research.

Denoth *et al.* (2002) analyzed the classical biological control record and found that in over 50% of multi-agent introductions against both weeds and arthropods, successful biological control was attributed to a single species. This result is consistent with the lottery model and supports the importance of species identity. Logistic regressions of biological control success (successful or not successful) on the number of agents released showed that increasing natural enemy species richness improved the biological control of weeds, but had no effect on arthropod pests. This result suggests that species complementarity (the cumulative stress model) may be more important for the classical biological control of weeds than it is for arthropods. To explain this, Denoth *et al.* (2002) hypothesized that weeds may be able to support multiple agents, e.g., foliage and seed feeders, while arthropods provide fewer opportunities for such resource partitioning. In other words, the link between taxonomic and functional diversity may be stronger for weed control agents than it is for arthropod control agents. This seems particularly likely given that multiple parasitoid species are often introduced for the biological control of arthropod pests. Unlike predators, parasitoids often attack and "kill" the same individuals (i.e., superparasitism, hyperparasitism). This is perhaps the clearest case of functional redundancy imaginable. Such functional redundancy may also explain why parasitoid species richness has no effect on total percent parasitism in grass-feeding chalcid wasps (Rodriguez and Hawkins 2000). Thus, the classical biological control record suggests that species identity, not richness, is the key to successful arthropod biocontrol. Whether this conclusion can be extended to CBC, where the available species pool is not limited to a single functional group, remains an open question.

Numerous BEF studies have attempted to distinguish between sampling and species complementarity effects to assess whether the conservation of species richness itself is warranted. In plant studies where primary productivity is the response variable, measures of overyielding are commonly used to provide evidence for species complementarity. Numerous indices of overyielding have been developed, primarily as a tool to serve agriculture by maximizing plant yields through intercropping (Vandermeer 1989, Petchey 2003). Recently, BEF workers have developed more sophisticated statistical procedures for partitioning sampling and species complementarity effects (Loreau and Hector 2001, Špaèková and Lepš. 2001, Petchey 2003). These procedures may be adapted by ecologists working with animals (Cardinale *et al.* 2002, Duffy *et al.* 2003). In addition to using statistical procedures to distinguish between sampling and species complementarity effects, researchers have also attempted to control experimentally for sampling effects to isolate the effects of species complementarity (e.g., Reich *et al.* 2004, Straub and Snyder 2006). By ensuring that the mean abundance of each species is equal across levels of species richness (as opposed to random or intentionally varied), differences between levels of richness cannot be caused by variation in the relative abundance of single species. This minimizes the potential for sampling effects. However, it is important to realize that controlling for species *abundance* is not the same as controlling for species *occurrence*. So while the mean abundance of any one species may be equal across levels of richness, it is still possible (and indeed likely) that a given species will be present in a greater proportion of high than low richness communities. Thus, species identity is not entirely controlled and the potential for sampling effects has not been completely eliminated. Nevertheless, controlling for species abundance is an important advance in experimental methodology and can strengthen investigations into the effects of increasing predator species richness.

4. THE IMPORTANCE OF EXTRAGUILD PREY DIVERSITY

In most communities, predators have multiple prey types to choose from. Wilby and Thomas (2002a) have argued that the value of predator diversity will depend on the diversity of prey types that are present. Specifically, they used a simulation model to show that suppression of exopterygote pests will not be affected by increasing predator biodiversity because there is little opportunity for resource partitioning among predator species when there are few phenotypic differences among prey life stages. Multiple predator species are thus functionally redundant, and of neutral value, in the case of exopterygotes. In contrast, endopterygotes, which have phenotypically diverse life stages (i.e., egg, larvae, pupae, adult), provide ample opportunity for resource partitioning among predator species and are thus more effectively controlled by a diverse predator assemblage. Here predator species fill different functional roles by attacking different prey life stages and, through resource partitioning, multiple predators act additively to suppress their prey. Ives *et al.* (2005) used a mathematical model to demonstrate an analogous situation. In their model predator species varied in their attack efficiencies on different prey species. Increasing predator diversity strengthened suppression of the entire prey community, but was of no value to the suppression of single prey species where there was no opportunity for resource partitioning among predator species. The hypothesis that the value of predator diversity depends on the diversity of the prey base has clear implications for biological control: increasing predator biodiversity will be important when the pests are endopterygotes and/or there are several target pest species. Surprisingly, few multiple-predator studies have included more than one prey species or life stage, leaving this simple hypothesis largely untested.

In addition to providing a substrate for resource partitioning and thus additivity among predator species, including multiple prey species can modify the interactions among predators and focal pests in important ways (e.g., Cardinale *et al.* 2003). For example, Cardinale *et al.* (2003) examined how *Coccinella* sp. (Coleoptera: Coccinellidae), *Nabis* sp. (Hemiptera: Nabidae), and the parasitoid *Aphidius ervi* (Hymenoptera: Braconidae) combine to control the pea aphid in alfalfa. Adding these natural enemies together led to synergistic suppression of the pea aphid. Interestingly, this outcome depended on the presence of a second herbivore. In addition to pea aphids, the study also included cowpea aphids, which commonly occur in alfalfa but achieve pest status only rarely. Apparently, the presence of the cowpea aphid caused the parasitoid to attack the pea aphid less often. When all three natural enemies were present, *Coccinella* reduced cowpea aphid densities, thus causing the parasitoid to attack pea aphids more often. This interaction modification led to a positive emergent effect such that pea aphid suppression was greater in the high diversity treatment than would be expected based on the individual impacts of each natural enemy when alone.

The takehome message, then, is that the value of increasing predator biodiversity may often depend on the diversity of the available prey base. Most agroecosystems have multiple pest species as well as herbivores that do not achieve pest status. To accurately evaluate the consequences of increasing predator biodiversity for biological control, these community members should be included in experiments whenever possible.

5. CONCLUSIONS

With predator biodiversity increasing in response to organic agriculture while global declines in biodiversity continue at an unprecedented rate, CBC researchers are now finding themselves at the intersection of biological control, sustainable agriculture, and biodiversity conservation. A deeper understanding of the relationship between predator biodiversity and biological control is needed to assess the compatibility of biodiversity and biocontrol and to improve the efficacy of CBC programs. BEF researchers have recognized that biodiversity is a complex concept that includes species richness, abundance, composition, and identity. While these components of biodiversity are often correlated, they may each be important to varying degrees. Thus, BEF researchers have developed an experimental framework that may be used to isolate their independent effects in an effort to identify which of these components should be prioritized for conservation. This approach can be easily extended to the higher trophic levels that CBC researchers work with. We see an important opportunity here, and provide several suggestions for how CBC researchers might proceed.

First, CBC researchers will want to extensively survey their agroecosystem, and perhaps the surrounding landscape, to identify the species pool from which they will work (see van Veen, this volume, for related discussion). Predator species that do not obviously impact target pests should not be ignored, because these species may still have important indirect effects (Wootton 1994). For the same reason, apparently unimportant herbivores and detritivores should be included as well. We are well aware of the logistical demands of collecting so many species, and appreciate that no study can account for all of the biocomplexity in the system. Further, we recognize that by including so much complexity, the underlying mechanisms will not be easily revealed. However, we cannot ignore this complexity if we want to accurately assess the effect of increasing predator diversity on biological control. Thus, a top-down approach that embraces complexity and compares communities with natural or elevated levels of species richness to less speciose predator communities should be used to measure the effect of species richness on biological control. As we have discussed in this essay, BEF-type experiments can be used to identify the components of predator biodiversity that are responsible for the observed pattern. And once this pattern is generated, hypotheses about the species interactions that are at work can be formulated and, in the tradition of much predator ecology research, we can create much simpler communities to examine these interactions more closely.

An important incongruence between CBC and BEF research is that the latter has generally ignored the effects of abundance, yet this component of biodiversity is at least as relevant to CBC as is species richness. Thus, experiments that manipulate both predator abundance and richness will be extremely valuable. Further, we see the integration of abundance and richness effects as a priority for CBC research, and as an exciting opportunity to make an important contribution to our understanding of biodiversity and ecosystem functioning more generally.

Finally, consistent differences between organic and conventionally managed fields in predator species richness, abundance, composition or identity provide an obvious platform for experimental work. Alternatively, researchers may want to use their agroecosystem as a model system to examine how increasing predator species richness affects biological control. Almost no experiments have properly isolated the effect of predator species richness on biological control, making this an important objective. And while logistically difficult, creating a gradient of predator species richness and measuring pest suppression and/or primary production (i.e., trophic cascades) would be particularly valuable. It is a reasonable hypothesis that, because predator species can both compete with *and* eat each other, they are fundamentally different from other consumers and will thus produce a very different diversity-function relationship than has been observed with plants (Fig. 1). Alternatively, as we have alluded to, interspecific interference may not be so severe and a similar diversity-function relationship may exist. Research on this problem is needed from the perspective of both CBC and BEF, and will help to evaluate the compatibility of biodiversity conservation and biological control.

6. ACKNOWLEDGEMENTS

We thank Deborah Finke, Shawn Steffan, and two anonymous reviewers for suggestions that greatly improved this essay. We also extend our thanks to Guy Boivin and Jacques Brodeur for the opportunity to contribute to this volume. During the writing of this chapter the authors were supported by the National Research Initiative of the USDA Cooperative Research, Education and Extension Service (CSREES), grant #2004-01215.

7. REFERENCES

Allison, G. W., 1999, The implications of experimental design for biodiversity manipulations, Am. Nat. 153: 26-45.

Altieri, M.A., 1994, Biodiversity and Pest Management in Agroecosystems, Food Products Press, an imprint of the Haworth Press, New York.

Aquilino, K. M., Cardinale, B. J., and Ives, A. R., 2005, Reciprocal effects of host plant and natural enemy diversity on herbivore suppression: an empirical study of a model tritrophic system, Oikos 108: 275-282.

Aukema, B. H., and Raffa, K. F., 2004, Does aggregation benefit bark beetles by diluting predation? Links between a group-colonisation strategy and the absence of emergent multiple predator effects, Ecol. Entomol. 29: 129-138.

Ball, G. E., and Bousquet, Y., 2001, Carabidae, In: American Beetles, Vol. 1, R. H. Arnett, Jr., and M. C. Thomas, Eds, CRC Press, Boca Raton, pp. 32-132.

Barbosa, P.,ed., 1998, Conservation Biological Control, Academic Press, San Diego.

Bengtsson, J., Ahnström, J., and Weibull, A., 2005, The effects of organic agriculture on biodiversity and abundance: a meta-analysis, J. Appl. Ecol. 42: 261-269.

Björkman, C., and Liman, A., 2005, Foraging behaviour influences the outcome of predator-predator interactions, Ecol. Entomol. 30: 164-169.

Cardinale, B. J., Harvey, C. T., Gross, K., and Ives, A. R., 2003, Biodiversity and biocontrol: emergent impacts of a multi-enemy assemblage on pest suppression and crop yield in an agroecosystem, Ecol. Lett. 6: 857-865.

Cardinale, B. J., Palmer, M. A., and Collins, S. L., 2002, Species diversity enhances ecosystem functioning through interspecific facilitation, Nature 415: 426-429.

Chalcraft, D. R., and Resetarits, W. J., 2003, Predator identity and ecological impacts: Functional redundancy or functional diversity?, Ecology 84: 2407-2418.

Chang, G. C., 1996, Comparison of single versus multiple species of generalist predators for biological control. Environ. Entomol. 25: 207-212.

Chang, G. C., and Kareiva, P., 1999, The case for indigenous generalists in biological control, In: Theoretical approaches to biological control, B. A. Hawkins and H. V. Cornell, eds, Cambridge University Press, New York, pp. 103-115.

Connolly, J., 1988, What is wrong with replacement series?, TREE 3(1): 24-26.

DEFRA, 2002, Action Plan to Develop Organic Food and Farming in England; http: //www.defra.gov.uk/farm/organic/actionplan/index.htm.

Denoth, M., Frid, L., and Meyers, J., 2002, Multiple agents in biological control: improving the odds? Biol. Cont. 24: 20-30.

Duffy, J. E., 2002, Biodiversity and ecosystem function: the consumer connection, Oikos 99: 201-219.

Duffy, J. E., 2003, Biodiversity loss, trophic skew and ecosystem functioning, Ecol. Lett. 6: 680-687.

Duffy, J. E., Richardson, J. P., and Canuel, E. A., 2003, Grazer diversity effects on ecosystem functioning in seagrass beds, Ecol. Lett. 6: 637-645.

Ehler, L. E., 1998, Conservation biological control: past, present, and future, In: Conservation Biological Control, P. Barbosa, ed., Academic Press, San Diego, pp. 1-8.

Environmental Protection Agency, 2003, Food Quality and Protection Act of 1996; http://www.epa.gov/opppsps1/fqpa/.

EU, 2002, Organic Farming in the EU: Facts and Figures; http://europa.eu.int/comm/agriculture/qual/ organic/facts_en.pdf.

Evans, E.W., 1991, Intra versus interspecific interactions of ladybeetles (Coleoptera: Coccinellidae) attacking aphids, Oecologia 87: 401-408.

Ferguson, K. I., and Stiling, P., 1996, Non-additive effects of multiple natural enemies on aphid populations. Oecologia 108: 375-379.

Finke, D. L., and Denno, R. F., 2005, Predator diversity and the functioning of ecosystems: the role of intraguild predation in dampening trophic cascades, Ecol. Lett. 8: 1299-1306.

Finke, D. L., and Denno, R. F., 2004, Predator diversity dampens trophic cascades, Nature 429: 407-410.

Hole, D. G., Perkins, A. J., Wilson, J. D., Alexander, I. H., Grice, P. V. and Evans, A. D., 2005, Does organic farming benefit biodiversity?, Biol. Conserv. 122: 113-130.

Hooper, D. U., Chapin, F. S., Ewel, J. J., Hector, A., Inchausti, P., Lavorel, S., Lawton, J. H., Lodge, D. M., Loreau, M., Naeem, S., Schmid, B., Setala, H., Symstad, A. J., Vandermeer, J., Wardle, D. A., 2005, Effects of biodiversity on ecosystem functioning: A consensus of current knowledge, Ecol. Mon. 75: 3-35.

Hurd, L. E., and Eisenberg, R. M., 1990, Arthropod community responses to manipulation of a bitrophic predator guild, Ecology 71: 2107-2114.

Huston, M. A., and McBride, A. C., 2002, Evaluating the relative strengths of biotic versus abiotic controls on ecosystem processes, In: Biodiversity and Ecosystem Functioning, M. Loreau, S. Naeem, and P. Inchausti, eds., Oxford University Press, New York, pp. 47-60.

Ives, A. R., Cardinale, B. J., and Snyder, W. E., 2005, A synthesis of subdisciplines: predator- prey interactions, and biodiversity and ecosystem functioning, Ecol. Lett. 8: 102-116.

Jolliffe, P. A., 2000, The replacement series, J. Ecol. 88: 371-385.

Kinzig, A. P., Pacala, S., and Tilman, G. D., (eds.), 2002, The Functional Consequences of Biodiversity: Empirical Progress and Theoretical Extensions, Princeton University Press, Princeton.

Koss, A. M., Jensen, A. S., Schreibner, A., Pike, K. S., and Snyder, W. E., 2005, A comparison of predator and pest communities in Washington potato fields treated with broad-spectrum, selective or organic insecticides. Environ. Entomol. 34: 87-95.

Krebs, J. R., Wilson, J. D., Bradbury, R. B., Siriwardena, G. M., 1999, The second silent spring?, Nature 400: 611-612.

Kruess, A., and Tscharntke, T., 1994, Habitat fragmentation, species loss, and biological-control, Science 264: 1581-1584.

Kruess, A., and Tscharntke, T., 2000, Species richness and parasitism in a fragmented landscape: experiments and field studies with insects on *Vicia sepium*, Oecologia 122: 129-137.

Loreau, M., and Hector, A., 2001, Partitioning selection and complementarity in biodiversity experiments, Nature 412: 72-76.

Loreau, M., Naeem, S., and Inchausti, P., eds, 2002, Biodiversity and ecosystem functioning: synthesis and perspectives, Oxford University Press, New York.

Loreau, M., Naeem, S., Inchaust, P., Bengtsson, J., Grime, J. P., Hector, A., Hooper, D. U., Huston, M. A., Raffaelli, D., Schmid, B., Tilman, D., and Wardle, D. A., 2001, Biodiversity and ecosystem functioning: current knowledge and future challenges, Science 294: 804-808.

Losey, J. E., and Denno, R. F., 1998, Positive predator-predator interactions: enhanced predation rates and synergistic suppression of aphid populations, Ecology 79: 2143-2152.

Myers, J.H., 1985, How many insect species are necessary for successful biocontrol of weeds? In: Proceedings of the 6th International Symposium on Biological Control of Weeds, Agriculture Canada, E. S. Delfosse (ed), Canadian Govt. Printing Office, Ottawa, pp. 77-82.

Myers, J.H., Higgins, C., Kovacs, E., 1989, How many insect species are necessary for the biological control of insects? Environ. Entomol. 18: 541-547.

Petchey, O. L., 2003, Integrating methods that investigate how complementarity influences ecosystem functioning, Oikos 101: 323-330.

Pimentel, D., 1961, Species diversity and insect population outbreaks, Ann. Entomol. Soc. Amer. 54: 76-86.

Polis, G. A., Myers, C. A., and Holt, R. D., 1989, The ecology and evolution of intraguild predation: potential competitors that eat each other, Annu. Rev. Ecol. Syst. 20: 297-330.

Polis, G. A. and Strong, D. R., 1996, Food web complexity and community dynamics, Am. Nat. 147: 813-846.
Reich, P. B., Tilman, D., Naeem, S., Ellsworth, D. S., Knops, J., Craine, J., Wedlin, D., and Trost, J., 2004, Species and functional group diversity independently influence biomass accumulation and its response to CO_2 and N, PNAS 101: 10101-10106.
Rodriguez, M.A., and Hawkins, B. A., 2000, Diversity, function and stability in parasitoid communities, Ecol. Lett. 3: 35-40.
Rosenheim, J. A., Kaya, H. K., Ehler, L. E., Marois, J. J., and Jaffee, B. A., 1995, Intraguild predation among biological-control agents: theory and practice, Biol. Cont. 5: 303-335.
Rosenheim, J. A., Wilhoit, L. R., and Armer, C. A., 1993, Influence of intraguild predation among generalist insect predators on the suppression of an herbivore population, Oecologia 96: 439-449.
Schmitz, O. J., and Sokol-Hessner, L., 2002, Linearity in the aggregate effects of multiple predators in a food web, Ecol. Lett. 5: 168-172.
Sih, A., Englund, G., and Wooster, D., 1998, Emergent impacts of multiple predators on multiple prey, TREE 13: 350-355.
Snyder, W. E., Chang, G. C., and Prasad, R. P., 2005, Conservation biological control: Biodiversity influences the effectiveness of predators, In: Ecology of Predator-Prey Interactions, P. Barbosa, ed., Oxford University Press, New York, pp. 324-343.
Snyder, W. E., and Ives, A. R., 2001, Generalist predators disrupt biological control by a specialist parasitoid, Ecology 82: 705-716.
Snyder, W. E. and Ives, A. R., 2003, Interactions between specialist and generalist natural enemies: parasitoids, predators, and pea aphid biocontrol, Ecology 84: 91-107.
Snyder, W. E., and Wise, D. H., 2001, Contrasting trophic cascades generated by a community of generalist predators, Ecology 82: 1571-1583.
Soluk, D. A., and Collins, N. C., 1988, Balancing risks? Responses and non-responses of mayfly larvae to fish and stonefly predators, Oecologia 77: 370-374.
Špaèková, I., and Lepš, J., 2001, Procedure for separating the selection effect from other effects in diversity-productivity relationship, Ecol. Lett. 4: 585-594.
Straub, C. S., and Snyder, W. E., 2006, Species identity dominates the relationship between predator biodiversity and herbivore suppression, Ecology, in press.
Strong, D. R., 1992, Are trophic cascades all wet? Differentiation and donor-control in speciose systems, Ecology 73: 747-754.
Symondson, W. O. C., Sunderland, K. D., and Greenstone, M. H., 2002, Can generalist predators be effective biocontrol agents?, Ann. Rev. Entomol., 47, 561-594.
Tilman, D., Fargione, J., Wolff, B., D'Antonio, C., Dobson, A., Howarth, R., Schindler, D., Schlesinger, W. H., Simberloff, D., and Swackhamer, D., 2001, Forecasting agriculturally driven global environmental change, Science 292: 281-292.
Tilman, D., Knops, J., Wedin, D., Reich, P., Ritchie, M., and Siemann, E., 1997, The influence of functional diversity and composition on ecosystem processes, Science 277: 1300-1302.
Tilman, D., Reich, P. B., Knops, J., Wedin, D., Mielke, T., and Lehman, C., 2001, Diversity and productivity in a long-term grassland experiment, Science 294: 843-845.
Vance-Chalcraft, H. D., Soluk, D. A., and Ozburn, N., 2004, Is prey predation risk influenced more by increasing predator density or predator species richness in stream enclosures? Oecologia 139: 117-122.
Vandermeer, J. H., 1989, The Ecology of Intercropping, Cambridge University Press, Cambridge.
Waage, J., and Mills, N. J., 1992, Biological control. In: Natural Enemies, M. J. Crawley, (ed), Blackwell, Oxford, pp. 412-430.
White, C, and Eigenbrode, S. D., 2000, Leaf surface waxbloom in *Pisum sativum* influences predation and intra-guild interactions involving two predator species, Oecologia 124(2): 252-259.
Wilby, A., and Thomas, M. B., 2002a, Natural enemy diversity and pest control: patterns of pest emergence with agricultural intensification, Ecol. Lett. 5: 353-360.

Wilby, A., and Thomas, M. B., 2002b, Are the ecological concepts of assembly and function of biodiversity useful frameworks for understanding natural pest control? Agric. For. Entomol. 4: 237-243.

Wilby, A., Villareal, S. C., Lan, L. P., Heong, K. L., and Thomas, M. B., 2005, Functional benefits of predator species diversity depend on prey identity. Ecol. Entomol. 30: 497-501.

Wootton, J. T., 1994, The nature and consequences of indirect effects in ecological communities. Annu. Rev. Ecol. Syst. 25: 443-466.

Index